Modelling and Management of Irrigation System

Modelling and Management of Irrigation System

Special Issue Editors

Juan Antonio Rodríguez Díaz
Rafael González Perea
Miguel Ángel Moreno

MDPI • Basel • Beijing • Wuhan • Barcelona • Belgrade • Manchester • Tokyo • Cluj • Tianjin

Special Issue Editors
Juan Antonio Rodríguez Díaz
University of Córdoba
Spain

Rafael González Perea
University of Castilla-La Mancha
Spain

Miguel Ángel Moreno
University of Castilla-La Mancha
Spain

Editorial Office
MDPI
St. Alban-Anlage 66
4052 Basel, Switzerland

This is a reprint of articles from the Special Issue published online in the open access journal *Water* (ISSN 2073-4441) (available at: https://www.mdpi.com/journal/water/special_issues/water_modlling_and_management_of_irrigation_system).

For citation purposes, cite each article independently as indicated on the article page online and as indicated below:

LastName, A.A.; LastName, B.B.; LastName, C.C. Article Title. *Journal Name* **Year**, *Article Number*, Page Range.

ISBN 978-3-03928-790-1 (Hbk)
ISBN 978-3-03928-791-8 (PDF)

Contents

About the Special Issue Editors

Juan Antonio Rodríguez Díaz has a M.S. in Agricultural Engineering (2000) and a Ph.D. in Irrigation Water Management (2004), both from the University of Córdoba. He is currently a professor of hydraulics and irrigation engineering in the Faculty of Agricultural and Forestry Engineering (ETSIAM). With 18 years' experience in research, education, and consultancy in the field of hydraulics and irrigation water resources management, his research fields include the optimization of water energy nexus in irrigation systems, renewable energy sources for irrigation water supply, benchmarking and performance indicators in irrigation districts, climate change impacts on irrigation demand, applications of artificial intelligence in water management, and precision irrigation. He is the author of 87 papers in scientific peer-reviewed journals (h-index of 22 in JCR and 23 in Scopus). He has participated in over 50 research and consultancy projects at the national and international levels.

Rafael González Perea has a M.S. in Agricultural Engineering (2012) and Ph.D. in Irrigation Water Management by Artificial Intelligence techniques. He is a researcher in the "uan de La Cierva-Formación program in the Faculty of Agricultural and Forestry Engineering (ETSIAM) at the University of Castilla La-Mancha. His research fields include the optimization of water energy nexus in irrigation systems, application of artificial intelligence in water management, big data, and precision irrigation. He is the author of 15 papers in scientific peer0reviewed journals (h-index of 6 in Scopus).

Miguel Ángel Moreno has a Ph.D. in Agricultural Engineering from the University of Castilla-La Mancha (2005). He was a postdoctoral researcher at Utah State University and The University of Arizona (2005 and 2006). His research lines focus on optimizing inputs in agriculture, mainly water and energy, but also fertilizers, herbicides, and pesticides. He leads a line of research dealing with the water–energy nexus in irrigation with the development of models to decrease water and energy consumption. His second line of research is in the framework of very-high-resolution remote sensing with UAVs for developing decision making systems within agriculture, environment, and forestry applications by exploiting 3D features of the canopy. He has contributed 57 JCR papers and has an h-index of 18.

Editorial

Modelling and Management of Irrigation System

Juan Antonio Rodríguez Díaz [1],*, Rafael González Perea [2] and Miguel Ángel Moreno [3]

[1] Department of Agronomy, University of Córdoba, Campus Rabanales, Edif. da Vinci, 14071 Córdoba, Spain
[2] Department of Plant Production and Agricultural Technology, School of Advanced Agricultural Engineering, University of Castilla-La Mancha, Campus Universitario, s/n, 0207 Albacete, Spain; rafael.gonzalezperea@uclm.es
[3] Institute for Regional Development, University of Castilla-La Mancha, Campus Universitario, s/n, 02071 Albacete, Spain; miguelangel.moreno@uclm.es
* Correspondence: jarodriguez@uco.es

Received: 18 February 2020; Accepted: 3 March 2020; Published: 4 March 2020

Abstract: Nowadays, irrigation is becoming an activity of precision, whereby it is necessary to combine information collected from various sources to manage resources in an optimal way. New management strategies, such as big data techniques, sensors, artificial intelligence, unmanned aerial vehicles (UAV) and new technologies in general, are becoming more relevant every day. Related to this, modeling techniques, both at the water distribution network and at farm level, will be essential to gather information from various sources and offer useful recommendations for decision-making processes. In this Special Issue, ten high-quality papers were selected that cover a wide range of issues that are relevant to the different aspects related to irrigation management: water source and distribution network, plot irrigation systems and crop water management.

Keywords: water-energy nexus; decision support systems; soil-water-plant-atmosphere models; optimization

---

1. Introduction

Agricultural irrigation is the main water user, accounting for more than 70% of all water withdrawals worldwide [1] and plays an essential role of feeding the world population since, representing 20% of the total cropped area, it produces 40% of the global food production (rainfed agriculture produces 60% with 80%). However, irrigated agriculture will face important challenges in the coming decades and must feed, in a climate change context, a growing population with less soil and water resources. For this reason, it will be increasingly important to use water as efficiently as possible.

Seeking to increase water use efficiency, in recent decades, many water conveyance networks and irrigation systems have evolved to pressurized ones, making energy another key resource for the irrigation sector, which represents a growing percentage of the total water costs and increases the carbon footprint of irrigation activities [2].

Nowadays, agriculture in general is in a technological revolution involving the use of sensors, unmanned aerial vehicles (UAV), satellite images, renewable energy sources and new technologies in general. This new technology offers a huge amount of information that must be processed, analyzed and made available to users in a friendly way. In all this, concepts such as big data, ICTs and, of course, decision support systems that facilitate irrigation management and efficient use of resources will play an increasingly important role.

Given this situation, it is essential to use modeling techniques that, collecting information from the wide variety of sources currently available, allow to analyze the different systems in depth and facilitate decision-making processes. Better control of the irrigation process, as well as better management of pressurized irrigation networks, are essential to convert irrigation to a precision activity, where the efficiency of the use of the involved resources is maximized. These facts highlight the need to improve efficiency in the water–energy nexus, which is essential for economic, social and environmental development of the sector [3].

However, irrigation systems modeling is somewhat complex given that many factors are involved, such as the water distribution network, the hydraulic configuration of the on-farm irrigation system and the soil-water-plant-atmosphere system. For this reason, it is necessary to model the different levels of the whole irrigation system since all of them are relevant for the efficient use of resources.

The selected papers included in this special number cover a wide range of issues that are relevant to the three abovementioned levels: the water distribution network, on-farm irrigation system engineering, and irrigation management at field level.

2. Summary of the Papers

2.1. Water Distribution Network

In this regard, four papers are presented that address topics such as the optimal allocation of resources, the optimal management of water distribution networks and the efficient use of energy in pressurized networks.

Ehteram et al. (2018) [4] presents a heuristic method based on an improved weed algorithm for reservoir operation aimed at reducing the irrigation deficits. The result is a useful tool for solving complex problems in water management.

The management of open channels water distribution networks is especially complex. It is necessary to analyze, on the one hand, the offtake and flowing discharge in the canal, and on the other hand, the water users and the peculiarities of the water demand. This topic is addressed by Luppi et al. (2018) [5] who combined agronomic and hydraulic aspects thanks to the tools IRRINET management Decisional Support System (DSS) and the SIC2 (Simulation and Integration of Control for Canals) hydraulic software. The methodology is applied in the open-canal Canale Emiliano Romagnolo (CER), which is one of the major irrigation infrastructures in Northern Italy.

The optimization of the water-energy is essential in pressurized irrigation systems. Related to this, two papers are included. Córcoles et al. (2018) [6] presents a decision Support System tool useful to reduce the energy consumption of water abstraction from wells. It is based on the installation of variable speed drives and results shows potential energy savings of up to 23% with payback periods ranging from 4.5 to 10 years.

Renewable energy sources will play an important role in the water supply. There are several options, such as photovoltaic energy or windmills, but another option is energy recovery from excesses of pressure. It is useful to recover part of the energy used in the pumping station but also to have energy in the farms when there is no other source available. This is especially important for the supply of programmers, fertigation equipment or filters. This aspect is addressed by Crespo Chacón et al. (2019) who analyses the potential of microhydropower and pumps working as turbines (PATs) in pressurized water distribution networks [7]. They developed a methodology for optimum selection of PATs that maximizes the energy production and minimizes the payback period. The methodology was tested in Sector VII of the right bank of the Bembézar River (BMD), in Southern Spain. Five potential sites for PAT installation were found with an energy recovery potential of 93.9 MWh and an annual energy index of 0.10 MWh year^{-1} ha^{-1}.

2.2. On-Farm Irrigation System Engineering

Two papers are presented that address on-farm irrigation system modelling. Robles Rovelo et al. (2019) [8] characterized and simulated the behavior of a Low-Pressure Rotator Spray Plate Sprinkler frequently used in center pivots. The experiments were performed under two pressures (69 kPa and 103 kPa) and in calm and windy conditions. The energy losses due to the impact of the out-going jet with the sprinkler plate were measured using an optical technique. They satisfactorily analyzed water distribution and estimated energy losses over 16,500 droplets.

Buono da Silva Baptista el al. (2019) [9] analyze also center pivots but from an energy efficiency point of view. They analyze the effect of variable speed drives to adjust the pumping pressure to that which is strictly needed. They developed the VSPM model (Variable Speed Pivot Model) to perform hydraulic and energy analyses of center pivot systems in an integrated manner. The application in a real case study of maize cropped in Albacete (Spain) offered reductions in energy consumption of 12% with annual economic savings of €2821.47.

2.3. Irrigation Management at Field Level

Finally, in the last section, agronomy and optimal irrigation scheduling are considered. In this sense, there are four papers that can be divided into two groups, on the one hand, the estimation of the real crop irrigation needs and on the other, in soil-water-plant-atmosphere models useful to facilitate irrigation scheduling. There are two papers in each subcategory.

Li and Ma (2019) [10] validated the dual crop coefficient approach experimentally for drip irrigated summer maize in Aksu, Xinjiang (China). Evapotranspiration and transpiration are validated with a root mean square error (RSME) of 10 mm and 20 mm respectively. They also compare the results with those obtained for partially mulched maize in terms of evapotranspiration and water consumption.

In the second paper about crop water requirements, Benson and Chen (2018) [11] quantify the daily water requirements of container-grown Calathea and Stromanthe produced in shaded greenhouse. To address water requirements of greenhouse-grown plants, this study adapts a canopy closure model and investigated actual evapotranspiration (ETA) of Calathea G. Mey. 'Silhouette' and Stromanthe sanguinea Sond. from transplanting to marketable sizes in a shaded greenhouse.

Regarding the soil-water-plant-atmosphere models, firstly we have a calibration and application of the well known Aquacrop model but for sugarbeet production in central Spain [12]. Using data from a single deficit irrigation experiment and from eight different commercial farms, the model is calibrated and validated for simulating canopy cover, biomass and final yield with accurate results (RMSE = 11.39%, 2.10 t ha^{-1}, and 0.85 t ha^{-1}, respectively). Finally, they use the validated model to simulate different irrigation water allocations in the two main production areas of sugar beet in Spain.

In the last paper, Alcaide Zaragoza et al. (2019) developed the REUTIVAR model for precision fertigation of olive orchards using reclaimed water [13]. This crop is particularly relevant in Southern Spain, it usually receives deficit irrigation but fertilization is commonly imprecise, which causes over-fertilization, especially nitrogen. This problem is aggravated when using reclaimed water, which carries a significant amount of nutrients. A new model for optimum precision fertigation scheduling is developed which combines weather information, both historical and forecast data, soil characteristics, hydraulic characteristics of the system, water allocation, tree nutrient status, and irrigation water quality.

3. Conclusions

The optimum management of irrigation systems is very complex given that many factors are involved, structured in clearly differentiated levels (water distribution, on-farm irrigation system and agronomy). To adequately address the optimal management of each of these levels, it is necessary to use a large amount of information, which is usually found in unfriendly formats that are not too useful to extract useful recommendations for the decision-making processes. In this context, the objective of this Special Issue was to show the latest advances in the modeling of irrigation systems at each of those levels. Thus, ten high-quality papers related to the modeling of irrigation systems were included corresponding to the three levels abovementioned. The results show the possibilities that the models offer for the optimal management of irrigation systems, focusing each of the works on a particular aspect of them.

These works establish a framework of options to continue advancing in the different aspects started here. Future works should be directed towards the integration of the different models included in each of these levels. Thus, new aspects such as advanced artificial intelligence, incorporation of new communication networks, integration of the models in friendly applications particularly in mobile devices, integration UAV and satellite images, renewable energy sources for water supply and the optimization of the water energy nexus in general should be integrated in the news tools that will be developed in the coming years.

Author Contributions: Conceptualization, J.A.R.D., R.G.P. and M.Á.M.; writing, J.A.R.D., R.G.P. and M.Á.M. All authors have read and agreed to the published version of the manuscript.

Funding: This work was supported by the Spanish Ministry of Economy and Competitiveness (AGL2017-82927-C3-1-R).

Acknowledgments: We acknowledge all the authors that has contributed to this special issue with their research works.

Conflicts of Interest: The authors declare no conflict of interest.

References

1. FAO. *The Future of Food and Agriculture: Trends and Challenges*; FAO: Rome, Italy, 2017; ISBN 9789251095515.
2. Daccache, A.; Ciurana, J.S.; Rodriguez Diaz, J.A.; Knox, J.W. Water and energy footprint of irrigated agriculture in the Mediterranean region. *Environ. Res. Lett.* **2014**, *9*, 124014. [CrossRef]
3. Tarjuelo, J.M.; Rodriguez-Diaz, J.A.; Abadía, R.; Camacho, E.; Rocamora, C.; Moreno, M.A. Efficient water and energy use in irrigation modernization: Lessons from spanish case studies. *Agric. Water Manag.* **2015**, *162*, 67–77. [CrossRef]
4. Ehteram, M.; Singh, V.P.; Karami, H.; Hosseini, K.; Dianatikhah, M.; Hossain, M.S.; Ming Fai, C.; El-Shafie, A. Irrigation Management Based on Reservoir Operation with an Improved Weed Algorithm. *Water* **2018**, *10*, 1267. [CrossRef]
5. Luppi, M.; Malaterre, P.O.; Battilani, A.; Di Federico, V.; Toscano, A. A Multi-disciplinary Modelling Approach for Discharge Reconstruction in Irrigation Canals: The Canale Emiliano Romagnolo (Northern Italy) Case Study. *Water* **2018**, *10*, 1017. [CrossRef]
6. Córcoles, J.I.; Gonzalez Perea, R.; Izquiel, A.; Moreno, M.Á. Decision Support System Tool to Reduce the Energy Consumption of Water Abstraction from Aquifers for Irrigation. *Water* **2019**, *11*, 323. [CrossRef]
7. Crespo Chacón, M.; Rodríguez Díaz, J.A.; García Morillo, J.; McNabola, A. Pump-as-Turbine Selection Methodology for Energy Recovery in Irrigation Networks: Minimising the Payback Period. *Water* **2019**, *11*, 149. [CrossRef]
8. Robles Rovelo, C.O.; Zapata Ruiz, N.; Burguete Tolosa, J.; Félix Félix, J.R.; Latorre, B. Characterization and Simulation of a Low-Pressure Rotator Spray Plate Sprinkler Used in Center Pivot Irrigation Systems. *Water* **2019**, *11*, 1684. [CrossRef]
9. Buono da Silva Baptista, V.; Córcoles, J.I.; Colombo, A.; Moreno, M.Á. Feasibility of the Use of Variable Speed Drives in Center Pivot Systems Installed in Plots with Variable Topography. *Water* **2019**, *11*, 2192. [CrossRef]
10. Li, F.; Ma, Y. Evaluation of the Dual Crop Coefficient Approach in Estimating Evapotranspiration of Drip-Irrigated Summer Maize in Xinjiang, China. *Water* **2019**, *11*, 1053. [CrossRef]
11. Beeson, R.C., Jr.; Chen, J. Quantification of Daily Water Requirements of Container-Grown *Calathea* and *Stromanthe* Produced in a Shaded Greenhouse. *Water* **2018**, *10*, 1194. [CrossRef]
12. Garcia-Vila, M.; Morillo-Velarde, R.; Fereres, E. Modeling Sugar Beet Responses to Irrigation with AquaCrop for Optimizing Water Allocation. *Water* **2019**, *11*, 1918. [CrossRef]
13. Alcaide Zaragoza, C.; Fernández García, I.; González Perea, R.; Camacho Poyato, E.; Rodríguez Díaz, J.A. REUTIVAR: Model for Precision Fertigation Scheduling for Olive Orchards Using Reclaimed Water. *Water* **2019**, *11*, 2632. [CrossRef]

Article

Irrigation Management Based on Reservoir Operation with an Improved Weed Algorithm

Mohammad Ehteram [1,*], Vijay P. Singh [2], Hojat Karami [1], Khosrow Hosseini [1], Mojgan Dianatikhah [1], Md. Shabbir Hossain [3], Chow Ming Fai [3] and Ahmed El-Shafie [4]

[1] Department of Water Engineering and Hydraulic Structures, Faculty of Civil Engineering, Semnan University, Semnan 3513119111, Iran; hkarami@semnan.ac.ir (H.K.); khhoseini@semnan.ac.ir (K.H.); Mojgan_dianatikhah@semnan.ac.ir (M.D.)

[2] Department of Biological and Agricultural Engineering, Zachry Department of Civil Engineering, Texas A&M University, 321 Scoates Hall, 2117 TAMU, College Station, TX 77843-2117, USA; vsingh@tamu.edu

[3] Department of Civil Engineering, University Tenaga National, Kajang 43000, Malaysia; mdshabbir@uniten.edu.my (M.S.H.); chowmf@uniten.edu.my (C.M.F.)

[4] Department of Civil Engineering, Faculty of Engineering, University Malaya, Kuala Lumpur 50603, Malaysia; elshafie@um.edu.my

* Correspondence: mohammdehteram@semnan.ac.ir; Tel.: +98-091-6228-1744

Received: 30 August 2018; Accepted: 13 September 2018; Published: 17 September 2018

Abstract: Water scarcity is a serious problem throughout the world. One critical part of this problem is supplying sufficient water to meet irrigation demands for agricultural production. The present study introduced an improved weed algorithm for reservoir operation with the aim of decreasing irrigation deficits. The Aswan High Dam, one of the most important dams in Egypt, was selected for this study to supply irrigation demands. The improved weed algorithm (IWA) had developed local search ability so that the exploration ability for the IWA increased and it could escape from local optima. Three inflows (low, medium and high) to the reservoir were considered for the downstream demands. For example, the average solution for the IWA at high inflow was 0.985 while it was 1.037, 1.040, 1.115 and 1.121 for the weed algorithm (WA), bat algorithm (BA), improved particle swarm optimization algorithm (IPSOA) and genetic algorithm (GA). This meant that the IWA decreased the objective function for high inflow by 5.01%, 5.20%, 11.65% and 12% compared to the WA, BA, IPSOA and GA, respectively. The computational time for the IWA at high inflow was 22 s, which was 12%, 18%, 24% and 29% lower than the WA, BA, IPSOA and GA, respectively. Results indicated that the IWA could meet the demands at all three inflows. The reliability index for the IWA for the three inflows was greater than the WA, BA, IPSOA and GA, meaning that the released water based on IWA could well supply the downstream demands. Thus, the improved weed algorithm is suggested for solving complex problems in water resources management.

Keywords: water resources management; Aswan High Dam; weed algorithm; irrigation demands

---

1. Introduction

Water resources management is an important area of hydrological sciences that deals with the sustainable allocation and utilization of water as a valuable resource [1]. Water scarcity has been identified as a major problem for water resource managers in different regions throughout the world. Technologies and innovative ideas need to be explored and embraced to resolve this practical challenge through strategic decisions made about water utilization [2,3]. Optimizing reservoir operation is one of the most important approaches to solving water scarcity. In this regard, a suitable optimization algorithm that allows core decision-markers to generate optimal operating rules for water utilization and to minimize perceived gaps between water release and demand is practical for water resources management [4]. A primary aim for controlling a reservoir's operation is to develop a set of rules that govern the controlled release of water for downstream users that lead to a reduction in water shortages [5]. It is also imperative to ensure that the amount of water stored in the reservoir is not less than the permissible range used in the dry season and during critical conditions [6]. Consequently, new measures implemented for the controlled release of water and to provide a sustainable water supply to the different water users (e.g., those engaged in industrial activities and in agricultural practices), must be identified through an appropriately selected optimization algorithm to resolve the issues faced in the operation of a reservoir [7,8].

Over the last two decades, evolutionary and meta-heuristic optimization methods have been developed and explored to solve complex engineering-based optimization problems. A major advantage of these approaches lies is their simplicity for adjusting the objective function and the system constraints within the method's mathematical process. On the other hand, attaining the global optima that allow a universal solution can be a time-consuming process and thus, it is undesirable for real-time operation purposes. It has been shown that evolutionary and meta-heuristic methods involve less computational time in solving a complex problem and therefore may be useful for core decision-makers who need to attain an optimal solution for real-time problem-solving [4].

1.1. Background

Fallah Mehdipour et al. [9] applied genetic programming for reservoir operation to reduce irrigation deficits. The released water for downstream was considered as decision variables and different arithmetic and mathematical operators were tested for the genetic programming. The results indicated that the irrigation demands were met with the highest reliability index compared to the genetic algorithm (GA) and particle swarm optimization (PSO). The decision variable for the study was the released water as an unknown value.

Ostadrahimi et al. [10] applied PSO for multi-reservoir operation to generate different rule curves to reduce irrigation deficits for a case study in Iran. The applied PSO acted based on modified inertia weight as new version of PSO algorithm to increase the convergence rate. The results were compared with the GA and nonlinear programming. The PSO had greater convergence velocity than the GA and based on PSO, the released water could meet irrigation demands.

Afsahr [5] optimized reservoir operation with a modified and adapted version of PSO with the aim of increasing the benefits of power generation. This version of PSO adapted with the hydrological and hydraulic boundary condition of problem so that the algorithm could model the reservoir operation during the drought years. The new version of PSO had average solutions that were close to the global solution of the problem and annual generated power was greater than that calculated by the GA and harmony search algorithm (HA).

Ant colony optimization (ACO) was used for the optimal operation of multi-reservoir systems and results indicated that the applied algorithm could meet irrigation demands with the least risk so that the average irrigation deficit was reduced to a probable minimum value [11]. Also, Moeini and Afshar [11] developed the ACO based on mutation operator to increase the diversity for the initial population of algorithm.

Zhang et al. [12] applied a new dynamic programming method and GA for the operation of a multi-reservoir system with the aim of increasing of power generation in the China. The GA had a better performance than dynamic programming although the power generation for the dry years accorded with the limitation so that a rationing process could be good strategy for dry years.

The water cycle algorithm (WCA) was used for reducing power deficits based on the movement of drops in the environment and with the positions of drops considered as the initial population [13]. The position of drops in the environment was considered as a decision variable and the algorithm acted based on movement of drops to the rivers. The computational time of the WCA was decreased compared to the GA and PSO and the averages of deficits were decreased by the WCA to the probable minimum value.

Bozorg-Hadad et al. [14] applied the bat algorithm (BA) with the aim of decreasing of power deficits for a case study in Iran. The algorithm is based on the received frequencies of bats from their surroundings. The position of bats were considered as decision variables and they were updated based on velocity and frequency value. The convergence velocity for the BA based on a random walk and local search process was more than the GA and PSO and the BA power generation process had a higher reliability index.

Different orders of nonlinear rule curves based on the GA were used for the operation of a reservoir [15]. Results indicated that the third order rule curve could meet irrigation demand based on the high value obtained for the reliability index so that the released water met demands well and irrigation deficits were minimized.

Biogeography based optimization (BBO) was used for decreasing power deficits for a case study in Iran [16]. The algorithm was based on environmental habitats and migration processes in the environment. Results indicated that the algorithm could generate solutions which were close to the global solution based on less iteration. The process of the algorithm was simple and based on an initial population of species in the environment.

Asghari et al. [17] applied the weed optimization algorithm (WA), based on weeds growing in the environment, for multi-reservoir operation with the aim of increasing power generation. The results were compared with the GA, PSO and WCA and indicated that the WA had the greatest convergence rate and highest values of the generated benefits.

Fixed length genetic programming (FLGP) was used for the operation of a multi-reservoir system with the aim of decreasing irrigation demands [18]. Results indicated that the demand for irrigation could be supplied based on FLGP, which had a higher reliability index than GP and a greater convergence velocity than other algorithms.

The gravity search (GS) algorithm was used for a multi-reservoir system with the aim of decreasing power generation [19]. The results indicated that the GS needs to the accurate sensitivity analysis for the random parameters in the algorithm. Annual power generation based on GS had a higher value and greater convergence velocity than the GA and PSO.

The shark algorithm (SA) based on shark life and rotational movement was applied to reservoir operation to reduce irrigation deficits [20]. The rotational movement was considered as having powerful ability to avoid trapping in the local optimums. Results indicated that the SA could supply the irrigation demands based on a higher reliability index compared to the PSO and GA. The SA also had a greater convergence speed than the GA and PSO.

Ehteram et al. [21] applied monarch butterfly optimization algorithm for the multi reservoir systems in the China. The algorithm acted based on migration operator for the butterflies. The results indicated that the annual average for power production based on monarch algorithm was 12% and 14% more than GA and PSO.

Mousavi et al. [22] applied the crow algorithm for the extraction of rule curves for the irrigation management for a case study. The algorithm acted based on ability of crows for the finding and hiding of food. The results indicated that the vulnerability index based on crow algorithm was decreased significantly by the crow algorithm.

Karami et al. [23] applied the krill algorithm for reservoir operation and irrigation management. The initial position of krill was considered as decision variable and the results indicated the average annual for irrigation deficits based on krill algorithm was 20% and 23% less than PSO and genetic algorithm.

The kidney algorithm (KA), based on the way a kidney performs in the body, was used for the operation of a reservoir with the aim of decreasing power generation [24]. The KA had a simpler optimization process than the GA, PSO and HA and the vulnerability index for power generation was decreased 10%, 12% and 14% by KA compared to the PSO, HA and GA, respectively.

Ehteram et al. [25] applied the spider monkey algorithm based on monkey life. The local and global leaders were considered as important future of these algorithms and the method was used for irrigation management for multi reservoir in Iran. The results indicated that spider monkey algorithm could the downstream demands based on higher reliability index compared to the genetic algorithm and particle swarm algorithm.

Karami et al. [26] applied the simple weed algorithm for irrigation management for a case study in Iran. The results indicated that the annual average of deficits based on WA was more than PSO and the algorithm did not have the good convergence velocity.

1.2. Innovation and Objectives

One of the known algorithms in the field of optimization is the weed algorithm (WA). This algorithm acts based on weed life and the processes of generating weeds and seeds in the environment. One of the main advantages of the WA is related to its simple architecture in the optimization process and the simple method of setting the random parameters. The WA has been used in different fields such as training controller robots [27], designing recommender systems [28], optimization of mathematical benchmarks [29], optimization of power plant performance [23] and optimization of reservoir performance [17]. Asghari et al. [17] applied the weed algorithm for reservoir operation based on increasing of power generation for downstream power plant. The results indicated that the weed algorithm needed to high computational time compared to the PSO. Karami et al. [23] applied the weed algorithm for the decreasing of irrigation deficits and the results indicated the annual average deficits for the WA was 20% PSO and also, the algorithm could not obtain the global solution for the benchmark with some local optimums.

One of the main weaknesses of the WA is a tendency to get trapped in local optima. Another challenge is related to the exploration ability, which allows the algorithm to search a large volume in the problem space and find the best solution (close to the global solution) in the least computational time [17]. Thus, it is necessary that the WA be improved based on new operators in the local search section of algorithm. The new algorithm is defined in this study based on an elite weed. One level and an operator are added to the algorithm to search the space around the elite weed within a specific radius [23]. This ability causes new weeds to be generated during the WA optimization so that the algorithm can exit from the local optima and the diversity of population is added with the improved WA (IWA). The increased population diversity and the improved ability of the WA search, results in better potential for finding the best solution based on a more accurate search around of global solution. Thus, the paper address these issues: (1) prepare improved weed algorithm based on new operators, (2) formulate the mathematical weed algorithm for the reservoir operation, (3) introduce comprehensive study based on comparing of new weed algorithm with other evolutionary algorithms, (4) examine the new weed algorithm based on different inflows to the reservoir with the aim of irrigation deficits. In fact, the IWA has more ability to avoid trapping in local optimum and experienced less computational time compared to the classical WA. These two advantages provide the IWA the ability to search for the global optima and better suitability over the other algorithms to be proposed for real-time application. In addition, the IWA has experienced good balance between exploration and exploitation ability.

The Aswan High Dam in Egypt, an important dam for increasing water storage for irrigation demand in areas downstream of the dam, is used as a case study. This dam plays an important role in Egyptian economics. Reports showed that the monetary benefit generated after dam construction was 255 million USD of which 180 USD was related to agricultural production and indicates the importance of the Aswan High Dam for meeting irrigation demands. In our study, the IWA is used for optimal operation of this dam with the aim of reducing irrigation deficits and results are compared with the weed algorithm (WA), bat algorithm (BA), improved particle swarm optimization algorithm (IPSOA) and the genetic algorithm (GA). The reliability index, vulnerability index and resiliency index are used to evaluate the new weed algorithm relative to the other evolutionary algorithms for its performance in meeting irrigation demands. The output of this paper will show the amount of water storage required at different inflows to meet water demands of the IWA compared to the other algorithms for this important problem in water resources management. The main innovation of this paper is related to the improvement of the WA and its application in water resources management.

2. Method

2.1. Weed Algorithm (WA)

The WA acts based on weed behavior in the environment and each weed is considered one solution. The initial weed population for the algorithm is NP_0. The weeds produce seeds with a normal distribution and these seeds can grow and become offspring weeds; the offspring weeds and the parent weeds generate the next population for the WA. The weed with the best value for the objective function is selected and the iteration cycle continues until the convergence criteria are satisfied. The following assumptions are considered for the algorithm [17]:

1. A limited number of seeds can grow in the environment.
2. The seeds can grow and become weeds to continue the next generation.
3. The growth process and generation of seeds and weeds continue until the number of seeds reaches a maximum number. The algorithm is considered based on the following levels:

2.1.1. Initialization

The size of the initial population equals NP_0. Each weed is considered as a solution candidate so that the accurate size of the population and the value of NP_0 are determined based on sensitivity analysis.

2.1.2. Reproduction

Reproduction is a process where the weeds generate seeds and the maximum and minimum number of seeds, N_0S_{max} and NS_{min}, are generated based on the quality of parent weeds. Weeds that live in one location are known as a colony [17]. Reproduction is important because although plants of lower quality have a smaller chance of continuing of their presence in the next generation, the process allows even the low-quality plants (weeds) to generate seeds because they may contain important information. Figure 1 shows the production level of the WA.

The number of generated seeds is computed based on the equation

$$Nseed_j = NS_{min} + \frac{Fit_j - Fit_{min}}{Fit_{max} - Fit_{min}} \times (NS_{max} - NS_{min}) \tag{1}$$

where, Fit_j is the jth objective function, Fit_{min} is the minimum value of the objective function, Fit_{max} is the maximum value of the objective function and $Nseed_j$ is the number of seeds in each level.

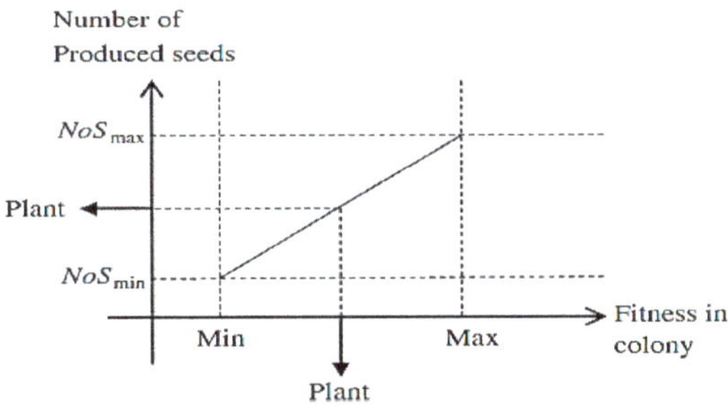

Figure 1. Production levels in the weed algorithm (WA).

2.1.3. Spatial Dispersal

The seeds generated by weeds are distributed randomly. The generated offspring weeds are distributed according to a mean 0 and a standard deviation. The standard deviation is considered a large positive value and it decreases as the number of iterations increases. This results in the combination of the global search and local search to get the optimal search space. The standard deviation is computed based on the equation

$$\sigma_{iter} = \frac{(It_{max} - Iter)^t}{It_{max}} \times \left(\sigma_{initial} - \sigma_{final}\right)^2 + \sigma_{final} \tag{2}$$

where σ_{iter} is the standard deviation for each level, It_{max} is the maximum number of iterations, $Iter$ is the number of iterations in each level, $\sigma_{initial}$ is the initial standard deviation, σ_{final} is the final standard deviation and t is the nonlinear modulus

2.1.4. Competitive Selection

Weeds do not go extinct as long as they can generate offspring weeds. There is a limit to the number of weeds that could live in the environment. When the weeds generate offspring weeds, they constitute the next generation for the population. If their number is not more than NP_0, there is not a specific problem; otherwise, the NP_0 weed with the better objective function based on a competition is selected for continuing the algorithm. The weed with the best value for the objective function is then selected in the new population; if it is better than the best weed computed so far, the new weed will replace the previous best weed.

2.2. Improved Weed Algorithm (IWA)

Local search ability is added to the WA to increase the searches of the problem space and give the algorithm more exploration ability to obtain optimal solutions. First, the weed with the best value for the objective function is selected as the elite weed. N_{es} seeds are then produced by the elite weed. The new seeds grow around the elite weed and generate new weeds, which grow within a radius S_r from the elite weed. Each new weed, as one of the decision variables, can be defined based on

$$x_{(new)s,i} = x_{s,i}.(rand() \times 2 \times S_r + (1 - S_r)) \tag{3}$$

where $x_{(new)s,i}$ is the new value of the weed for the ith variable, $x_{s,i}$ is the value of ith weed and S_r is the radius.

Then, the computed value for the weeds is checked with the upper and lower values of the decision variables

$$x_{s,i} = \left[\begin{array}{l} x_{min} \leftarrow if(x_{s,i}) < x_{min} \\ x_{max} \leftarrow if(x_{s,i}) > x_{max} \end{array} \right] \tag{4}$$

where x_{min} is the lower limit for the decision variable and x_{max} is the upper limit for the decision variable.

Figure 2 shows the IWA process.

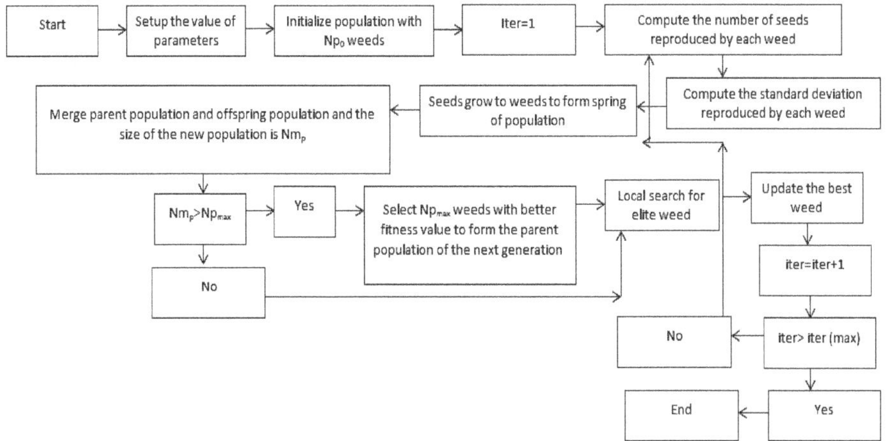

Figure 2. Improved weed algorithm (IWA) process.

2.3. Bat Algorithm (BA)

Bats have a powerful ability to receive sounds from their surroundings. They generate sounds, which are echoed back and allow the bats to distinguish obstacles from food based on the received frequencies. The BA acts based on the following assumptions [21]:

1. The echolocation ability is used by all bats so that they can identify an obstacle in their surroundings.
2. Bats have random velocity (v_l) at a random position (y_l) and the frequency, wavelength and loudness for the BA are f_l, λ and A, respectively.
3. The loudness varies for the bats from a large positive to a small positive.

The sounds generated by the bats have a pulsation rate (r_l), which is between 0 (minimum) and 1 (maximum). The position, pulsation rate and frequency are updated based on the equations

$$f_l = f_{\min} + (f_{\max} - f_{\min}) \times \beta \tag{5}$$

$$v_l(t) = [y_l(t) - Y_*] \times f_l \tag{6}$$

$$y_l(t) = y_l(t-1) + v_l(t) \times t \tag{7}$$

where f_{min} is the minimum, f_{max} is the maximum frequency, Y_* is the best solution, $v_l(t)$ is the velocity, t is time interval, $y_l(t-1)$ is position at time $(t-1)$, β is the random vector and f_l is the frequency.

The random walk is used as a local search strategy

$$y(t) = y(t-1) + \varepsilon A(t) \tag{8}$$

where ε is a random value between -1 and 1 and $A(t)$ is loudness.

The loudness and pulsation rate are updated for each level. When the bats find prey, the pulsation rate increases and the loudness decreases and also, there is this condition vice versa. The pulsation rate is updated based on

$$r_l^{t+1} = r_l^0[1 - \exp(-\gamma t)] A_l^{t+1} = \alpha A_l^t \tag{9}$$

where α and γ are constant parameters. The BA process is shown in Figure 3.

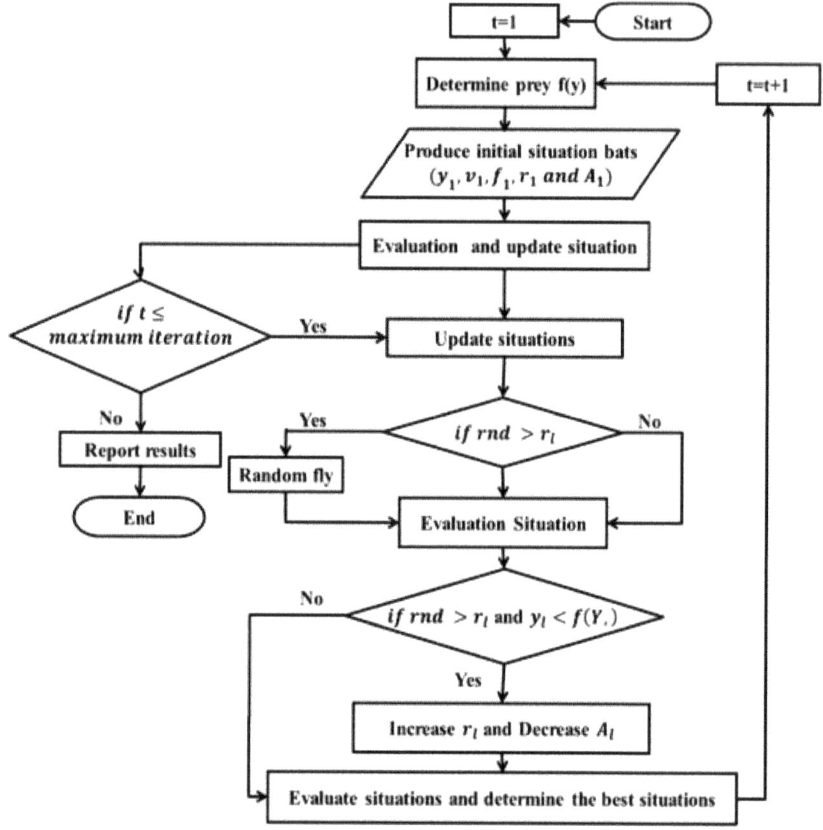

Figure 3. Bat algorithm (BA) process.

2.4. Improved Particle Swarm Optimization Algorithm (IPSOA)

The optimization process in this algorithm starts with the collection of particles. Each particle is considered as a candidate solution of the optimization problem and contains three vectors: the current position of the particle (x_i), the best-obtained position of each particle in the previous iteration (y_i) and the velocity vector. A central aim of each cycle in the algorithm is the identification of the best position of each particle [18]. Following this, the best position of the particle (X_i^{iter+1}) is considered as the new position of the new particle for the purpose of continuance in such a way that it yields two main equations [23]

$$V_i^{iter+1} = wV_i^{iter} + c_1 rnd\left(Y_i^{iter} - X_i^{iter}\right) + c_2 rnd\left(Y_*^{iter} - X_i^{iter}\right) \tag{10}$$

$$X_i^{iter+1} = X_i^{iter} + V_i^{iter+1} \tag{11}$$

where V_i^{iter+1} is the new velocity vector for each particle, c_1 is the personal learning coefficient, c_2 is the global learning coefficient, rnd is a random value between 0 and 1, Y_*^{iter} is the current best solution and w is the inertial coefficient.

Past experimental results have shown that it is better to consider high values for the inertia coefficient at the start of the process because the algorithm is able to search the solution space with a higher accuracy and its value can be reduced linearly. In this study, an improved version of the PSOA (denoted as IPSOA) was used for solving the optimization problem where the following equation shows the linear reduction in the inertial coefficient

$$w^{iter+1} = w^{iter} \times w_{damp} \qquad (12)$$

where w_{damp} is a reduction (damping) coefficient bounded by (0.9–1).

In order to optimize the given problem, the particles in the algorithm spread in the search space and they are improved by the best solutions, which are found in each iteration of the search space. The second term in Equation (10) is the internal knowledge of each particle, which compares its current position to the previous best position. The third term in this equation shows the social interaction among the particles, which is used to measure the difference between the current and the best position of each particle.

2.5. Genetic Algorithm (GA)

Decision variables in the GA are defined based on a random population and its chromosomes. A particular code is considered for each variable and the fitness function is then computed based on the initial value of each chromosome assigned to the problem of interest [23]. Each chromosome has an objective function and the operator selection chooses the best chromosome. These superior chromosomes are considered for crossover operators to generate new chromosomes based on the parents' characteristics. Following this, the mutation operator is used for the change of one or more genes for the generation of better chromosomes. Figure 4 shows the schematic diagram of the GA.

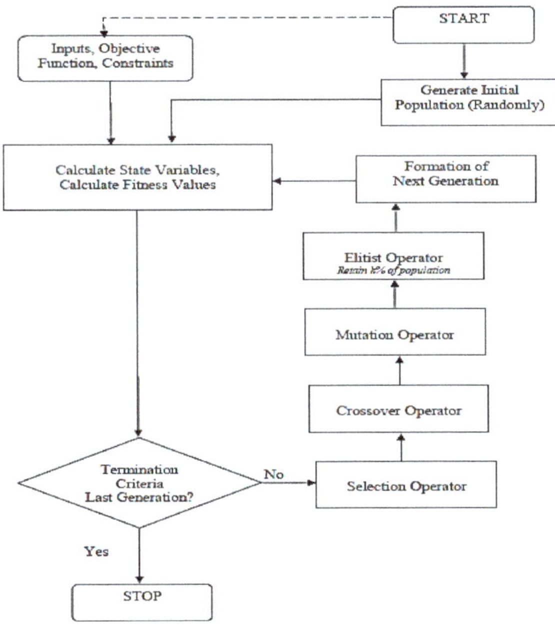

Figure 4. Genetic algorithm (GA) process.

3. Case Study

The case study site, Aswan High Dam (AHD), located on the Nile River, is the highest dam in Egypt (Figure 5).

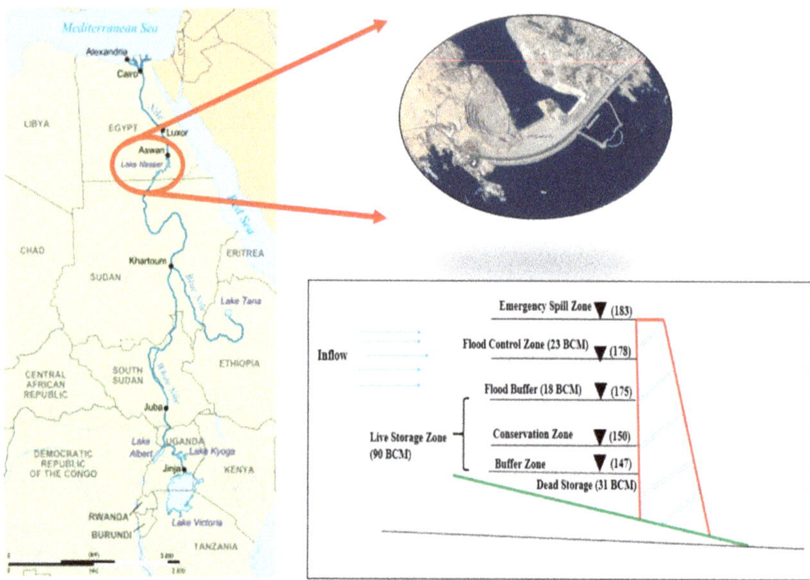

Figure 5. Location and characteristics of Aswan High Dam.

Minimizing the water deficit downstream from the dam is considered as an objective function in terms of the optimization problem that needs to be solved for core decision-making and for optimal water resources management. The AHD is one of the largest dams in the world and has a relatively large reservoir, namely Lake Nasser that lies between latitudes 22°00″ and 23°58″ N and longitudes of 31°07″ and 33°15″ E. Lake Nasser has a maximum width of 60 km and a mean width of 10 km. Data of monthly inflow to the reservoir is extracted from 1989 to 1999 because there is a real water scarce for this period. Inflow is classified based on the United State Geological Survey (USGS). A value on a scale of 100, known as a percentile, indicates inflow distribution. For example, when river discharge on a specific day is in the 70th percentile, it means that this discharge is equal to or greater than 70% of discharge recorded on this day of the year during all years that measurements have been made. The USGS suggests that a percentile of more than 75 is high inflow and a percentile below 25 is low inflow. Normal inflow means the percentile is between 25 to 75. Table 1 shows the classification of average monthly inflows into the reservoir during 1989 to 1999.

Table 1. Classification of monthly inflow to the reservoir (BCM) (1989–1999).

Month	High	Medium	Low	Demand
	Billion Cubic Meters (BCM)			
January	4.8	3.15	1.90	3.5
February	3.7	1.95	0.80	3.8
March	3.5	1.7	0.55	4.4
April	2.7	1.15	0.30	4.9
May	2.5	1.35	0.65	5.1
June	2.8	1.65	0.90	5.2
July	7.7	4.75	2.80	5.8
August	27.5	20.4	15.50	5.1
September	31	24.05	18.55	4.5
October	21.2	15.6	11.3	3.9
November	10.9	7.30	4.75	3.2
December	6.5	4.30	2.7	2.9

Several parameters of the AHD reservoir were limited to satisfy the optimization model. Eleven values were considered for the initial reservoir storage and the model was then executed for each storage value and three inflow scenarios (i.e., high, medium and low). This yielded a total of 36 rule curves for a one-year period. The required information was extracted from the study of El-Shafie et al. [30] where the seepage value for the reservoir was set at 0.08 billion cubic meters (BCM) and the water level was to be maintained over the range 147–183 m. The permissible range in storage was approximately 32–162 BCM and the maximum value for water release was approximately 7.5 BCM. For this dam, the water level at the end of July was expected to be less than 175 m (122 BCM). The objective function for this reservoir is

$$minimize Z = \sum_{t=1}^{1} (D_t - R_t)^2 \tag{13}$$

where Z is the objective function, D_t is the water demand (BCM) and R_t is the water release (BCM).

The constraints for this problem are explained in the following equations.

Continuity equation:

$$S_{t+1} = S_t + I_t - R_t - L_t \tag{14}$$

where S_{t+1} is the storage at the end of the period t, S_t is the storage at the beginning of the period t, I_t is the inflow to the reservoir, R_t is the release volume and L_t is the evaporation loss from the reservoir.

Storage constraint (for all months except *July*):

$$32 BCM \leq S_t \leq 162 BCM \tag{15}$$

$$S_{July} \leq 122 \tag{16}$$

Storage constraint for *July* is:

If the above-mentioned constraints are not satisfied, the penalty functions are considered to be:

$$penalty1 = \begin{bmatrix} 0 \leftarrow if(S_t) > S_{min} \\ C_1(S_{min} - S_t)^2 \leftarrow if(S_t) < S_{min} \end{bmatrix} \tag{17}$$

$$penalty2 = \begin{bmatrix} 0 \leftarrow if(S_t) < S_{max} \\ C_2(S_{max} - S_t)^2 \leftarrow if(S_t) > S_{max} \end{bmatrix} \tag{18}$$

$$penalty3 = \begin{bmatrix} 0 \leftarrow if\left(S_{July}\right) < 122 \\ C_3\left(S_{July} - 122\right)^2 \leftarrow if\left(S_{July}\right) > 122 \end{bmatrix} \tag{19}$$

where S_{max} is the maximum storage, S_{min} is the minimum storage, S_{July} is the storage in *July* and C_1 is the penalty coefficient.

Finally, the objective function can be written as

$$minimization(Z) = Z + penalty1 + penalty2 + penalty3 \tag{20}$$

Model Performance Indicators

Different indices were used to investigate the ability of the various algorithms to simulate the dam–reservoir operation.

Volumetric Reliability: This index aimed to explain the ratio of the released water volume over the total period compared to the total demand values and it can be stated as [23]

$$\alpha_V = \frac{\sum\limits_{i=1}^{N} \sum\limits_{t=1}^{T} R_{t,i}}{\sum\limits_{i=1}^{N} \sum\limits_{t=1}^{T} D_{i,t}} \times 100 \tag{21}$$

where α_V is volumetric reliability. A higher percent for this index shows the demands are well supplied based on released water.

Vulnerability: This index aimed to explain the maximum ratio of the generated failures of the system during the total period and can be stated based on [23]

$$\lambda = Max_{i=1}^{N}\left(Max_{t=1}^{T}\left(\frac{D_{i,t} - R_{i,t}}{D_{i,t}} \right) \right) \tag{22}$$

where λ is the vulnerability index. A low value for this index indicates a low intensity of failure occurrences in the system based on the difference between released water and demands.

Resiliency: This index aims to investigate the exit speed of the system from a failure based on

$$\gamma = \frac{N_{t=1}^{T}(D_{t+1} \pounds R_{t+1} | D_t > R_t)}{N_{t=1}^{T}(D_t > R_t)} \tag{23}$$

where $N_{t=1}^{T}(D_{t+1} \leq R_{t+1} | D_t > R_t)$ is the number of reservoir consecutives to meet the demands well and $N_{t=1}^{T}(D_t > R_t)$: number of periods in which the demands are not met. When the system has higher value for the resiliency index, the system the system can rapidly recover from failures. The mathematical model is considered for reservoir operation based on the following levels:

1. The released water is considered as a decision variable and is defined based on the initial population of weeds.
2. The continuity equation is computed and the reservoir storage is computed for each release decision variable.
3. The constraints for each variable are computed and the penalty functions are computed if the constraints are not satisfied.
4. The objective function is computed for each member.
5. The number of seeds generated by each weed is computed.
6. The standard deviation for spatial dispersal is calculated.
7. A local search for an elite weed is carried out.
8. The best weed is updated
9. If the convergence criterion is satisfied, the algorithm finishes; otherwise, the algorithm returns to the second step.

4. Results and Discussion

4.1. Sensitivity Analysis

The random parameters in the evolutionary algorithms have an important effect on the final results of the optimization process. An accurate determination of parameter values is therefore required, based on a sensitivity analysis. Sensitivity analysis means that the variation of parameter values is considered versus the variation of the objective function value. For example, the objective function of this problem was related to the minimization of irrigation deficits and was computed for the different intervals of the value of each parameter. When the objective function reached the minimum value versus the value of the random parameter, that value was considered the best value for that parameter. For example, the results of the sensitivity analysis for all the algorithms tested are shown in Tables 2–5. The initial size (NP_{min}) of the IWA population is 10 because the objective function at this size had the least probable value. NP_{max} the maximum size of the population, was 50 because at this size the objective function was 0.985 BCM, the lowest value among other population sizes. Following this logic, NS_{max}, the maximum number of seeds was 10 and NS_{min} was 2. For the BA (Table 3), the optimum population size at high inflow was 50 with an objective function of 1.039 BCM. The maximum frequency was 5 Hz, the minimum frequency was 2 Hz and the maximum loudness was 0.60 Db. However, the sensitivity analysis based on this section shows the variations of value of parameters for the accurate of value of parameters. The optimum population size for the IPSOA was 30 for high inflow (Table 4) with an objective function of 1.115. At the same objective function value; the best value of $c_1 = c_2$ equaled 2 and the best value of w was 0.6. For the GA (Table 5), population size at high inflow was 30 with an objective function value of 1.121. Also, the mutation probability and the crossover probability were 0.6 and 0.4, respectively, with the lowest objective function at 0.121.

Table 2. Sensitivity analysis for IWA.

			High Inflow				
NP_{min}	Objective Function (BCM)	NP_{max}	Objective Function (BCM)	NS_{min}	Objective Function (BCM)	NS_{max}	Objective Function (BCM)
5	0.994	25	0.993	1	0.998	5	0.993
10	0.985	50	0.985	2	0.985	10	0.985
15	0.989	75	0.992	3	0.989	15	0.987
20	0.998	100	0.999	4	0.991	20	0.991
			Medium Inflow				
5	1.122	25	1.110	1	1.14	5	1.098
10	1.021	50	1.021	2	1.100	10	1.021
15	1.098	75	1.076	3	1.021	15	1.078
20	1.111	100	1.085	4	1.045	20	1.079
			Low Inflow				
5	1.141	25	1.161	1	1.128	5	1.090
10	1.121	50	1.141	2	1.021	10	1.121
15	1.124	75	1.121	3	1.155	15	1.157
20	1.135	100	1.124	4	1.179	20	1.178

Table 3. Sensitivity analysis for BA.

			High Inflow				
Population Size	Objective Function (BCM)	Maximum Frequency	Objective Function (BCM)	Minimum Frequency	Objective Function (BCM)	Maximum Loudness	Objective Function (BCM)
10	1.055	3	1.078	1	1.091	0.20	1.067
30	1.045	5	1.039	2	1.086	0.40	1.055
50	1.039	7	1.042	3	1.039	0.60	1.039
70	1.042	9	1.055	4	1.045	0.80	1.067
			Medium Inflow				
10	1.124	3	1.145	1	1.147	0.20	1.135
30	1.119	5	1.112	2	1.112	0.40	1.112
50	1.112	7	1.118	3	1.116	0.60	1.118
70	1.132	9	1.121	4	1.121	0.80	1.122
			Low Inflow				
10	1.149	3	1.148	1	1.143	0.20	1.147
30	1.132	5	1.132	2	1.132	0.40	1.139
50	1.156	7	1.139	3	1.138	0.60	1.141
70	1.170	9	1.154	4	1.145	0.80	1.145

Table 4. Sensitivity Analysis for IPSOA.

High Inflow							
Population Size	Objective Function (BCM)	$c_1 = c_2$	Objective Function (BCM)	W	Objective Function (BCM)	w_{damp}	Objective Function (BCM)
10	1.118	1.6	1.122	0.4	1.131	0.60	1.123
30	1.115	1.8	1.115	0.60	1.125	0.70	1.117
50	1.121	2.0	1.117	0.80	1.115	0.80	1.115
70	1.130	2.2	1.120	1.0	1.119	0.90	1.121
Medium Inflow							
10	1.145	1.6	1.142	0.4	1.134	0.60	1.141
30	1.133	1.8	1.128	0.60	1.128	0.70	1.128
50	1.128	2.0	1.134	0.80	1.136	0.80	1.135
70	1.135	2.2	1.140	1.0	1.140	0.90	1.140
Low Inflow							
10	1.167	1.6	1.169	0.4	1.165	0.60	1.169
30	1.155	1.8	1.155	0.60	1.155	0.70	1.155
50	1.159	2.0	1.161	0.80	1.159	0.80	1.157
70	1.163	2.2	1.168	1.0	1.163	0.90	1.161

Table 5. Sensitivity analysis for GA.

High Inflow					
Population Size	Objective Function (BMC)	Crossover Probability	Objective Function (BMC)	Mutation Probability	Objective Function (BMC)
10	1.135	0.20	1.133	0.20	1.134
30	1.121	0.40	1.121	0.40	1.129
50	1.129	0.60	1.125	0.60	1.121
70	1.136	0.80	1.129	0.80	1.128
Medium Inflow					
10	1.154	0.20	1.167	0.20	1.165
30	1.142	0.40	1.141	0.40	1.142
50	1.145	0.60	1.155	0.60	1.149
70	1.149	0.80	1.167	0.80	1.54
Low Inflow					
10	1.197	0.20	1.199	0.20	1.192
30	1.185	0.40	1.185	0.40	1.185
50	1.189	0.60	1.191	0.60	1.187
70	1.191	0.80	1.195	0.80	1.189

4.2. Analysis of 10 Random Results for Different Algorithms

Table 6 shows ten random results for different algorithms for high inflow. The average solution for IWA was 0.985 while it is 1.037, 1.040, 1.115 and 1.121 for the WA, BA, IPSO and GA, respectively. The IWA decreased the objective function by 5.01%, 5.20%, 11.65% and 12% compared to the WA, BA, IPSOA and GA, respectively. The computational time for IWA was 22 s, which was 12%, 18%, 24% and 29% less than the WA, BA, IPSOA and GA, respectively. The variation coefficient for the IWA was also less than the GA, IPSOA, BA and WA. Table 7 shows the average of 10 random results for medium inflow. The average solution for IWA was 1.021, which was 8.01%, 8.20%, 9.4% and 10.5% less than the WA, BA, IPSO and GA, respectively. Computational time for the IWA was 22 s and was 3, 5, 7 and 9 s less than the WA, BA, IPSO and GA, respectively and the variation coefficient was also lowest for the IWA. Table 8 shows the low inflow results. The average solution for the IWA was 1.121, which was 0.30%, 0.90%, 2.9% and 5.4% less than the WA, BA, IPSO and GA, respectively. Computational time for the IWA was 21 s and was 3, 2, 6 and 7 s less than the WA, BA, IPSO and GA, respectively. The variation coefficient for the IWA was less than other evolutionary algorithms which proves that one computer run can be reliable because the variation coefficient has a small value. The average value of the objective function for the high inflow was 0.985, which was 3.5% and 12% less than the medium and low inflows, respectively. This suggested that the square of the difference of released water and demand based on high inflow was less than that based on low and medium inflow and that the reservoir based on the IWA and high inflow had the best performance. Similar results were found for the other evolutionary algorithms.

Table 6. Ten random results for high inflow.

Run	IWA	WA	BA	IPSOA	GA
1	0.985	1.035	1.039	1.115	1.121
2	0.985	1.037	1.045	1.118	1.125
3	0.987	1.037	1.039	1.115	1.121
4	0.985	1.037	1.039	1.115	1.121
5	0.985	1.037	1.039	1.115	1.121
6	0.985	1.037	1.039	1.115	1.121
7	0.985	1.037	1.039	1.115	1.121
8	0.985	1.037	1.039	1.115	1.121
9	0.985	1.037	1.039	1.115	1.121
10	0.985	1.037	1.039	1.115	1.121
Average	0.985	1.037	1.040	1.115	1.121
Computational Time(s)	22	25	27	29	31
Variation Coefficient	0.0003	0.0006	0.001	0.0008	0.001

Table 7. Ten random results for medium inflow.

Run	IWA	WA	BA	IPSOA	GA
1	1.021	1.110	1.112	1.128	1.142
2	1.024	1.110	1.114	1.128	1.146
3	1.022	1.114	1.112	1.128	1.142
4	1.021	1.110	1.112	1.133	1.142
5	1.021	1.110	1.114	1.128	1.142
6	1.021	1.110	1.112	1.128	1.142
7	1.021	1.110	1.112	1.128	1.142
8	1.021	1.110	1.112	1.128	1.142
9	1.021	1.110	1.112	1.128	1.142
10	1.021	1.110	1.112	1.128	1.142
Average	1.021	1.110	1.112	1.128	1.142
Computational Time (s)	22	25	28	30	32
Variation Coefficient	0.0007	0.0010	0.0008	0.0010	0.0008

Table 8. Ten random results for low inflow.

Run	IWA	WA	BA	IPSOA	GA
1	1.121	1.125	1.132	1.155	1.187
2	1.122	1.129	1.132	1.155	1.88
3	1.123	1.125	1.136	1.155	1.185
4	1.121	1.125	1.132	1.159	1.185
5	1.121	1.125	1.132	1.155	1.185
6	1.121	1.125	1.132	1.155	1.185
7	1.121	1.125	1.132	1.155	1.185
8	1.121	1.125	1.132	1.155	1.185
9	1.121	1.125	1.132	1.155	1.185
10	1.121	1.125	1.132	1.155	1.185
Average	1.121	1.125	1.132	1.155	1.185
Computational Time (s)	21	24	26	27	29
Variation Coefficient	0.0005	0.001	0.001	0.001	0.0009

4.3. Convergence Curves for Different Algorithms

Figures 6–8 shows the convergence curves for the different algorithms and inflows. At high inflow (Figure 6), after 1800 iterations, the IWA had converged while the WA, BA, IPSOA and GA converged after 2000, 2500, 3000 and 3000 iterations, respectively, indicating the IWA has the fastest convergence process among all the tested algorithms. The IWA converged after 3900 iterations for medium inflow while the WA, BA, IPSOA and GA converged after 4200, 4380, 4600 and 4800 iterations, respectively, (Figure 4). Figure 4 shows the fastest convergence was also for the IWA at low inflow.

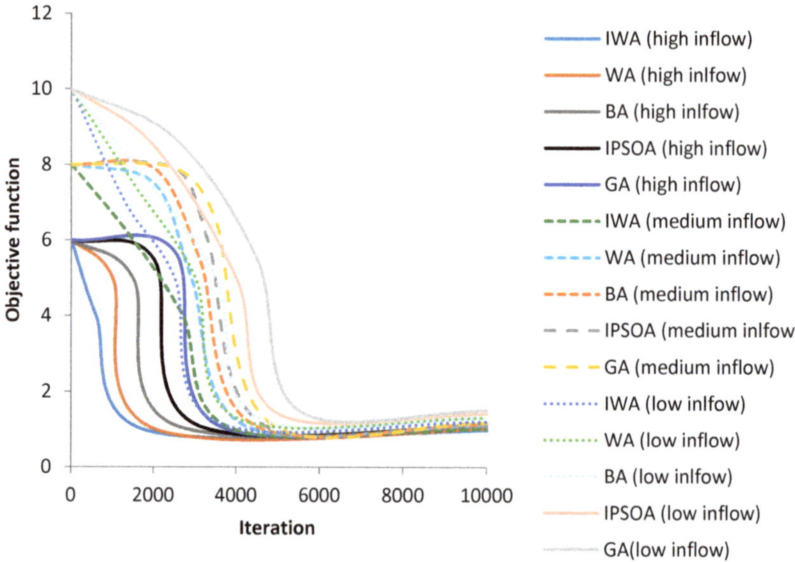

Figure 6. Convergence curve for low inflow, high inflow and medium inflow.

4.4. Analysis of Monthly Rule Curves for Reservoir Operation

Table 1 shows the average monthly inflows and demand during 1989 to 1999. July is the month with the greatest demand and therefore a critical month for analysis and avoiding of repetition of the same results. Figure 7 shows the average released water versus reservoir storage in July for 1989 to 1999 at high inflow. When the reservoir storage was 41 BCM, the released water could meet the demand (5.8 BCM) based on the IWA while the reservoir storage was 46, 47, 50 and 55 BCM for the WA, BA, IPSOA and GA, respectively, to meet the same demand. The IWA could therefore decrease reservoir storage at high inflow by about 10%, 14%, 18% and 25% compared to the WA, BA, IPSOA and GA, respectively. When the reservoir storage equaled 110 BCM at the high inflow using the IWA, the released water was greater than the demand (5.8 BCM) and could be used for the other needs such as power generation. At medium inflow (Figure 8), the IWA indicated that released water could meet demands with storage of 47 BCM, while the storage was 49, 52, 56 and 58 BCM for the WA, BA, IPSOA and GA, respectively. The IWA reduced storage for the demand supply by about 4.08%, 9.9%, 16% and 18% compared to the WA, BA, IPSOA and GA, respectively. In fact, the IWA can meet the demands earlier than WA, BA, IPSOA and GA. When storage calculated by the IWA at medium inflow was 112 BCM, the released water was greater than demand. For low inflow (Figure 9), the IWA indicated storage of 52 BCM was required to meet demands while the storage was 54, 55, 56 and 60 for the WA, BA, IPSOA and GA, respectively. The IWA reduced the storage for water supply at low inflow by about 3.7%, 5.4%, 7.6% and 14% compared to the WA, BA, IPSOA and GA, respectively. Figure 10 shows the released water for the different algorithms during 1989–1999 and the results indicated that the IWA met the demands more frequently than the GA, IPSOA, BA and WA. The reservoir storage for high inflow using the IWA was 41 BCM while for low and medium inflows it was 52 and 47 BMC, respectively. Thus, the storage should be increased for the low and medium inflows to satisfy the demands. A similar trend was noted for the other algorithms. Table 9 shows the root mean square error (RMSE) between average released water with the demand value based on values in Table 1 for each month during the study period. Table 9 shows an RMSE for the low inflow using the IWA of 12.382 BCM, which was 4.1%, 18.21%, 18.69% and 28.3% lower compared to the WA, BA, IPSOA and GA, respectively. For the medium inflow (Table 9), the RMSE for the IWA was 11.38 BCM, which was 4.51%, 24.44%, 25.71% and 29.16% lower than the WA, BA, IPSOA and GA, respectively. Table 9 shows the same trend for high inflow. In other words, Figure 10 shows that the operation rule generated using IWA experienced more match between water released and demand.

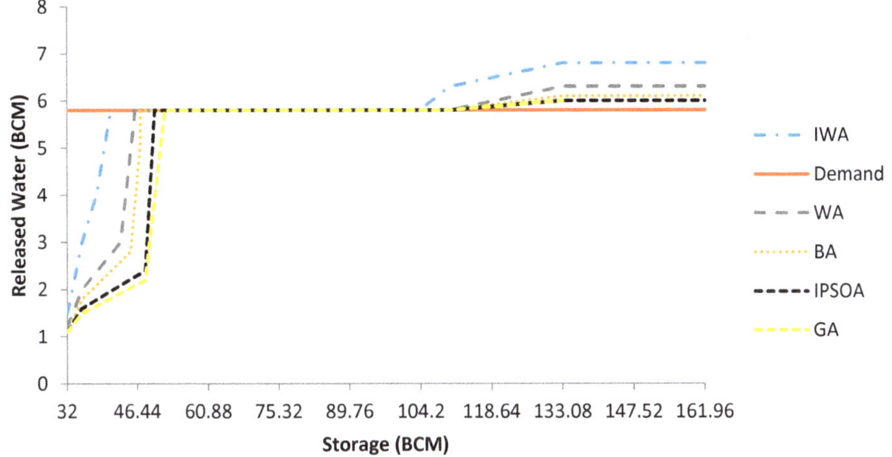

Figure 7. Released water for high inflow.

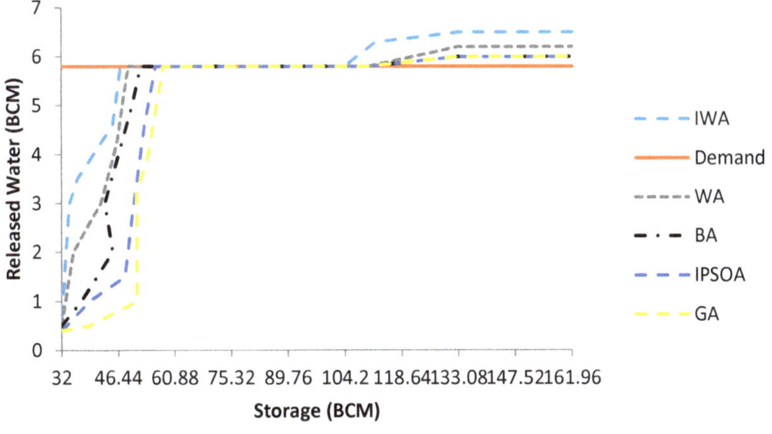

Figure 8. Released water for medium inflow.

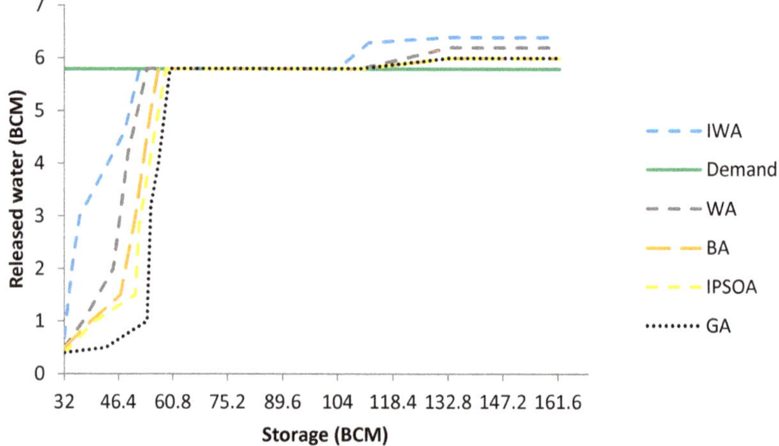

Figure 9. Released water for low inflow.

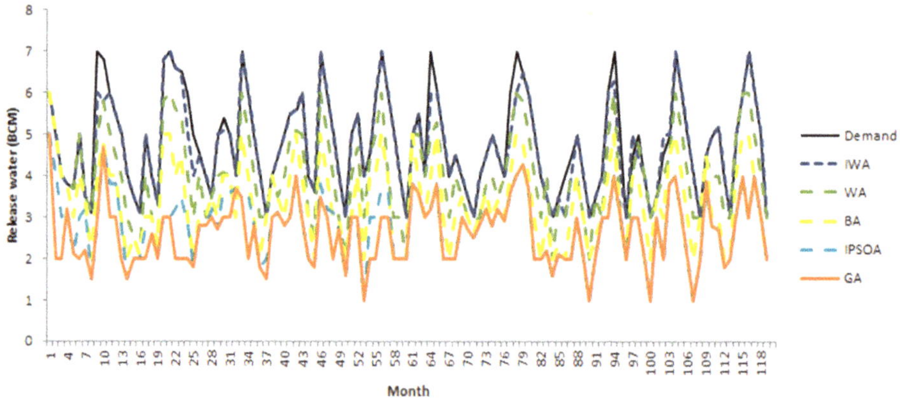

Figure 10. Released water for the total of operation periods.

Table 9. Computed RMSE for low inflow.

$RMSE = \sqrt{\frac{(R_i - D_i)^2}{T}}, T : Number(of)periods$ (BCM) (Low Inflow)					
Month	**IWA**	**WA**	**BA**	**IPSOA**	**GA**
January	1.32	1.43	2.12	2.18	2.30
February	1.21	1.31	2.11	2.15	2.55
March	1.12	1.14	1.65	1.78	1.79
April	1.14	1.16	1.78	1.82	1.75
May	0.98	1.02	1.22	1.45	1.57
June	0.91	0.93	0.98	1.14	1.21
July	1.56	1.57	1.59	1.61	1.52
August	1.11	1.21	1.14	1.16	1.18
September	1.12	1.14	1.15	1.17	1.21
October	0.76	0.78	0.82	0.85	0.87
November	0.012	0.015	0.017	0.019	0.020
December	1.14	1.21	1.31	1.32	1.30
Total	12.382	12.915	15.14	16.649	17.27

Table 9. *Cont.*

Medium Inflow				
Month	**IWA**	**WA**	**BA**	**IPSOA**	**GA**
---	---	---	---	---	---
January	1.12	1.33	2.10	2.15	2.28
February	1.20	1.22	1.99	2.12	2.45
March	1.09	1.11	1.45	1.72	1.68
April	1.11	1.14	1.59	1.65	1.75
May	0.77	1.01	1.32	1.34	1.52
June	0.82	0.91	0.98	1.10	1.11
July	1.54	1.47	1.61	1.54	1.32
August	1.10	1.11	1.24	1.14	1.19
September	1.10	1.10	1.17	1.12	1.21
October	0.74	0.77	0.84	0.78	0.87
November	0.012	0.014	0.017	0.019	0.020
December	1.12	1.20	1.35	1.25	1.30
Total	11.83	12.51	15.657	15.925	16.7

Low Inflow				
Month	**IWA**	**WA**	**BA**	**IPSOA**	**GA**
---	---	---	---	---	---
January	0.958	1.31	2.10	2.12	2.17
February	1.12	1.12	1.89	1.95	2.29
March	1.09	1.01	1.32	1.55	1.65
April	0.99	1.12	1.29	1.65	1.72
May	0.71	1.00	1.12	1.34	1.51
June	0.81	0.90	0.98	0.91	1.10
July	1.52	1.26	1.31	1.42	1.29
August	1.02	1.25	1.14	1.14	1.19
September	0.99	1.10	1.57	1.11	1.11
October	0.72	0.75	0.84	0.78	0.87
November	0.012	0.014	0.017	0.019	0.020
December	1.11	1.19	1.25	1.23	1.28
Total	11.05	12.024	14.827	15.390	16.18

4.5. Analysis of Different Algorithms for Reservoir Operation during Total Operation Periods

Table 10 shows the performance of the different algorithms for reservoir operation using the reliability, vulnerability and resiliency indices. The highest value for the reliability index for high inflow was 95% based on the IWA, which was 4%, 7%, 17% and 20% more than the WA, BA, IPSOA and GA, respectively. This indicated that the IWA could better supply the volume of demand compared to the other evolutionary algorithms. The reliability index for medium and low inflows based on the IWA were 90% and 89%, respectively, suggesting that at high inflow and based on the IWA, the reservoir could supply more of the demand.

For the IWA at high inflow, the vulnerability index was 12% while it was 14%, 15%, 17% and 19% for WA, BA, IPSOA and GA, respectively (Table 10). This meant that the intensity of failures for the reservoir operation based on the IWA was less than the WA, BA, IPSOA and GA and the system faced lower average deficits during the operational period (1989–1999). The system faced more deficits based on the IWA during the periods with low and medium inflows. Although the vulnerability index of the IWA was 2%, 4%, 5% and 6% lower for the medium inflow compared to the WA, BA, IPSOA and GA, respectively and 1%, 4%, 5% and 6% lower for low inflow, the vulnerability index based on IWA for medium and low inflow (14% and 16%) is 2% and 4% more than high inflow. Decreasing the inflow led to decreasing the volume of released water and thus, the system would face more deficits.

The highest value for the resiliency index at high inflow was for the IWA and it was 2%, 4%, 5% and 7% more than WA, BA, IPSOA and GA, respectively (Table 10). The resiliency index shows the model's ability to recover from a critical period (failure). It is based on the ratio of the number of periods when reservoir water supply meets the demand to the number of periods when the demand is not supplied; thus, a high percent for this index is desirable. Although, the IWA acted better than the other evolutionary algorithms based on the resiliency index at all flows, the index declined from the high to lower flows. A similar trend was observed for the other algorithms. Figure 8 shows the released water for the different algorithms during 1989–1999 and the results indicated that the IWA met the demands more frequently than the GA, IPSOA, BA and WA.

Table 10. Computed different indexes for different algorithms

Algorithm	Reliability Index (%)	Vulnerability Index (%)	Resiliency Index
High Inflow			
IWA	95	12	54
WA	91	14	52
BA	88	15	50
IPSO	78	17	49
GA	75	19	47
Medium Inflow			
IWA	90	14	51
WA	89	16	50
BA	87	18	49
IPSO	74	19	48
GA	72	20	45
Low Inflow			
IWA	89	16	50
WA	87	17	47
BA	85	20	44
IPSO	82	21	42
GA	70	22	40

5. Conclusions

The present study investigated the performance of an improved weed algorithm (IWA) and compared it to other evolutionary algorithms: the weed algorithm (WA), bat algorithm (BA), improved particle swarm optimization algorithm (IPSOA) and the genetic algorithm (GA). The operation of the reservoir of the Aswan High Dam, one of the most important dams in Egypt, was selected for this study. The IWA based on a local search operator attempted to improve the simple weed algorithm. Low, medium and high inflows to the reservoir were considered. The average solution by the IWA for low inflow was 1.121 and it was 0.30%, 0.90%, 2.9% and 5.4% less than the WA, BA, IPSO and GA, respectively. The computational time for the IWA at low inflow was 21s and was 3, 2, 6 and 7 s less than the WA, BA, IPSO and GA, respectively. Similar results were found for all three inflows. Based on extracted rule curves, the IWA could meet water demand earlier than the other algorithms. Using the IWA, reservoir storage was 41 BCM for the high inflow and the released water could meet the demand (5.8 BCM); the reservoir storage required was 46, 47, 50 and 55 BCM for the WA, BA, IPSOA and GA, respectively, to meet the demand at high inflow. The value of RMSE based on released water and demand was computed for each month and was less for the IWA than the other evolutionary algorithms. The vulnerability index for the IWA at high inflow was 12% while it was 14%, 15%, 17% and 19% for the WA, BA, IPSOA and GA, respectively. The reliability and resiliency indices were also better for the IWA at all three inflow rates compared to the other algorithms. The new weed algorithm has a high potential for solving different water resources management problems and it is suggested it can be used for multi-reservoir operation and multi-propose systems in future research. Although the method was used for Aswan High Dam but it can be used for other dams with the different applications because the method can supply the irrigation demands under different inflow conditions and thus, the method can be used for the dams with the aim of power generation on other demands. Some limitations for the methods are related to the accurate determination of random parameters that needs further research to accurate estimated utilizing advanced sensitivity analysis.

Water **2018**, *10*, 1267

In addition, even though the proposed model could be generalized to be applied for different case studies, it is recommended to the adaptation for the new case study constraints with the procedure of the IWA.

Author Contributions: H.K., K.H., M.E. and V.P.S. initiated the research point of the hydrological problem and supervised the study. M.D. collected the data and performed the modeling. M.S.H. and C.M.F. handled the writing up of the introduction and methodology. The analysis of the modeling outcomes has been handled by A.E.-S., M.D. and M.S.H. organized the whole manuscript and managed the paper submission and revision.

Funding: This research was funded by the University of Malaya Research Grant (UMRG) coded RP017C-15SUS and RP025A-18SUS.

Acknowledgments: The authors would like to appreciate the financial support received from University of Malaya research grants coded UMRG (RP017C-15SUS) and RP025A-18SUS.

Conflicts of Interest: The authors declare no conflict of interest.

References

1. Srinivasan, K.; Kumar, K. Multi-objective simulation-optimization model for long-term reservoir operation using piecewise linear hedging rule. *Water Resour. Manag.* **2018**, *32*, 1901–1911. [CrossRef]
2. Ehteram, M.; Mousavi, S.F.; Karami, H.; Farzin, S.; Singh, V.P.; Chau, K.W.; El-Shafie, A. Reservoir operation based on evolutionary algorithms and multi-criteria decision-making under climate change and uncertainty. *J. Hydroinform.* **2018**, *20*, 332–355. [CrossRef]
3. Ahmadianfar, I.; Adib, A.; Salarijazi, M. Optimizing multireservoir operation: Hybrid of bat algorithm and differential evolution. *J. Water Resour. Plan. Manag.* **2015**, *142*, 05015010. [CrossRef]
4. Cheng, C.T.; Wang, W.C.; Xu, D.M.; Chau, K.W. Optimizing hydropower reservoir operation using hybrid genetic algorithm and chaos. *Water Resour. Manag.* **2008**, *22*, 895–909. [CrossRef]
5. Afshar, M.H. Large scale reservoir operation by constrained particle swarm optimization algorithms. *J. Hydro Environ. Res.* **2012**, *6*, 75–87. [CrossRef]
6. Chau, K.W. A split-step particle swarm optimization algorithm in river stage forecasting. *J. Hydrol.* **2007**, *346*, 131–135. [CrossRef]
7. Afshar, M.H. Extension of the constrained particle swarm optimization algorithm to optimal operation of multi-reservoirs system. *Int. J. Electr. Power Energy Syst.* **2013**, *51*, 71–81. [CrossRef]
8. Ehteram, M.; Karami, H.; Mousavi, S.F.; El-Shafie, A.; Amini, Z. Optimizing dam and reservoirs operation based model utilizing shark algorithm approach. *Knowl. Based Syst.* **2017**, *122*, 26–38. [CrossRef]
9. Fallah-Mehdipour, E.; Haddad, O.B.; Mariño, M.A. Real-time operation of reservoir system by genetic programming. *Water Resour. Manag.* **2012**, *26*, 4091–4103. [CrossRef]
10. Ostadrahimi, L.; Mariño, M.A.; Afshar, A. Multi-reservoir operation rules: Multi-swarm PSO-based optimization approach. *Water Resour. Manag.* **2012**, *26*, 407–427. [CrossRef]
11. Moeini, R.; Afshar, M.H. Extension of the constrained ant colony optimization algorithms for the optimal operation of multi-reservoir systems. *J. Hydroinform.* **2013**, *15*, 155–173. [CrossRef]
12. Zhang, Z.; Zhang, S.; Wang, Y.; Jiang, Y.; Wang, H. Use of parallel deterministic dynamic programming and hierarchical adaptive genetic algorithm for reservoir operation optimization. *Comput. Ind. Eng.* **2013**, *65*, 310–321. [CrossRef]
13. Haddad, O.B.; Moravej, M.; Loáiciga, H.A. Application of the water cycle algorithm to the optimal operation of reservoir systems. *J. Irrig. Drain. Eng.* **2014**, *141*, 04014064. [CrossRef]
14. Bozorg-Haddad, O.; Karimirad, I.; Seifollahi-Aghmiuni, S.; Loáiciga, H.A. Development and application of the bat algorithm for optimizing the operation of reservoir systems. *J. Water Resour. Plan. Manag.* **2014**, *141*, 04014097. [CrossRef]
15. Bolouri-Yazdeli, Y.; Haddad, O.B.; Fallah-Mehdipour, E.; Mariño, M.A. Evaluation of real-time operation rules in reservoir systems operation. *Water Resour. Manag.* **2014**, *28*, 715–729. [CrossRef]
16. Haddad, O.B.; Hosseini-Moghari, S.M.; Loáiciga, H.A. Biogeography-based optimization algorithm for optimal operation of reservoir systems. *Water Resour. Plan. Manag.* **2015**, *142*, 04015034. [CrossRef]
17. Asgari, H.R.; Bozorg Haddad, O.; Pazoki, M.; Loáiciga, H.A. Weed optimization algorithm for optimal reservoir operation. *J. Irrig. Drain. Eng.* **2015**, *142*, 04015055. [CrossRef]

18. Akbari-Alashti, H.; Haddad, O.B.; Mariño, M.A. Application of fixed length gene genetic programming (FLGGP) in hydropower reservoir operation. *Water Resour. Manag.* **2015**, *29*, 3357–3370. [CrossRef]

19. Bozorg-Haddad, O.; Janbaz, M.; Loáiciga, H.A. Application of the gravity search algorithm to multi-reservoir operation optimization. *Adv. Water Resour.* **2016**, *98*, 173–185. [CrossRef]

20. Ehteram, M.; Allawi, M.F.; Karami, H.; Mousavi, S.F.; Emami, M.; Ahmed, E.S.; Farzin, S. Optimization of chain-reservoirs' operation with a new approach in artificial intelligence. *Water Resour. Manag.* **2017**, *31*, 2085–2104.

21. Ehteram, M.; Karami, H.; Mousavi, S.F.; Farzin, S.; Kisi, O. Optimization of energy management and conversion in the multi-reservoir systems based on evolutionary algorithms. *J. Clean. Prod.* **2017**, *168*, 1132–1142. [CrossRef]

22. Mousavi, S.F.; Vaziri, H.R.; Karami, H.; Hadiani, O. Optimizing reservoirs exploitation with a new crow search algorithm based on a multi-criteria decision-making model. *JWSS* **2018**, *22*, 279–290.

23. Karami, H.; Mousavi, S.F.; Farzin, S.; Ehteram, M.; Singh, V.P.; Kisi, O. Improved krill algorithm for reservoir operation. *Water Resour. Manag.* **2018**, *32*, 3353–3372. [CrossRef]

24. Ehteram, M.; Karami, H.; Farzin, S. Reservoir optimization for energy production using a new evolutionary algorithm based on multi-criteria decision-making models. *Water Resour. Manag.* **2018**, *32*, 2539–2560. [CrossRef]

25. Ehteram, M.; Karami, H.; Farzin, S. Reducing irrigation deficiencies based optimizing model for multi-reservoir systems utilizing spider monkey algorithm. *Water Resour. Manag.* **2018**, *32*, 2315–2334. [CrossRef]

26. Karami, H.; Ehteram, M.; Mousavi, S.F.; Farzin, S.; Kisi, O.; El-Shafie, A. Optimization of energy management and conversion in the water systems based on evolutionary algorithms. *Neural Comput. Appl.* **2018**, 1–4. [CrossRef]

27. Roshanaei, M.; Lucas, C.; Mehrabian, A.R. Adaptive beamforming using a novel numerical optimisation algorithm. *IET Microw. Antennas Propag.* **2009**, *3*, 765–773. [CrossRef]

28. Rad, H.S.; Lucas, C. A recommender system based on invasive weed optimization algorithm. In Proceedings of the 2007 IEEE Congress on Evolutionary Computation, Singapore, 25–28 September 2007; IEEE: Piscataway, NJ, USA, 2007; pp. 4297–4304.

29. Chakraborty, P.; Roy, G.G.; Das, S.; Panigrahi, B.K. On population variance and explorative power of invasive weed optimization algorithm. In Proceedings of the 2009 World Congress on Nature & Biologically Inspired Computing (NaBIC), Coimbatore, India, 9–11 December 2009; IEEE: Piscataway, NJ, USA, 2009; pp. 227–232.

30. El-Shafie, A.; Jaafer, O.; Akrami, S.A. Adaptive neuro-fuzzy inference system based model for rainfall forecasting in Klang River, Malaysia. *Int. J. Phys. Sci.* **2011**, *6*, 2875–2888.

Article

A Multi-disciplinary Modelling Approach for Discharge Reconstruction in Irrigation Canals: The Canale Emiliano Romagnolo (Northern Italy) Case Study

Marta Luppi [1], Pierre-Olivier Malaterre [2], Adriano Battilani [3], Vittorio Di Federico [4] and Attilio Toscano [1,*]

[1] Department of Agricultural and Food Sciences, University of Bologna, Viale Giuseppe Fanin 50, 40127 Bologna, Italy; marta.luppi2@unibo.it
[2] UMR G-eau, IRSTEA, 361 rue Jean-François Breton, 34196 Montpellier, France; pierre-olivier.malaterre@irstea.fr
[3] Consorzio del Canale Emiliano Romagnolo (CER), via Ernesto Masi 8, 40137 Bologna, Italy; battilani@consorziocer.it
[4] Department of Civil, Chemical, Environmental and Materials Engineering, University of Bologna, Viale Risorgimento 2, 40136 Bologna, Italy; vittorio.difederico@unibo.it
[*] Correspondence: attilio.toscano@unibo.it; Tel.: +39-051-2096179

Received: 26 June 2018; Accepted: 25 July 2018; Published: 31 July 2018

Abstract: Agriculture is the biggest consumer of water in the world, and therefore, in order to mitigate the effects of climate change, and consequently water scarcity, it is important to reduce irrigation water losses and to improve the poor collection of hydraulic status data. Therefore, efficiency has to be increased, and the regulation and control flow should be implemented. Hydraulic modelling represents a strategic tool for the reconstruction of the missing hydraulic data. This paper proposes a methodology for the unmeasured offtake and flowing discharge estimation along the open-canal Canale Emiliano Romagnolo (CER), which is one of the major irrigation infrastructures in Northern Italy. The "multi-disciplinary approach" that was adopted refers to agronomic and hydraulic aspects. The tools that were used are the IRRINET management Decisional Support System (DSS) and the SIC^2 (Simulation and Integration of Control for Canals) hydraulic software. Firstly, the methodology was developed and tested on a Pilot Segment (PS), characterized by a simple geometry and a quite significant historical hydraulic data availability. Then, it was applied on an Extended Segment (ES) of a more complex geometry and hydraulic functioning. Moreover, the available hydraulic data are scarce. The combination of these aspects represents a crucial issue in the irrigation networks in general.

Keywords: lined irrigation open-canal; unmeasured discharges estimation; hydraulic modelling; irrigation DSS

---

1. Introduction

Counting on the intensive exploitation of the water resources, many works of the last decades have addressed agricultural water management practices towards the productivity strengthening and the defeating poverty [1–3]. Nowadays, the water scarcity, combined with the rising food demand, has involved a gradual switch of the objectives [1,3] to the following: Resource preservation (quantitatively, qualitatively, and ecologically) in relation to agricultural production (crop irrigation, animal rearing, and on-farm operations) [4–6], rural realities economy improvement [7,8], and facing climate change [9].

The sustainable development resulted from these key components is promoted by the Water Framework Directive (WFD/2000/60CE) [10] and policies that are closely related to the EU2020 program [11–13]. At the regional scale, the water management practices for irrigation are identified as a primary challenge because of their socio-economic implications [13]. They consist in the improvement of the irrigation consumption knowledge at the field scale and the increase in the efficiency and the discharge regulation at conveyance system scale [14].

Despite the evolution of irrigation infrastructures tends to be focused mainly on pressurized systems, many districts are often fed by dense canal networks that have remained basically unchanged since they were constructed decades ago. They are characterized by significant water losses and irrecoverable outflow at their end [15–18]. The irrigation systems performances can be improved through hardware (physical/structural) changes, such as the canal lining or the installation of sophisticated control structures [19,20], or through software (operational) techniques, such as appropriate delivery rules and an effective communication between water supply agencies and water users [21].

A common flaw in irrigation delivery systems that are characterized by open canals and by many users is the absence of a proper information system that ensures and collects measured and monitored data about hydraulic status [22–25]. When considering that the total water consumption for irrigation is projected to increase by 10% by 2050 [26], it will represent a central issue in the near future [27]. Generally, the only known quantities are measured water levels at specific locations, often with limited precision and possible failures [19].

Hydraulic modelling emerges as a strategic tool for: 1) the reconstruction of unmeasured data, such as discharges or water levels at other locations, unknown perturbations (inflows and outflows) [28,29], and hydraulic variables (friction coefficients and hydraulic device discharge coefficients) [19,30]. 2) the visualization and control of the flow at several structures [15,31].

In parallel, irrigation Decisional Support Systems (DSS) can characterize the crops that are served by a specific irrigation delivery system, and also, can indirectly monitor their hydraulic status. In the last few decades, DSS underwent many changes [32,33] ranging from the prevention of extreme events (droughts and floods) and pollution [32] to the irrigation scheduling [34–39]. The latter is based on the integration of several models, processes, and factors (i.e., meteorological and soil conditions and types of crops) [40,41].

This study presents a tool for the reconstruction of unmeasured discharges along a specific irrigation delivery canal. The combination of hydraulic modelling and irrigation DSS can solve the problem that was created by the poor hydraulic data collection. The multi-disciplinary approach that is proposed in this paper reflects the merging of hydraulic engineering and agronomy aspects. It was developed on one of the most important irrigation canals in Northern Italy: The Canale Emiliano Romagnolo (CER) [42]. The methodology was developed on a simple geometry 7 km long Pilot Segment (PS) and over a more complex 22 km long Extended Segment (ES).

2. Materials and Methods

2.1. Description of the CER

The CER starts in Salvatonica di Bondeno (Ferrara, Italy) on the right bank of the Po River and it provides the irrigation supply for an area of about 3000 km^2. That area represents the 93% of the irrigated and the 22% of the agricultural land in the Emilia Romagna Region. The agricultural land covers the 60% of the regional territory [43], where different cultures are irrigated, among which extensive crops, vegetables, and orchards [44]. To convey and to distribute water, the CER hydraulic system uses seven pumping stations (the main one on the Po River) and 165 km of canal networks (Figure 1).

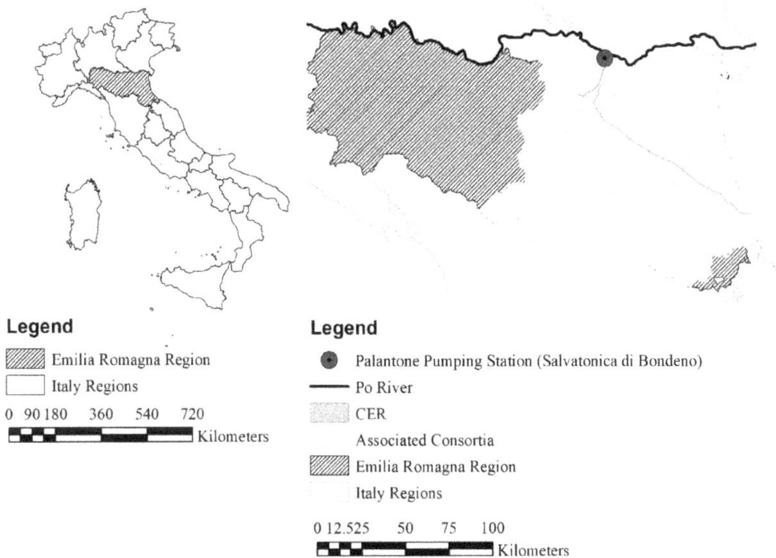

Figure 1. The Emilia Romagna Region, the Associated Consortia and the CER.

The main reach is 133 km long and its first 104 km are characterized by 60–17.6 m width at the top and 6.0–6.4 m at the bottom of the canal. The side slopes are 3:1 and 1.5:1 or 1.75:1 for composite trapezium sections (first 37 km) and 2:1 for the simple ones. The cross section of the canal later becomes narrower with a rectangular shape: open (width range: 6.8–5.6 m, elevation range: 3–2.7 m) or closed (width range: 6.4–5.6 m, elevation range: 2.1–1.9 m) and made of reinforced concrete. The CER receives no inflow from surface runoff, drainage, or different types of discharges, but it has several offtakes. From the canal, the water is offtake using pumps or gates, and it is conveyed to the irrigated fields through secondary channels that are managed by Associated Consortia. The irrigation offtakes have a seasonal variability, and therefore the maximum permitted discharge at the main pumping station varies from 68 m^3/s (from May to September) to 25 m^3/s (the rest of the year). Moreover, discharges are also affected by the meteorological issues (e.g., long dry seasons), the type of cultivated crops, and the irrigation practices. The Consortium of the CER is in charge of: (1) maintenance operations (geometric and functioning repairs, periodic cleanings); (2) collection of quantitative and qualitative measurements; and, (3) supply of irrigation services to farmers (by means of several irrigation Associated Consortia that distributes water to final users).

2.2. Investigation Period and Available Data

This study focuses on the period of full operation of the CER i.e., the irrigation season (June–August) that is characterized by the highest water demand and irrigation frequency. The irrigation period selected comprises 73 days (20 June–31 August) of the years from 2012 to 2015. These years were characterized by different average daily rainfall. For example, 2013 (1.30 mm/day) and 2015 (0.94 mm/day) had daily rainfall that was close to the decennial (2005–2015) average value (1.1 mm/day), while 2014 (2.22 mm/day) and 2012 (0.13 mm/day) were especially rainy and dry, respectively.

The main available data for this study are: (1) water volumes at offtakes (calculated indirectly); (2) crop water requirements (estimated); and, (3) water levels at the main canal (measured); (4) functioning data of pumping stations along the CER (measured).

In particular, for each irrigation offtake, calculated and estimated water amounts were provided. The former refers to monthly cumulated volumes indirectly calculated by the Associated Consortia on the basis of flow rates and working times of offtakes pumps or the opening gate area, the opening time, and the water level at offtakes manual gates.

On the other hand, estimated water volumes were based on the crop water requirements provided by the IRRINET management DSS, which was developed by the Consortium of the CER [45]. IRRINET is identified as the reference tool for the estimation of irrigation volumes in the Emilia Romagna Region [46], and it provides to farmers a day-by-day information on how much and when to irrigate crops [47]. It is based on a daily water balance of soil-plant-atmosphere system. IRRINET processes a huge quantity of information related to: areas (meteorological, water table depth and soil data) and farms (types of irrigated crops, start and stop crops dates). Since 2012, at the end of every irrigation period, the Consortium of the CER has collected daily optimum crop water requirement (*CWR*) values for all the crops that are served by IRRINET. For every type of crop (*i*) and for every day, these values are averaged; afterwards, they are cumulated on a decadal time scale giving CWR_i (Section 2.4.1).

Along the CER, the only hydraulic measurements available are water levels. In total, forty cross-sections are equipped with ultrasonic level transmitters (The Probe PL-517, Terry Ferraris &C. S.p.A., Milan, Italy). These instruments are generally located near two types of infrastructures: (a) culverts (passing under different rivers; in total, 29 instruments); and (b) pumping stations (in suctions and/or delivery tanks; in total 11 instruments). After direct field surveys, the measurement accuracy of both types of transmitters was estimated to be lower than the original instrument accuracy (± 0.02 m), in particular, ± 0.05 m and ± 0.10 m, respectively. The transmitters located near culverts serve for management purposes, and their accuracy was probably affected by flow disturbances (sediment build up and depressions next to the edges of culverts entrances due to velocity changes) [48]. On the other hand, the transmitters near pumping stations are used for operational purposes and they are strongly influenced by the pumps functioning.

At each of the 40 cross-sections, the water level value is transmitted and is registered with a time step of 30 min. Because of the offtake data time scale (monthly or decadal) and because of the general water level series incompleteness, the 30 min available measures were averaged on a daily time scale.

Finally, the daily measured functioning data at one pumping station (Pieve di Cento) were investigated (Section 2.6). Every time that the installed pumps would turn on or turn off the following parameters were measured: voltage (V), electric current (A), functioning time (h), discharge (m^3/s), volume (m^3), suction, and delivery tanks water level (m).

Table 1 provides a summary of all the available data used in the present application.

Table 1. The available data and their characteristics.

Available Data	Type	Unit	Time Step	Source
Offtake Volumes	Indirectly calculated	m^3	Monthly (cumulated values)	Associated Consortia
CWR_i	Estimated	mm	Decadal (cumulated values)	IRRINET
Water Levels	Measured	m	Daily (average values)	CER
Water Levels at Suction/Delivery Tanks	Measured	m	Pumps on/off (single values)	CER

2.3. Description of the Pilot Segment (PS)

The multi-disciplinary modelling approach was developed on a 7 km long Pilot Segment (PS) of the CER.

The PS extremities coincide with two concrete culverts called Culv_1 (upstream) and Culv_2 (downstream) (Figure 2). They are characterized by rectangular flow sections of 36 m^2 and of 31.5 m^2, respectively, and by submerged entrances and surface or/and piped-flow conditions. PS has three different trapezium cross sections with width ranges of 22.8–25.8 m (at the top) and 3.3–7 m (at the bottom). The side slopes are 3:1 and 1.5:1 for the first composite cross section and 2:1 for the other two simple sections. For the first 700 m along the segment, the bed altimetry goes from 12.81 m to 13.74 m above the sea level. After that part, the canal has a constant slope with a final value of 13.32 m above sea level.

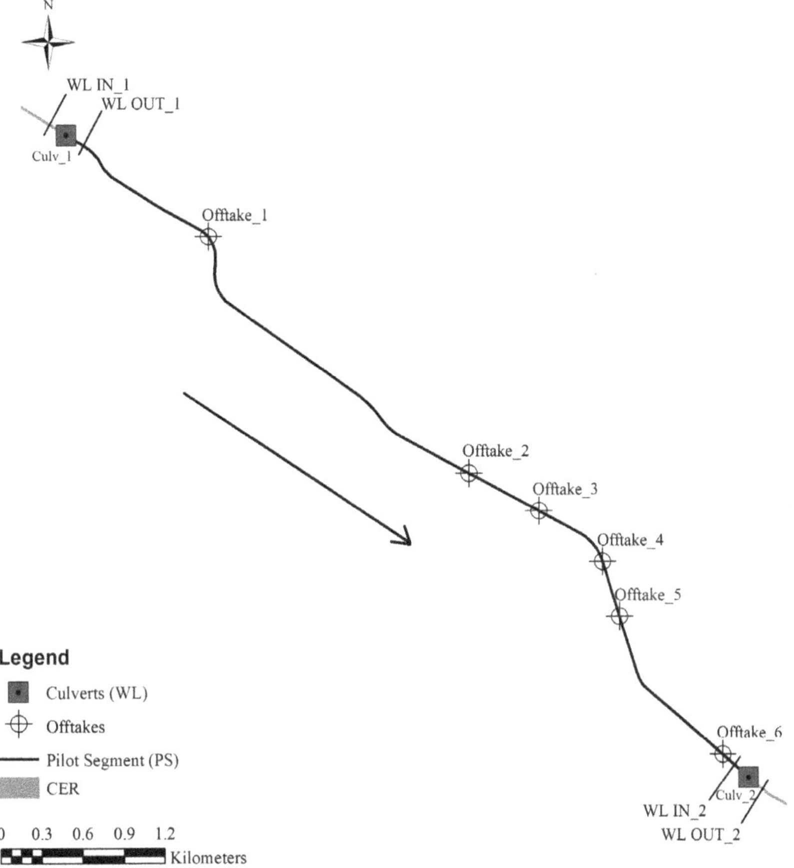

Figure 2. The scheme of Pilot Segment (PS): The six irrigation offtakes, the two culverts passing under the Idice River (Culv_1) and the Quaderna River (Culv_2), the four water gauges (WL) at the IN and OUT of both the culverts.

Six offtakes of PS serve a large irrigated area (8385 ha) through a network of not-pressurized irrigation channels. Over the four years of analysis, the biggest amount of water diverted from the segment (70% of the total offtake), was always diverted by the same three offtakes out of the six mentioned. The water gauges present at the segment are four: two at Culv_1 and two at Culv_2. They are located a few meters away from the entrance and the exit of both culverts (Figure 2).

2.4. Elaboration of the Multi-Disciplinary Modelling Approach on PS

The methodology was developed on a 7 km long Pilot Segment (PS), which was characterized by a simple geometry and a quite significant availability of water level measurements. The offtake discharges were estimated and verified also while considering the daily optimum *CWR* at field scale [46], that was estimated by the IRRINET, a regional irrigation DSS [45]. Combining hydraulic modelling of the CER with the optimization process of the hydraulic variables (Manning's coefficient and gate discharge coefficient) allowed for determining the flowing discharges. The simulations were run under steady flow conditions using the hydraulic software SIC^2 (5.38c, UMR G-eau IRSTEA, Montpellier, France) [49].

The methodology developed on PS was later applied on a 22 km Extended Segment (ES), that apart from a more complex geometry and hydraulic functioning (especially because of the presence of four culverts), also has a lower hydraulic data availability and lower accuracy when compared to PS. The methodology was tested on this particular segment, since it was characterized by different issues that are common in irrigation networks [19].

2.4.1. Reconstruction of the Unmeasured Offtake Discharges

The offtakes that were not measured were reconstructed using the indirectly calculated and the estimated data provided by the Associated Consortia and by the IRRINET service, respectively. In the following description, in order to distinguish these two data sources, different indexes are used: D for the former (Associated Consortia) and T for the latter (IRRINET). The index C indicates the results that were obtained by calculations done by the authors with the available data. The T-data aim to refine the time scale of the D-data and to verify them by a comparison with agronomic values, such as crop water requirements. Therefore, the obtained C-results (Equations (1)–(3)) have a decadal time scale instead of a monthly one; moreover, their values include agronomic aspects (e.g., optimum crop water requirement), the intensity, and the efficiency of the irrigation practices (Equation (4)).

During the decade n, the discharge exiting from a generic offtake k, q_{kCn} (m^3/s) can be written as:

$$q_{kCn} = q_{rDm}\, w_{kCn} \tag{1}$$

where q_{rDm} (m^3/s) is the average discharge diverted from the reference offtake during the month m (m = 1, 2, 3), and w_{kCn} is the weight of the offtake k during the decade n (n = 1, ..., 7).

The reference offtake was identified every year as the one diverting the greatest irrigation water volume. q_{rDm} was calculated as:

$$q_{rDm} = \frac{V_{rDm}}{D_m} \tag{2}$$

where V_{rDm} (m^3) is the indirectly calculated cumulated volume of the reference offtake for the month m, while D_m (s) is the duration of the month m.

The weight was obtained comparing the offtake k and the reference offtake in volumetric terms. The approach considered w_{kCn}, as follow:

$$w_{kCn} = \frac{(w_{kDm} + w_{kTn})}{2};\ w_{kDm} = \frac{V_{kDm}}{V_{rDm}};\ w_{kTn} = \frac{V_{kTn}}{V_{rTn}} \tag{3}$$

where w_{kDm} (-) and w_{kTn} (-) are the weights of the offtake k obtained using the D-data and the T-data, respectively, V_{kDm} (m^3) is the indirectly calculated volume of the offtake k during the month m, V_{kTn} (m^3), and V_{rTn} (m^3) are the volumes of the offtake k and of the reference offtake, respectively, calculated during the decade n using IRRINET.

In particular, for the decade n, the calculated volume of the generic offtake k (V_{kTn}) was determined by the expression [14]:

$$V_{kTn} = \left[\sum_{i=1}^{n} \left(\frac{CWR_i\, A_i\, II_i}{EI_i}\right)\right] \frac{1}{ED} \tag{4}$$

where A_i (m^2) is the area covered by the crop i per each year, CWR_i (mm) is the decadal cumulated optimum water requirement for the crop i, II_i (-) is the irrigation intensity of the crop i, EI_i (-) is the efficiency of the irrigation method for the crop i, and ED (-) is the efficiency of the delivery system.

If the generic offtake k is the reference offtake, the Equation (4) gives the quantity V_{rTn}.

The CWR values were provided by the Consortium of the CER, as already said in Section 2.2 for extensive cultivations (maize, soy, and alfa-alfa), for vegetables (beet, onion, melon, potato, and tomato), and for orchards (pear-tree, peach-tree, and vine).

The coefficient II indicates the intensity of irrigation, in other words, the ratio between the irrigated area and the area that potentially could be irrigated [50–52]. Its values were determined through field studies at the regional scale [53–56]. In particular, for the involved case-study crops, II ranges from 0.25 to 1, as shown in Table 2.

The coefficient EI indicates the efficiency of the irrigation method [57]. In Emilia Romagna, the considered value ranges are: 0.85–0.90 for drip irrigation and 0.70–0.80 for sprinkling irrigation [58]. In Table 2, the values of 0.85 and 0.75 were adopted for crops that were under the former and the latter irrigation efficiency, respectively.

The coefficient ED indicates the efficiency of the system that conveys water from the offtakes on the banks of the CER to the fields. For the present case-study, it was considered to be 0.50 [59,60]. In the area, in fact, 1122 km of channels (for both irrigation and drainage) and only 235 km of pipes provide water for crops. In particular, non-lined channels realize the 88% of the irrigation distribution [61].

Table 2. The values of the coefficients intensity of irrigation *(II)* and efficiency of the irrigation method *(EI)* for the irrigated crops served by PS and extended segment (ES).

Irrigated Crops	II_i (-)	EI_i (-)
Extensive crops		
Maize	0.75	0.75
Soy	0.50	0.75
Alfa-Alfa	0.25	0.75
Vegetables		
Beet	0.60	0.75
Onion	1.00	0.75
Melon	1.00	0.85
Potato	1.00	0.75
Tomato	1.00	0.85
Orchards		
Pear	1.00	0.85
Peach	1.00	0.85
Vine	0.50	0.85

2.4.2. Reconstruction of the Unmeasured Flowing Discharges

The hydraulic modelling combined with hydraulic variables optimization processes allowed for reconstructing the unmeasured flowing discharges along the segment.

SIC^2 (Simulation and Integration of Control for Canals) was selected as the most appropriate irrigation canal modelling software. It has been developed at IRSTEA (previously CEMAGREF, Montpellier, France) [62] and it enables describing the dynamics of rivers, drainage networks, and irrigation canals [63]. For the latter, devices (i.e., sills and gates) and irrigation offtakes can be specified in geometric and functioning terms [49]. SIC^2 can run steady flow computations under boundary conditions for discharge and/or water level [64]. In fact, it can consider several combinations of settings for devices and offtakes. The software provides the water level and the discharge profiles along the analyzed hydraulic system [29]. SIC^2 models also unsteady flow for initial conditions that were obtained from steady state computations [64] in discharge and water level terms. It can be used for water demand and control operations [19,65]. SIC^2 describes the dynamic behavior of water (discharge and water level) with the complete one-dimensional (1-D) Saint Venant equations in a bounded system [49]. This is the case of the CER in which the flow can be considered as mono-dimensional with a direction sufficiently rectilinear.

The 1-D Saint Venant equations are mathematically expressed as [66]:

$$\frac{\partial Q}{\partial x} + \frac{\partial S}{\partial t} = 0 \tag{5}$$

$$\frac{\partial Q}{\partial t} + \frac{\partial (Q^2/S)}{\partial x} + g\,S\,\frac{\partial Z}{\partial x} + g\,S\,J = 0 \tag{6}$$

where Q (m^3/s) is the discharge, S (m^2) is the wetted area, g (m/s^2) is the acceleration due to gravity, Z (m) is the water level, J (m/m) is the friction slope, x (m) is the longitudinal abscissa, and t (s) is the time.

The friction slope is obtained by the Manning-Strickler formula:

$$J = \frac{n^2\,Q^2}{S^2\,R^{4/3}} \tag{7}$$

where n (m$^{1/3}$/s) is the Manning's coefficient and R (m) is the hydraulic radius.

The continuity (Equation (5)) and the momentum (Equation (6)) equations are completed by boundary conditions for which SIC^2 provides a large range of options. They can be imposed in discharge, elevation, or rating curve terms. Lateral inflows and weir and gate equations can also be inserted. For example, the flow through a gate structure can be expressed by several classical or advanced equations, such as the submerged flow equation:

$$Q = C_d\,\sqrt{2\,g}\,L\,u\,\sqrt{Z_{up} - Z_{dn}} \tag{8}$$

where C_d (-) is the gate discharge coefficient, L (m) is the gate width, u (m) is the gate opening, Z_{up} (m), and Z_{dn} (m) are the water levels at the upstream and at the downstream of the gate, respectively.

The Saint Venant equations are non-linear partial differential equations and an analytical solution is restricted to problems of simple geometry. For all other cases, implicit finite difference approximations and a Preissmann scheme are used, as in the case of SIC^2 [66–68].

After the PS geometry entry, several hydraulic aspects were evaluated in SIC^2. The hydraulic variables values were set according to the literature: The Manning's coefficient presented a constant value of 0.013 (m$^{1/3}$/s) along the segment and within the two culverts [68] and the gate discharge coefficient that characterizes the entrances of each culvert was 0.6 [16,49,69]. The offtakes were modelled as "nodes" and they were characterized in discharge terms. In particular, the q_{kCn} values were inserted and were linearly interpolated in time.

For the year y, the vectors $Z1_{obs,y}$, $Z2_{obs,y}$, $Z3_{obs,y}$, and $Z4_{obs,y}$ can be defined. They contain the daily measured water levels at the four gauges: WL IN_1, WL OUT_1, WL IN_2, and WL OUT_2, respectively (Figure 2).

$$Z1_{obs,y} = \begin{pmatrix} Z1_{obs_1} \\ Z1_{obs_2} \\ Z1_{obs_j} \\ .. \\ Z1_{obs_e} \end{pmatrix} ; Z2_{obs,y} = \begin{pmatrix} Z2_{obs_1} \\ Z2_{obs_2} \\ Z2_{obs_j} \\ .. \\ Z2_{obs_e} \end{pmatrix} ; Z3_{obs,y} = \begin{pmatrix} Z3_{obs_1} \\ Z3_{obs_2} \\ Z3_{obs_j} \\ .. \\ Z3_{obs_e} \end{pmatrix} ; Z4_{obs,y} = \begin{pmatrix} Z4_{obs_1} \\ Z4_{obs_2} \\ Z4_{obs_j} \\ .. \\ Z4_{obs_e} \end{pmatrix} \tag{9}$$

where j is the index for the examined day of the year y ($j = 1, ..., e$).

The software SIC^2 can compute the values of discharge and water level along PS under two boundary conditions only in water level terms; for PS they were represented by $Z1_{obs,y}$, and $Z4_{obs,y}$. The daily simulated water level values at WL OUT_1, and WL IN_2 ($Z2_{sim,y}$ and $Z3_{sim,y}$) were compared to those that were measured ($Z2_{obs,y}$ and $Z3_{obs,y}$) in order to demonstrate the reliability and accuracy of the hydraulic model, and therefore, of the computed discharge values. The vectors $Z2_{sim,y}$ and $Z3_{sim,y}$ can be defined as:

$$Z2_{sim,y} = \begin{pmatrix} Z2_{sim_1} \\ Z2_{sim_2} \\ Z2_{sim_j} \\ .. \\ Z2_{sim_e} \end{pmatrix} \; ; \; Z3_{sim,y} = \begin{pmatrix} Z3_{sim_1} \\ Z3_{sim_2} \\ Z3_{sim_j} \\ .. \\ Z3_{sim_e} \end{pmatrix} \tag{10}$$

where j is the index for the examined day of the year y (j = 1, ..., e).

The simulations can be run under steady or unsteady state. The use of the former can be justified by the slow dynamics in the CER and the time and CPU (Central Processing Unit) memory saving. In particular, SIC^2 allows implementing a series of steady state simulations. The year 2015 was examined as a first test. The hydraulic model was run under a series of one-day steady state simulations and under one-day and 10-min unsteady state simulations.

A refined hydraulic model can be obtained after an optimization process. It allows for minimizing the differences in water level terms at WL OUT_1 and WL IN_2 playing on the values of the hydraulic variables and of a scaling factor for the offtakes; they were set as parameterized variables.

The optimization process consisted in a set of parameters to be evaluated, a criterion to be minimized, and a minimization function; it was based on the dialogue between SIC^2 and Matlab$^{\circledR}$ (version 9.1, The MathWorks, Inc., Natick, MA, USA).

In SIC^2, the parameterized hydraulic variables were explicit $Cd1$ and $Cd2$, gate discharge coefficients of Culv_1 and Culv_2; n, $n1$ and $n2$, Manning's coefficients along PS, within Culv_1 and Culv_2.

In Matlab$^{\circledR}$, this hydraulic set was recalled and the scaling factor Cq allowed multiplying the offtake discharge values from Section 2.4.1. In the math code, the criterion and the minimization function were implemented.

The vectors $diff2_y$ and $diff3_y$ can be defined as:

$$diff2_y = Z2_{sim,y} - Z2_{obs,y} \qquad and \qquad diff3_y = Z3_{sim,y} - Z3_{obs,y} \tag{11}$$

Therefore, the criterion to be minimized J was expressed as:

$$J = \sqrt{\sum_{j=1}^{e} \left[\frac{(diff2_y)^2}{\sigma 2_y^2} + \frac{(diff3_y)^2}{\sigma 3_y^2} \right]} \tag{12}$$

where j is the index for the examined day of the year y (j = 1, ..., e), $\sigma 2_y$, and $\sigma 3_y$ are the vectors containing the weights (values of 10 or 1), indicating whether a measure is affected by errors or not.

The iterative play on the parameterized hydraulic variables influenced the elements of the vectors $diff2_y$ and $diff3_y$, and consequently, the criterion J.

The minimization function considered was based on the Nelder-Mead simplex direct search algorithm, already implemented in Matlab$^{\circledR}$ [70]. In Figure 3, the iterations on J are shown for the year 2015.

At the end of the process, the minimization function identified parameterized hydraulic variables values that represent real minimum for the criterion (Figure 4).

For every year, these values were used for running the hydraulic simulations in SIC^2. The obtained model was called "optimized" and it returned the simulated discharges and water levels along PS. Finally, the optimization process was characterized by the cost of J that indicated the criterion value at the end of the iterations.

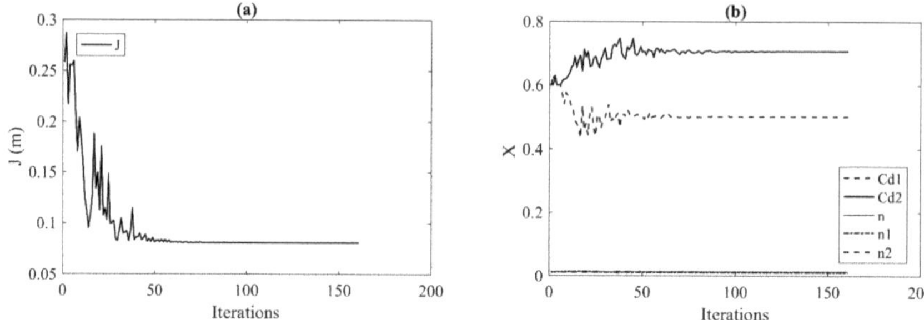

Figure 3. For 2015, the values of J (**a**) and of the parameterized hydraulic variables contained in the vector X (**b**) during the iterations of the optimization process that resulted in the minimization of the criterion.

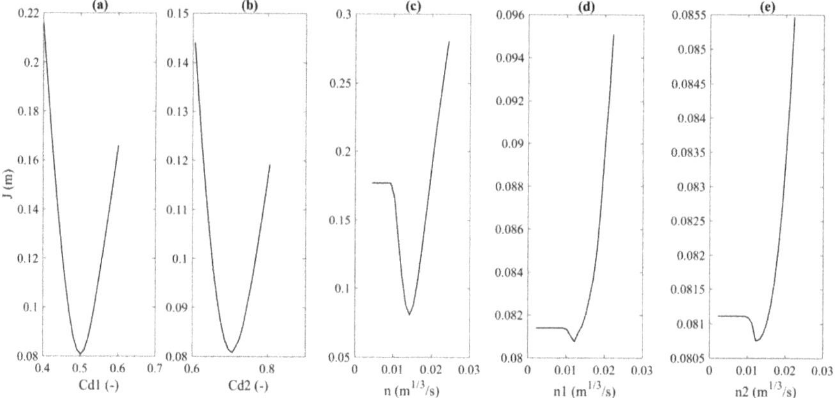

Figure 4. For 2015, $Cd1$ (**a**), $Cd2$ (**b**), n (**c**), $n1$ (**d**), $n2$ (**e**), the cuts of J around the minimum values of the parameterized hydraulic variables. In all cases they represent real minimum for the criterion.

Within the overall methodology, the measurement reliability represented a significant issue. The measures that are probably affected by errors (called "suspicious measures") can be contained in WL IN_1 and WL OUT_2 (boundary conditions), as in WL OUT_1 and WL IN_2 (optimization conditions) data series. The former affected the hydraulic model, while the latter the optimization process.

The days that are affected by suspicious measures were weighted in the optimization process through the elements of $\sigma2_y$ and $\sigma3_y$. In particular, if a day j is affected by a suspicious measure, the weight $(\sigma2_j; \sigma3_j)$ was set as 10; otherwise, it was equal to 1.

A detection method was elaborated considering the vectors $Z1_{obs,y}$, $Z2_{obs,y}$, $Z3_{obs,y}$, $Z4_{obs,y}$, $Q2_{sim,y}$, and $Q3_{sim,y}$. The latter two contained simulated values of discharge (output of the optimized hydraulic model) at the Culv_1 and Culv_2, respectively.

They can be expressed as:

$$Q2_{sim,y} = \begin{pmatrix} Q2_{sim_1} \\ Q2_{sim_2} \\ Q2_{sim_j} \\ .. \\ Q2_{sim_e} \end{pmatrix} ; \ Q3_{sim,y} = \begin{pmatrix} Q3_{sim_1} \\ Q3_{sim_2} \\ Q3_{sim_j} \\ .. \\ Q3_{sim_e} \end{pmatrix} \tag{13}$$

where j is the index for the examined day of the year y ($j = 1, ..., e$).

The method was based on the vectors:

$$delta_y = Z2_{obs,y} - Z3_{obs,y};$$

$$delta1_y = Z1_{obs,y} - Z2_{obs,y}; \tag{14}$$

$$delta2_y = Z3_{obs,y} - Z4_{obs,y};$$

For the day j, their elements represented the differences in water level terms along the segment and at the Culv_1 and Culv_2, respectively. The plots of $delta_y$-$delta1_y$, and $delta_y$-$delta2_y$ were used to evaluate in which vector the suspicious measures were located. The outliers of the data linear fitting were investigated. If the element j of $delta_y$ results as an outlier in both plots, a suspicious measure was in $Z2_{obsj}$ or in $Z3_{obsj}$. If the element j of $delta_y$ results as an outlier in the first plot but not in the second, the suspicious measure was in $Z1_{obsj}$. If the element j of $delta_y$ is an outlier in the second plot but not in the first, the suspicious measure was in $Z4_{obsj}$. To evaluate if a suspicious measure is in $Z2_{obsj}$ or $Z3_{obsj}$, $Q2_{sim,y}$-$delta1_y$, and $Q3_{sim,y}$-$delta2_y$ were plotted. For both, a data quadratic fitting of data was considered. If the j-th element of $delta1_y$ results as an outlier, the suspicious measure was in $Z2_{obsj}$ while if the element j results as an outlier of $delta2_y$, the suspicious measure was in $Z3_{obsj}$.

The most significant results obtained are given in Section 3.1.

2.5. Description of the Extended Segment (ES)

The multi-disciplinary modelling approach was then applied over a 22 km Extended Segment (ES) of the CER (Figure 5). Its downstream corresponds to WL IN_1 and its upstream is located in a delivery tank, few meters away from the pumping station Pieve di Cento exit. The latter counts seven pumps with a maximum capacity of 50 m^3/s and a maximum head of 4.5 m. For the first 33 m along the segment, the trapezium cross section top width is higher (85 m) and the bed altimetry varies from 10.79 m to 13.50 m above the sea level. Later, ES presents three different composite trapezium cross sections (top width from 26.4 m to 22.8 m; bottom width from 5.0 m to 3.3 m; side slope 3:1 and 1.5:1) and a constant slope (bed altimetry from 13.50 m to 12.81 m above the sea level). Four culverts under passing two roads (Road crossing_1 and Road crossing_2), the Navile Canal (Culv_3), and the Savena River (Culv_4) are characterized by a rectangular flow section of 36 m^2 (Figure 5). The road crossings present a modest length (about 20 m), while Culv_3 and Culv_4 are about 63 m and 86 m, respectively. The 12 occurring offtakes serve a total irrigated area of about 12,580 ha. The water gauges involved are only two at the ES extremities: WL OUT_0 (at the upstream) and WL IN_1 (at the downstream).

2.6. Application of the Multi-disciplinary Modelling Approach on ES

ES was characterized by a high complexity in geometric and functioning terms (Section 2.5). Moreover, the hydraulic data availability was poor; in fact, only two locations were equipped with water gauges. The multi-disciplinary modelling approach was applied over this segment in order to test its validity in more difficult conditions, representing a typical configuration in irrigation networks and with a significant lack of hydraulic measurements [19].

The offtake discharges decadal values were estimated as in Section 2.4.1. Due to the lack of available data, the PS flowing discharge resulted at WL OUT_1, was used to calculate the flowing discharges over ES, to run the optimized hydraulic model, and to compare the simulated and measured water level values at WL OUT_0. It was considered to be reliable due to the values of the parameterized hydraulic variables, of the linear interpolation parameters, and of the RMSE (Section 3.1.3). In particular, the PS flowing discharge values were used to define the ES upstream boundary conditions.

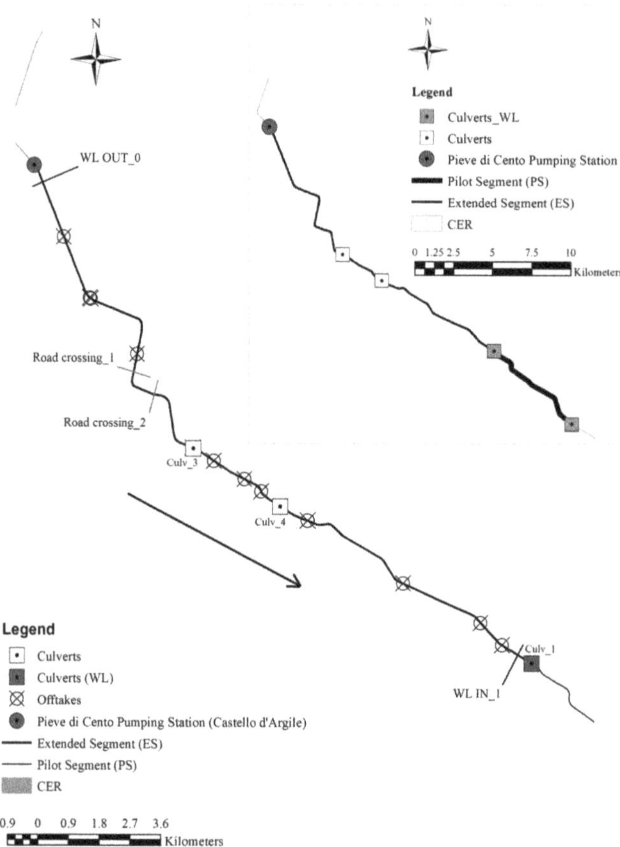

Figure 5. The scheme of ES: The 12 irrigation offtakes, the three culverts under passing the Idice River (Culv_1), the Navile Canal (Culv_3), and the Savena River (Culv_4), the two water gauges (WL OUT_0 and WL IN_1) at the OUT of the pumping station Pieve di Cento and at the IN of Culv_1, the two road crossings.

For the year y, the vector containing the calculated discharge values of a generic offtake k (Section 2.4.1) can be expressed as:

$$q_{kC,y} = \begin{pmatrix} q_{kCn_1} \\ q_{kCn_2} \\ q_{kCn_j} \\ .. \\ q_{kCn_e} \end{pmatrix} \tag{15}$$

where j is the index for the examined day of the year y ($j = 1, ..., e$).

Defining as $q_{totC,y}$, the total offtake discharges vector:

$$q_{totC,y} = \begin{pmatrix} q_{totC_1} \\ q_{totC_2} \\ q_{totC_j} \\ .. \\ q_{totC_e} \end{pmatrix} \tag{16}$$

Its element q_{totCj} was calculated as:

$$q_{totC_j} = \sum_{k=1}^{12} q_{kCn_j} \tag{17}$$

where k is the index of the generic offtake ($k = 1, \dots , 12$).

For the year y, the vector $Q0_y$ represented the ES upstream boundary conditions. It was obtained as:

$$Q0_y = Q2_{sim,y} + q_{totC,y} \tag{18}$$

whereas, the $Z1_{obs,y}$ values reported were used as the downstream boundary conditions. The hydraulic model was implemented under a series of one-day steady state simulations. For every year, the vector $Z0_{obs,y}$ contains the daily measured water levels values at WL OUT_0. They were used for testing the model performances and for evaluating the optimization process. $Z0_{obs,y}$ can be defined as:

$$Z0_{obs,y} = \begin{pmatrix} Z0_{obs_1} \\ Z0_{obs_2} \\ Z0_{obs_j} \\ .. \\ Z0_{obs_e} \end{pmatrix} \tag{19}$$

where j is the index for the examined day of the year y ($j = 1, ..., e$).

The optimized parameterized hydraulic variables set was larger than that of PS. It consisted in $Cd3$, $Cd4$, $Cd5$, and $Cd6$, gate discharge coefficients of Culv_3 and Culv_4 and of the two road crossings; n, $n3$, $n4$, $n5$, and $n6$, Manning's coefficients along ES, within the two culverts and the two road crossings. The significant uncertainty that affects the measured water levels at WL OUT_0 (Section 2.2) was reflected in the larger parameterized hydraulic variables set size. The high degree of freedom allowed for obtaining physically possible values of the parameters and the lower cost of J at the optimization process end. The gate discharge coefficients values could not be imposed as those of PS because the geometric and functioning characterization difference. Moreover, if the Manning's coefficients are imposed, the optimization process gives higher gate discharge coefficients values (>1) that are not physically correct. The offtake discharges scaling factor was not considered, as explained in Section 3.1.3.

For the year y, the vector $Z0_{sim,y}$ contained the daily simulated water levels at WL OUT_0:

$$Z0_{sim,y} = \begin{pmatrix} Z0_{sim_1} \\ Z0_{sim_2} \\ Z0_{sim_j} \\ .. \\ Z0_{sim_e} \end{pmatrix} \tag{20}$$

where j is the index for the examined day of the year y ($j = 1, ..., e$).

The optimization criterion was based on the definition of the vectors $diff0_y$ and $\sigma0_y$. The former contained the values of the daily differences between simulated and measured water levels at WL OUT_0, as:

$$diff0_y = Z0_{obs,y} - Z0_{sim,y} \tag{21}$$

The vector $\sigma0_y$ weighted the measures probably affected by errors ("suspicious") located in $Z0_{obs,y}$. The detection involved the Pieve di Cento pumps functioning data. In particular, for the year y, the vectors $Z0_{pmax,y}$ and $Z0_{pmin,y}$ contained the daily maximum and minimum values of the delivery tank water level that is registered by the pumps functioning, as:

$$Z0_{pmax,y} = \begin{pmatrix} Z0_{pmax_1} \\ Z0_{pmax_2} \\ Z0_{pmax_j} \\ .. \\ Z0_{pmax_e} \end{pmatrix} ; \ Z0_{pmin,y} = \begin{pmatrix} Z0_{pmin_1} \\ Z0_{pmin_2} \\ Z0_{pmin_j} \\ .. \\ Z0_{pmin_e} \end{pmatrix} \tag{22}$$

where j is the index for the examined day of the year y ($j = 1, ..., e$).

For every day j, the functioning range $Z0_{pmaxj}$-$Z0_{pminj}$ was identified. If $Z0_{obsj}$ do not belong to it, it is defined as a suspicious measure.

The expression of the criterion J was:

$$J = \sqrt{\sum_{j=1}^{e} \frac{(diff0_y)^2}{\sigma0_y^2}} \tag{23}$$

The minimization function is that of PS (Nelder-Mead simplex direct search algorithm). The most significant results obtained are given in Section 3.2.

3. Results and Discussion

3.1. Pilot Segment (PS)

3.1.1. Unmeasured Offtake Discharges

For every year, the values of w_{kDm}, w_{kTn}, and w_{kCn} were calculated, as in Section 2.4.1. Out of these weights, the first one resulted generally higher than the second one; the V_{rDm}-V_{kDm}, in fact, differed considerably from V_{rTn}-V_{kTn}. When considering the year 2015 as an example, the maximum values were 23.04×10^4 m^3 and 13.68×10^4 m^3, respectively. For the same year, Figure 6a underlines the monthly variability of w_{kDm} as compared to the decadal one of w_{kTn} for two offtakes: Offtake$_1$ (reference offtake) and Offtake$_5$ (Figure 2). The weights w_{kCn} were obtained averaging D-data and T-data according to Equation (3), and they were reported in Figure 6b. The averaging of those values was needed to minimize the possible measurement errors in D-data, and also, to take into account that CWR from IRRINET are "optimal requirements", when considering that water was always fully available.

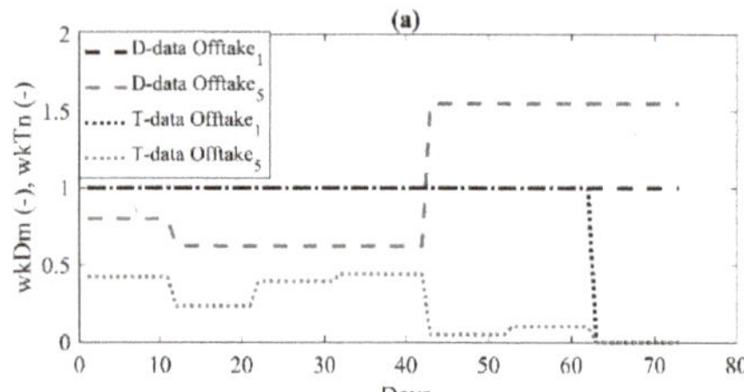

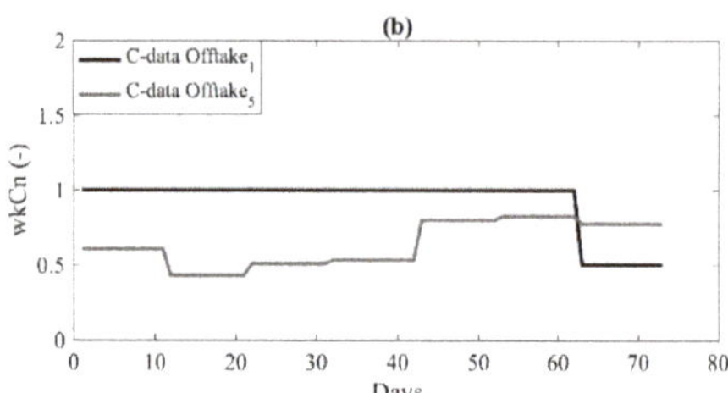

Figure 6. For the year 2015, the values of the weights w_{kDm} and w_{kTn} (**a**) and w_{kCn} (**b**) for the reference offtake (Offtake$_1$) and for a generic one (Offtake$_5$).

Over the four years of analysis, the trend of the offtake discharge values (q_{kCn}) was mainly coherent with the yearly meteo-climatic conditions (i.e., average daily rainfall). The reference offtake discharge values ranged from 0 m^3/s to 0.24 m^3/s. q_{kCn} of all other offtakes varied from 0 m^3/s to 0.17 m^3/s.

Figures 7a and 7b show that the two offtakes (Offtake$_1$ and Offtake$_5$) had the lowest values in 2014 (mean values of 0.021 m^3/s and 0.032 m^3/s, respectively) and the highest mainly in 2012 (mean values of 0.137 m^3/s and 0.082 m^3/s, respectively). If the month of July is considered, the discharge values of the reference offtake were lower in 2012 than in 2013 and 2015. This can be explained not only by meteo-climatic conditions (that resulted in crop stress), but also by insufficient machine, manpower, or energy availability at the field. Moreover, among the years that were analysed, the cultivated crops differ.

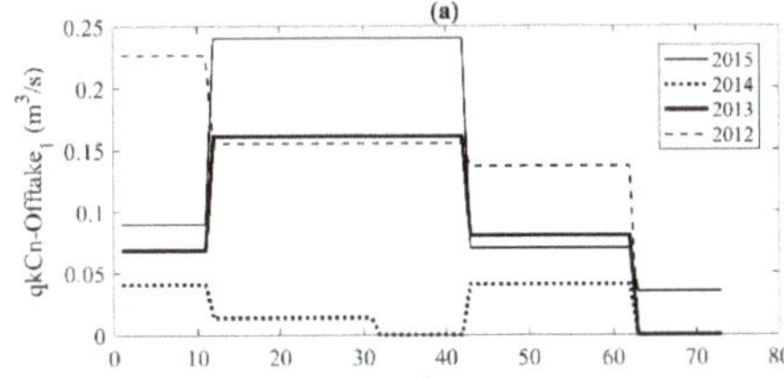
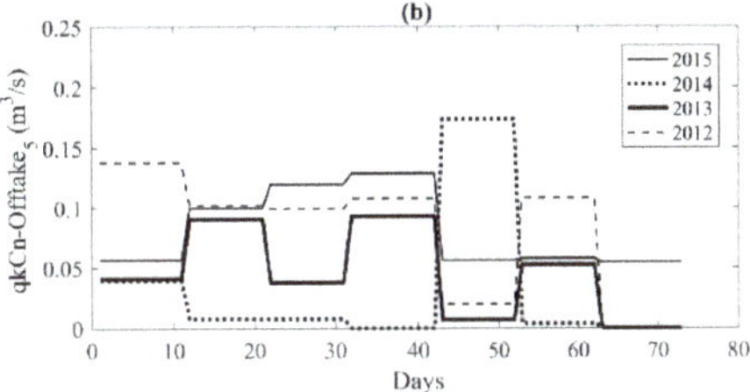

Figure 7. For every year, the variability of the diverted discharges for the reference offtake (Offatake$_1$) (**a**) and for a generic offtake (Offtake$_5$) (**b**).

3.1.2. Steady State Flow Condition

To evaluate if the hydraulic models should be run under steady or unsteady state conditions, the results of the year 2015 were analysed. They consisted in discharge and water level values at WL OUT_1 and WL IN_2. The hydraulic model of PS was run under a series of one-day steady state (Steady-1d) simulations, and under one-day (Unsteady-1d), and 10-min (Unsteady-10mn) unsteady state simulations.

The vectors *Z2sim*-2015 and *Z3sim*-2015 for Steady-1d and Unsteady-1d were completely overlaid. The differences obtained by comparing these vectors for Steady-1d and Unsteady-10mn reported mean values of 0.285 m and 5.347 $\times$ 10^{-4} m, respectively.

For Steady-1d and Unsteady-1d the vectors *Q2sim*-2015 and *Q3sim*-2015, so as the simulated water levels, were completely overlaid. If the simulations of steady state and those of unsteady state with time step 10 min are compared, the resulted maximum and mean differences were 3.843 m^3/s and 0.279 m^3/s, respectively.

For the optimization of the hydraulic model, the results in water level and discharge terms can be considered to be approximatively identical for the three flow conditions that are considered.

The series of one-day steady simulations was adopted for running the hydraulic models of both PS and ES. This assumption was justified by the slow dynamics occurring in the CER, and it is also coherent with the time scale of calculated offtake discharges (decadal) and of measured water level (daily) data. The use of steady state saves time and CPU memory that is an important point, since this hydraulic calculation is embodied into an optimization loop. Using only one run, SIC^2 computes 73 steady state simulations; one for every day of the irrigation period. The hydraulic variables on a daily basis are not function of time.

3.1.3. PS optimized Model

The optimized hydraulic model returned the flowing discharges along the PS. For example, in Figure 8, the $Q2_{sim,y}$ values are reported (values at the upstream of PS). For every year, they are grouped into two vectors: $Q2_{simc}$ (output from measured water levels not affected by errors) and $Q2_{sims}$ (output from measured water levels probably affected by errors, Section 2.4.2).

The lowest values of flowing discharge were calculated for the rainier year (2014) and they were 17.560 m^3/s ($Q2_{sim,2014}$) and 16.660 m^3/s ($Q3_{sim,2014}$), with standard deviations of 2.694 m^3/s and 3.204 m^3/s, respectively. When considering $Q2_{sim,y}$ as an example, the years 2013 and 2015 were characterized by higher flowing discharge mean values (23.930 m^3/s and 21.710 m^3/s) and standard deviation values (4.230 m^3/s and 4.841 m^3/s) as compared to those of the year 2012 (20.700 m^3/s and 2.538 m^3/s, respectively). This can be justified by the limiting factors that are mentioned in Section 3.1.1. Therefore, the years with extreme climatic conditions (2014 and 2012) presented less variability in relation to flowing discharge mean values when compared to the years 2013 and 2015 characterized by the alternation of dry and rainy intervals.

The values of the flowing discharge were the result of many factors: offtake discharges (that followed characterization, as explained in Section 3.1.1), the modelled functioning of culverts and the measured water levels.

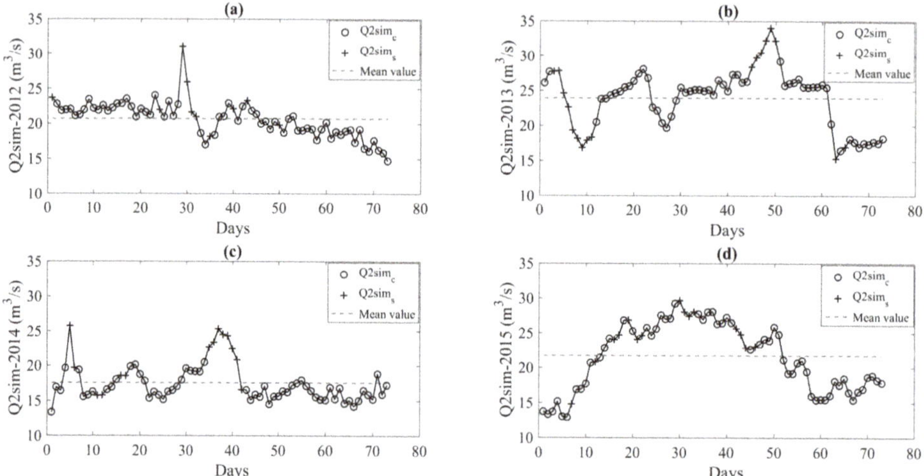

Figure 8. For every year: 2012 (**a**); 2013 (**b**); 2014 (**c**) and 2015 (**d**), the values of $Q2_{sim}$ obtained from the optimized model.

For the year y, the optimized hydraulic model performances were evaluated through the values of the parameterized variables and of the differences between water levels simulated and measured at WL OUT_1 and WL IN_2. At the end of the optimization process, the values of the hydraulic variables should be physically correct and coherent with literature [68].

For Cq, the yearly values that were obtained resulted close to 1. If the optimization process were cut around these values, they would not represent real minimum. The offtake discharges impact on the water levels at WL OUT_1 and WL IN_2 around their nominal values was less than the measurement accuracy considered ($\pm$0.05 m); the offtake discharges represented small rates if compared to flowing discharges. When considering the year 2015 as an example, the flowing discharge maximum and minimum values were 29.66 m^3/s and 12.88 m^3/s, respectively, while the reference offtake discharge ranged from 0.24 m^3/s (0.81% of the flowing discharge maximum) to 0.07 m^3/s (0.24% of the flowing discharge minimum). Cq cannot be considered as one of the parameters for the optimization process since it did not have any influence on it.

The parameterized hydraulic variables and the cost of the criterion are reported in Table 3 for every year of analysis. The results that were obtained with the suspicious measures weights are discussed in the following paragraphs.

Table 3. The values of the five parameterized variables and the cost of the criterion obtained from the optimization process: Without (above) and with the weights of suspicious measures (below).

Year	Parameterized Hydraulic Variables					Cost of the Criterion
	$Cd1$ (-)	$Cd2$ (-)	n (m$^{1/3}$/s)	$n1$ (m$^{1/3}$/s)	$n2$ (m$^{1/3}$/s)	J Cost (m)
	Without Suspicious Measures Weights					
2012	0.37	0.64	0.014	0.015	0.015	0.1742
2013	0.68	0.76	0.015	0.009	0.011	0.2799
2014	0.39	0.82	0.016	0.015	0.008	0.2331
2015	0.49	0.69	0.015	0.012	0.013	0.1036
	With Suspicious Measures Weights					
2012	0.37	0.65	0.014	0.014	0.015	0.1460
2013	0.71	0.74	0.016	0.009	0.011	0.1480
2014	0.44	0.80	0.015	0.013	0.010	0.1101
2015	0.50	0.71	0.014	0.012	0.012	0.0808

The gate discharge coefficients ($Cd1$ and $Cd2$) refer to submerged flow for both culverts. The $Cd2$ values were coherent with the range 0.60–0.85 that was reported in literature [69,71–73]. For all years, surface flow occurred within Culv_2. For some years, the $Cd1$ values significantly differed from the literature range, and it can be explained by applying Equation (8) to the two gates at Culv_1 and Culv_2. For example, when Equation (8) was applied on the year 2012 ($Cd1$ = 0.37, $Cd2$ = 0.64), for Culv_1, the term (Z_{up}-Z_{dn}) reported maximum and minimum values of 0.13 m and 0.03 m, respectively. They were higher than those at Culv_2 that were 0.07 m and 0.01 m, respectively. Due to the modest impact of the offtakes, the values of the discharges at the two culverts were similar, and therefore the gate discharge coefficient at Culv_1 has to be lower than the one at Culv_2. Within Culv_1, both flow types (surface and piped) occurred. The years 2012 and 2015 were characterized by 39 days of surface and 34 days of piped flows. The years 2013 and 2014, on the other hand, presented mainly surface flow (57 and 59 days, respectively).

The n values that were obtained were coherent with the reported literature range for concrete canals (0.010–0.020 m$^{1/3}$/s) [68]. Over the last four years, the mean value was 0.0147 m$^{1/3}$/s and the maximum difference attested was 0.002 m$^{1/3}$/s (2012–2013). The $n1$ and $n2$ values were coherent with the literature range for concrete culverts (0.010–0.014 m$^{1/3}$/s) [74] and both had a mean of 0.012 m$^{1/3}$/s. When considering the analysis period, the maximum difference among the years was 0.005 m$^{1/3}$/s (between 2012 and 2013 for $n1$, and between 2012 and 2014 for $n2$). The Manning's coefficient is the result of many factors: Basic value (roughness of the material that was used to line the canal), irregularities of the canal bed, cross sections variations, obstacles, vegetation growth, and meandering [75,76]. Along the PS, the Manning's coefficient was stable; its variations can be related to the presence of obstacles (debris, downed plants, and dropped obstacles) and algae growth. For the culverts, it showed more variability and it was the result of many possible factors, such as the grids at the culverts entrances, which involve head losses, the gates modelling approximations, and the additional head losses due to the change of geometry between open and closed flow cross sections.

For every year, the performances of the optimized hydraulic model were evaluated through the elements contained in $Z2_{sim,y}$ and $Z3_{sim,y}$ (Figure 9).

The differences between simulated and measured water levels at WL OUT_1 and at WL IN_2 affected the cost of the criterion and the linear interpolation parameters. The former was also influenced by the $\sigma2_y$ and $\sigma3_y$ vectors, as shown in Table 3. The maximum difference for the cost of the criterion was 0.1319 m (0.2799–0.1480 m) for 2013.

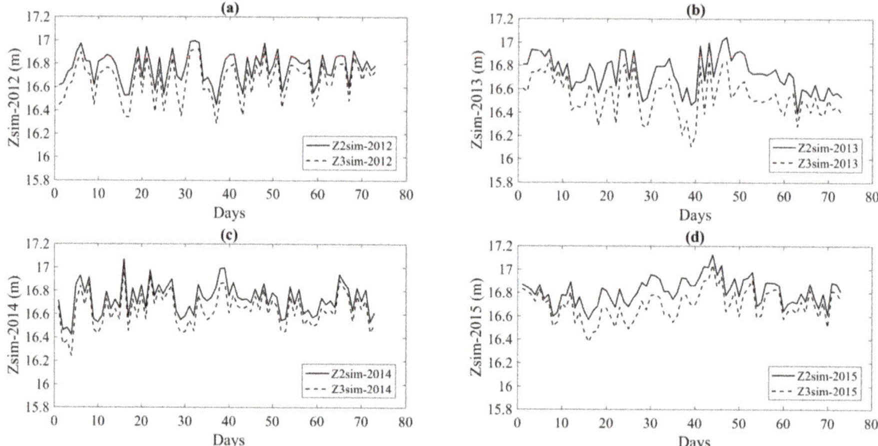

Figure 9. For every year: 2012 (**a**); 2013 (**b**); 2014 (**c**) and 2015 (**d**), the simulated water level values contained in $Z2_{sim}$ and $Z3_{sim}$.

For every year, in order to compare simulated and observed water level values, the former were plotted in the X-axis, while the latter in the Y-axis [77–81]. In this plot format, the points on the Y = X line represent the perfect correspondence between model-predicted and measured values; therefore, the intercept and the slope are 0 and 1, respectively [82]. Points below or above that line indicate over or under-estimations of the model [77]. In Figure 10, the elements of the vector $Z2_{sim,y}$ were plotted versus those of the vector $Z2_{obs,y}$. The former were reported for optimized (Opt) and non-optimized (Non-Opt) models.

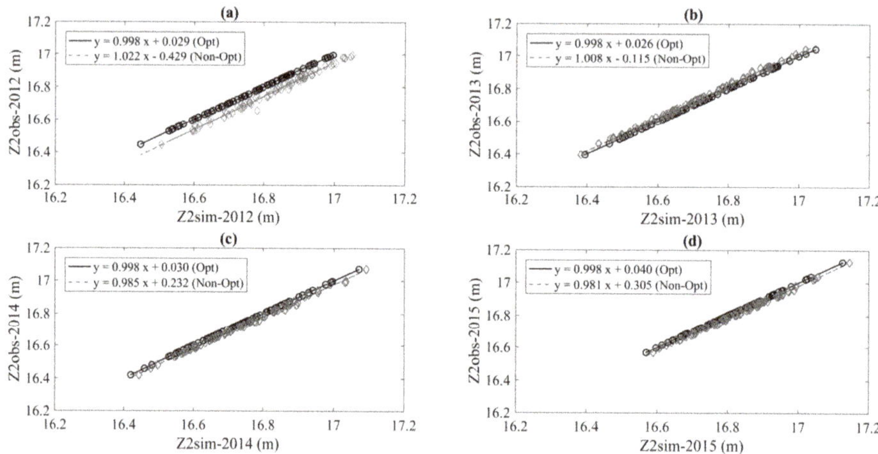

Figure 10. For every year: 2012 (**a**); 2013 (**b**); 2014 (**c**) and 2015 (**d**), the linear interpolation of $Z2_{obs}$ and $Z2_{sim}$ for both optimized and non-optimized models.

The validity of the optimized model was verified because of the line interpolation parameters values were closer to the optimum ones, especially in line intercept terms (i.e., 0.029 instead of 0.429 for $Z2sim$-2012). Over the four years, the mean values of intercept and slope line were 0.031 and 0.998, respectively. The same evaluation method was applied to $Z3_{sim,y}$ and $Z3_{obs,y}$ (Figure 11). Also, in this case, the optimized model shows an excellent fit, reporting mean values of intercept and slope line of 0.105 and 0.994, respectively.

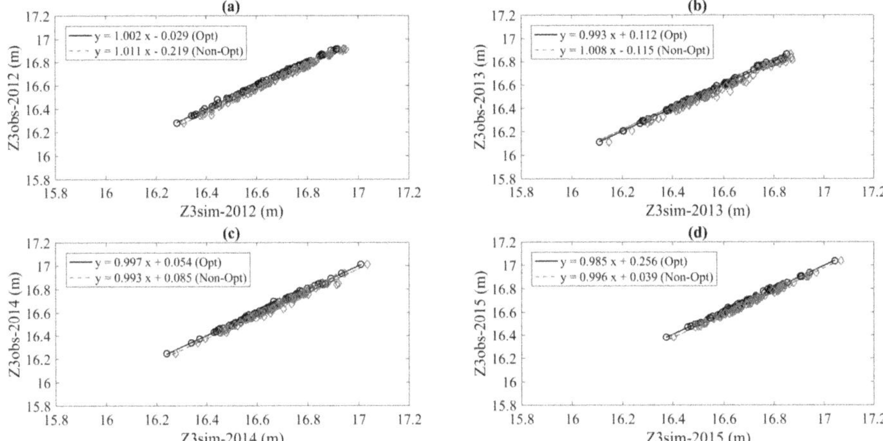

Figure 11. For every year: 2012 (**a**); 2013 (**b**); 2014 (**c**) and 2015 (**d**), the linear interpolation of $Z3_{obs}$ and $Z3_{sim}$ for both optimized and non-optimized models.

The performances of the optimization process have been also evaluated in terms of the root mean square error (RMSE) (Table 4). For the optimized model, the RMSE was calculated at WL OUT_1 and at WL IN_2 reporting mean values of 4.661×10^{-4} m and 8.150×10^{-3} m, respectively. They significantly differ from those of the non-optimized one (mean value of 0.0302 m at WL OUT_1 and 0.0285 m at WL IN_2).

Table 4. The root mean square error (RMSE) values for both optimized and non-optimized models at WL OUT_1 and WL IN_2.

Year	RMSE (m)	
	Non-Optimized Hydraulic Model	**Optimized Hydraulic Model**
WL OUT_1		
2012	0.0586	2.9×10^{-4}
2013	0.0220	6.1×10^{-4}
2014	0.0216	5.8×10^{-4}
2015	0.0185	3.9×10^{-4}
WL IN_2		
2012	0.0318	6.1×10^{-3}
2013	0.0340	11.2×10^{-3}
2014	0.0316	8.3×10^{-3}
2015	0.0265	7.0×10^{-3}

Overall, the comparison among simulations highlighted the fact that the optimized model achieved excellent results, which are very close to the measured values. In the RMSE terms, the differences between the two models (non-optimized vs optimized) had maximum value of 0.0583 m, that was recorded at WL OUT_1 for the dry year (2012). Moreover, the mean differences were 0.0297 m and 0.0228 m at WL OUT_1 and WL IN_2, respectively. When considering the measurements accuracy order of magnitude (± 0.05 m), the optimization process significantly improved the obtained results.

3.2. Extended Segment (ES)

The ES offtake discharges were calculated as in Section 2.4.1, and the q_{kCn} obtained were mainly coherent with the yearly meteo-climatic condition. In particular, the ES reference offtake reported minimum (0.02 m^3/s) and maximum (1.17 m^3/s) values during the rainy and the dry years, respectively. Moreover, q_{kCn} for all other offtakes varied from 0 m^3/s (2014) to 0.87 m^3/s (2012). This range was larger than those of PS (0–0.24 m^3/s for the reference offtake and 0–0.17 m^3/s for all other offtakes). In fact, the ES irrigated land supplied is 1.5 times larger (12,580 ha) than that of PS.

At WL OUT_0, the flowing discharges calculated with the Equation (18) are reported in Figure 12. For every year, the $Q0$ elements were grouped in $Q0_s$ and $Q0_c$ vectors in order to distinguish the flowing discharge values that are based on $Q2_{sims}$ and $Q2_{simc}$, respectively.

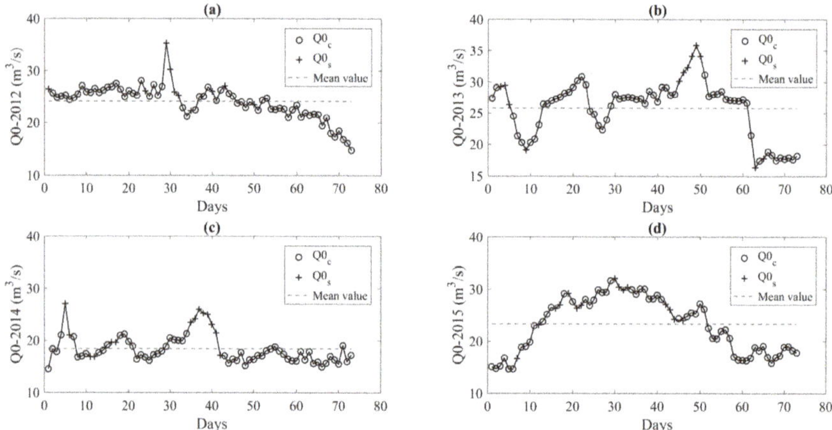

Figure 12. For every year: 2012 (**a**); 2013 (**b**), 2014 (**c**) and 2015 (**d**), the values of discharge ($Q0$) calculated at WL OUT_0.

The lowest values of flowing discharges resulted for the rainy year with a mean value of 18.46 m^3/s (standard deviation of 2.70 m^3/s); the highest values were related to 2012 (24.13 m^3/s on average with a standard deviation of 3.22 m^3/s) and 2013 (25.81 m^3/s on average, standard deviation of 4.569 m^3/s).

For the year y, the performances of the ES optimized hydraulic model were evaluated through the values of the parameterized hydraulic variables and the differences between simulated and measured water levels at WL OUT_0. When considering the hydraulic variables, the optimization process returned physically possible values while only using a larger set of parameters. In particular, four gate discharge coefficients and five Manning's coefficients were investigated to characterize ES (in roughness terms) and every culvert (in roughness and head loss terms). For three years, the values of the parameterized hydraulic variables obtained are reported in Table 5.

Table 5. The values of the nine parameterized variables and of the cost of the criterion obtained from the optimization process.

Year	Parameterized Hydraulic Variables									Cost of the Criterion
	$Cd3$ (-)	$Cd4$ (-)	$Cd5$ (-)	$Cd6$ (-)	n (m$^{1/3}$/s)	$n3$ (m$^{1/3}$/s)	$n4$ (m$^{1/3}$/s)	$n5$ (m$^{1/3}$/s)	$n6$ (m$^{1/3}$/s)	J Cost (m)
2012	0.60	0.45	0.62	0.52	0.020	0.020	0.013	0.011	0.010	0.5057
2013	0.58	0.60	0.58	0.58	0.014	0.013	0.013	0.013	0.013	0.4667
2015	0.42	0.59	0.43	0.45	0.011	0.019	0.019	0.019	0.010	0.3465

For every year of analysis, the optimization process was run in order to obtain the parameterized hydraulic variables values. For the year 2014, it could not end and it tended to minimized the criteria assigning negative values to the Manning's coefficients and high values (>1) to the gate discharge coefficients. So, for this year, the optimization loop was not finalized.

All gate discharge coefficients referred to submerged flow. The values obtained presented less variability than those of PS (Table 3). They were around 0.60, except for the year 2015 (mean value of 0.47). In 2012, the four culverts were mainly characterized by piped flow (as for Culv_2 of PS), while the year 2013, except for Culv_4, presented mainly free flow conditions.

As for PS, the n values that were obtained were coherent with the literature range reported for concrete canals 0.010–0.020 m$^{1/3}$/s [68]. Over the three years, the mean value was 0.015 m$^{1/3}$/s (very similar to PS n) and the maximum difference of 0.009 m$^{1/3}$/s was between the years 2012 and 2015 (0.002 m$^{1/3}$/s in PS). The $n3$, $n4$, $n5$ and $n6$ values were coherent with the range 0.010–0.014 m$^{1/3}$/s for concrete culverts [74], except for the year 2015, for which the values were higher (0.019 m$^{1/3}$/s maximum). As said before, the Manning's coefficient differences can be attributed to several factors, such as geometric irregularities or variations of the canal bed and of cross sections, obstacles, vegetation growth, and meandering. Moreover, the field survey estimated the accuracy of $Z0_{obs,y}$ values (±0.10 m) to be lower than those in PS.

When considering the same year y, the ES J cost quite significantly differed from PS, and the values of the elements contained in $diff2_y$, $diff3_y$ and $diff0_y$ can justify these results. In fact, considering the year 2015 as an example, the maximum absolute values (only for days not affected by suspicious measures) are 0.015 m at WL OUT_1 ($diff2_y$) and 0.011 m at WL IN_2 ($diff3_y$), while at WL OUT_0 ($diff0_y$), it was much higher and very close to the accuracy threshold (0.102 m). In any case, Figure 13 shows that all the single elements of $Z0_{sim,y}$ are within the $Z0_{obs,y}$ accuracy range (±0.10 m), except for few days (eight for 2012, nine for 2013, and two for 2015) that are related to $Z0_{obsj}$ or $Q2_{simj}$ and that were probably affected by errors.

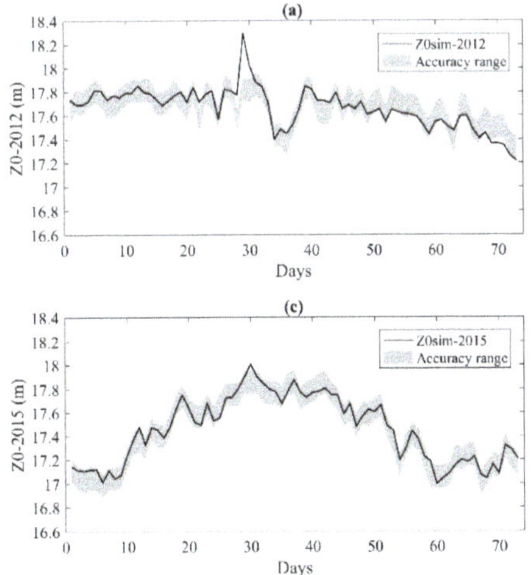
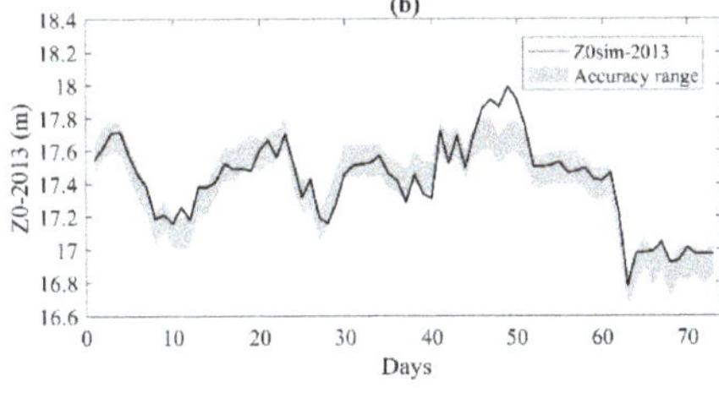

Figure 13. The simulated water level values contained in $Z0_{sim}$ for the years 2012 (**a**), 2013 (**b**) and 2015 (**c**).

As for PS, the vectors $Z0_{sim,y}$ and $Z0_{obs,y}$ were plotted (Figure 14) in order to detect the modelling impacts of the $diff0_y$ elements. The intercept and the slope of the linear correlation were evaluated, and they were compared with the optimum values (i.e., perfect fitting) and with those reported for PS.

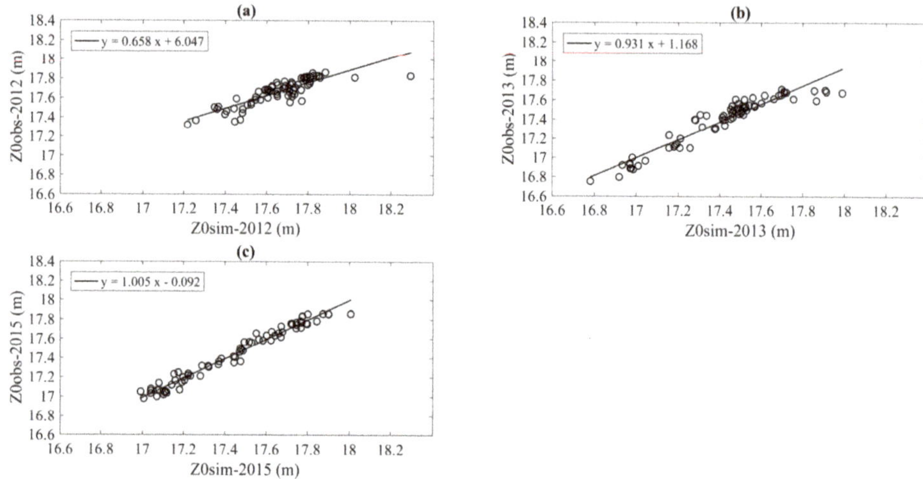

Figure 14. The linear interpolation of $Z0_{obs}$ and $Z0_{sim}$ for the optimized model for the years 2012 (**a**), 2013 (**b**) and 2015 (**c**).

The results can be considered excellent for the years 2013 and 2015, with RMSE values of 0.09 m and 0.05 m, respectively. For 2012, especially the intercept of the linear interpolation (6.047) was significantly different from the optimal value (0). As already reported by Mesplé [83], the modelling overestimation/underestimation was probably combined with the proportionality of the gap between the measured and simulated values. Therefore, the RMSE for 2012 (0.09 m) was similar to the other years.

Overall, the RMSE values in ES simulations were higher than those reported in PS, indicating that the model worked better in a segment that was characterized by simpler geometry and with higher availability and reliability of measured hydraulic data. When the model was tested on a more complex reality i.e., ES, it had to face two critical aspects: A scarce number and a lower accuracy of the hydraulic measured data. The latter affected the optimization process, especially for the years with extreme climatic conditions. In fact, the dry 2012 was characterized by an intense functioning of pumps with a maximum daily difference in water level terms of 1.23 m. On the contrary, the rainy 2014 presented lower irrigation demands and therefore the functioning of pumps was more intermittent. This implies that, due to the slope of ES (3.8×10^{-5}), a backwater flow occurred affecting the optimization process and leading to a poor representation of the reality. The multi-disciplinary modelling approach developed in this study presented satisfying results for the two remaining years (2013 and 2015).

4. Conclusions

A low availability of hydraulic data can seriously affect efficient management of irrigation canals. Therefore, this paper presents a novel approach that can be applied to reconstruct the missing hydraulic data by combining hydraulic modelling and an irrigation DSS (that was developed at a regional scale).

The approach was developed on a Northern Italian canal, more specifically, on its 7 km long segment (PS), which is characterized by a quite simple geometry and full availability of water levels, and it gave very good results. Its application on a more complex segment (ES) with a poor data availability and accuracy, confirmed that the approach can be successfully used to reconstruct data for years with standard meteo-climatic conditions, while years with extreme climatic conditions are more difficult to be simulated. It was found that the measuring point and consequently instrument accuracy are key factors for obtaining a model that can well represent the reality.

Moreover, the results showed that the offtake discharges can be estimated on the base of crop water delivery schedules and combining them with measured water levels could enable calculating the discharges that are flowing through the use of an optimized hydraulic model.

However, this approach was developed on a lined concrete canal. Therefore, its application on secondary channels, often on earth with considerable infiltration losses, have to be further studied in order to optimize the hydraulic model and to increase its relevance.

Since the approach proposed allows quantifying discharges and water levels along an irrigation canal, it can be integrated with water qualitative analysis (e.g., microbiological aspects), thus widening its multi-disciplinarity.

Author Contributions: Conceptualization, M.L., P.-O.M., A.B. and A.T.; Methodology, M.L., P.-O.M., A.B. and V.D.F.; Validation, M.L., P.-O.M., A.B. and A.T.; Formal Analysis, M.L. and V.D.F.; Investigation, M.L.; Resources, P.-O.M. and A.B.; Data Curation, M.L., P.-O.M. and A.B.; Writing-Original Draft Preparation, M.L. and P.-O.M.; Writing-Review & Editing, M.L., P.-O.M., A.B. and A.T.; Visualization, M.L.; Supervision, A.T.; Project Administration, A.T.; Funding Acquisition, A.T.

Funding: This research received no external funding.

Acknowledgments: Other data resources are the partners of the project: The consortium of the canale emiliano romagnolo (CER), the associated consortium bonifica renana and ARPAER (the local agro-meterological service).

Conflicts of Interest: The authors declare no conflict of interest.

Abbreviations

The following abbreviations are used in this manuscript.

Along the CER

$Culv\_1$, $Culv\_2$	Culverts of Pilot Segment passing under rivers
$Culv\_3$, $Culv\_4$	Culverts of Extended Segment passing under rivers
ES	Extended Segment
PS	Pilot Segment
$WL\ OUT\_0$	Water gauge at the exit of the pumping station Pieve di Cento
$WL\ IN\_1$	Water gauge at the entrance of Culv_1
$WL\ OUT\_1$	Water gauge at the exit of Culv_1
$WL\ IN\_2$	Water gauge at the entrance of Culv_2
$WL\ OUT\_2$	Water gauge at the exit of Culv_2

Measured data

$Z0_{obs,y}$	Vector containing daily water levels at WL OUT_0 for the year y
$Z0_{pmax,y}$	Vector containing maximum daily water levels from the functioning of Pieve di Cento pumps
$Z0_{pmin,y}$	Vector containing minimum daily water levels from the functioning of Pieve di Cento pumps
$Z1_{obs,y}$	Vector containing daily water levels at WL IN_1 for the year y
$Z2_{obs,y}$	Vector containing daily water levels at WL OUT_1 for the year y
$Z3_{obs,y}$	Vector containing daily water levels at WL IN_2 for the year y
$Z4_{obs,y}$	Vector containing daily water levels at WL OUT_2 for the year y

Offtakes

Ai	Irrigable area; area covered by the crop i
C-$data$	Calculated data
CWR_i	Decadal cumulated optimum crop water requirement for the crop i
D-$data$	Declared data provided by the Associated Consortia
D_m	Duration of the month m
ED	Coefficient of the efficiency of the delivery system CER-irrigable area
EI_i	Coefficient of the efficiency of the irrigation method of the crop i
II_i	Coefficient of irrigation intensity of the crop i
q_{kCn}	Calculated discharge exiting from the offtake k during the decade n
$q_{kC,y}$	Vector containing daily calculated discharge values of the offtake k for the year y
q_{rDm}	Discharge value exiting from the reference offtake during the month m
$q_{totC,y}$	Vector containing daily calculated offtake discharges from the segment (i.e., ES) for the year y
T-$data$	Estimated data provided by IRRINET
V_{kDm}	Monthly cumulated volume of the offtake k from D-$data$

V_{kTn}	Decadal cumulated volume of the offtake k from *T-data*
V_{rDm}	Monthly cumulated volume of the reference offtake from the *D-data*
V_{rTn}	Decadal cumulated volume of the reference offtake from the *T-data*
w_{kCn}	Weight of the offtake k during the decade n
w_{kDm}	Weight of the offtake k during the month m from *D-data*
w_{kTn}	Weight of the offtake k during the decade n from *T-data*

Optimization

$Cd1, Cd2$	Gate discharge coefficients at the entrances of Culv_1 and Culv_2
$Cd3, Cd4$	Gate discharge coefficients at the entrances of Culv_3 and Culv_4
$Cd5, Cd6$	Gate discharge coefficients at the entrances of 2 road crossings (ES)
Cq	Scaling factor of the offtake discharges
J	Criteria to be minimized
n	Manning's coefficient on the CER open-flow sections (along PS or ES)
$n1, n2$	Manning's coefficients within Culv_1 and Culv_2
$n3, n4$	Manning's coefficients within Culv_3 and Culv_4
$n5, n66$	Manning's coefficients within the 2 road crossings
$Q0_y$	Vector containing daily calculated flowing discharges at WL OUT_0 for the year y
$Q2_{sim,y}$	Vector containing daily simulated flowing discharges at WL OUT_1 for the year y
$Q3_{sim,y}$	Vector containing daily simulated flowing discharges at WL IN_2 for the year y
$Z0_{sim,y}$	Vector containing daily simulated water levels at WL OUT_0 for the year y
$Z2_{sim,y}$	Vector containing daily simulated water levels at WL OUT_1 for the year y
$Z3_{sim,y}$	Vector containing daily simulated water levels at WL IN_2 for the year y
$\sigma0_y$	Vector containing the daily weights of the suspicious measures located in $Z0_{obs,y}$
$\sigma2_y$	Vector containing the daily weights of the suspicious measures located in $Z1_{obs,y}$, $Z2_{obs,y}$ and $Z4_{obs,y}$
$\sigma3_y$	Vector containing the daily weights of the suspicious measures located in $Z1_{obs,y}$, $Z3_{obs,y}$ and $Z4_{obs,y}$

References

1. Sun, H.; Wang, S.; Hao, X. An Improved Analytic Hierarchy Process Method for the evaluation of agricultural water management in irrigation districts of north China. *Agric. Water Manag.* **2017**, *179*, 324–337. [CrossRef]
2. Chen, S.; Ravallion, M. *Absolute Poverty Measures for the Developing World, 1981–2004*; PNAS: Washington, DC, USA, 2007.
3. European Commission. *Roadmap to a Resource Efficient Europe*; European Environment Agency: Brussels, Belgium, 2011; pp. 1–24.
4. Gordon, L.J.; Finlayson, C.M.; Falkenmark, M. Managing water in agriculture for food production and other ecosystem services. *Agric. Water Manag.* **2010**, *97*, 512–519. [CrossRef]
5. De Fraiture, C.; Molden, D.; Wichelns, D. Investing in water for food, ecosystems, and livelihoods: An overview of the comprehensive assessment of water management in agriculture. *Agric. Water Manag.* **2010**, *97*, 495–501. [CrossRef]
6. European Environmental Agency (EEA). *Water Resources Across Europe-Confronting Water Scarcity and Drought*; European Environment Agency: Copenhagen, Denmark, 2009; pp. 1–55.
7. Molden, D. *Water for Food, Water for Life: A Comprehensive Assessment of Water Management in Agriculture*; Earthscan, IWMI: London, UK, 2007; pp. 1–39.
8. Namara, R.E.; Hanjra, M.A.; Castillo, G.E.; Ravnborg, H.M.; Smith, L.; Koppen, B.V. Agricultural water management and poverty linkages. *Agric. Water Manag.* **2010**, *97*, 520–527. [CrossRef]
9. Barker, R.; Molle, F. Evolution of irrigation in South and Southeast Asia, 2004. Available online: https://www.eea.europa.eu/publications/water-resources-across-europe (accessed on 27 July 2018).
10. European Commission. Directive 2000/60/EC of the European Parliament and of the Council Establishing a Framework for the Community Action in the Field of Water Policy. Available online: https://www.eea.europa.eu/policy-documents/directive-2000-60-ec-of (accessed on 27 July 2018).
11. European Commission DG ENV. *Water Saving Potential in Agriculture in Europe: Findings from the Existing Studies and Applications*; BIO Intelligence Service: Paris, France, 2012; pp. 1–232.
12. Masseroni, D.; Ricart, S.; de Cartagena, F.; Monserrat, J.; Gonçalves, J.; de Lima, I.; Facchi, A.; Sali, G.; Gandolfi, C. Prospects for Improving Gravity-Fed Surface Irrigation Systems in Mediterranean European Contexts. *Water* **2017**, *9*, 20. [CrossRef]

13. Levidow, L.; Zaccaria, D.; Maia, R.; Vivas, E.; Todorovic, M.; Scardigno, A. Improving water-efficient irrigation: Prospects and difficulties of innovative practices. *Agric. Water Manag.* **2014**, *146*, 84–94. [CrossRef]

14. Ministero delle Politiche Agricole Alimentari e Forestali (MiPAAF). *Approvazione delle linee guida per la regolamentazione da parte delle Regioni delle Modalità di quantificazione dei volumi idrici ad uso irriguo*; MiPAAF: Rome, Italy, 2015.

15. Litrico, X.; Fromion, V.; Baume, J.-P.; Arranja, C.; Rijo, M. Experimental validation of a methodology to control irrigation canals based on Saint-Venant Equations. *Control Eng. Pract.* **2005**, *13*, 1425–1437. [CrossRef]

16. Lozano, D.; Arranja, C.; Rijo, M.; Mateos, L. Simulation of automatic control of an irrigation canal. *Agric. Water Manag.* **2010**, *97*, 91–100. [CrossRef]

17. Cantoni, M.; Weyer, E.; Li, Y.; Ooi, S.K.; Mareels, I.; Ryan, M. Control of Large-Scale Irrigation Networks. *Proc. IEEE* **2007**, *95*, 75–91. [CrossRef]

18. Ooi, S.K.; Weyer, E. Control design for an irrigation channel from physical data. *Control Eng. Pract.* **2008**, *16*, 1132–1150. [CrossRef]

19. Jean-Baptiste, N.; Malaterre, P.-O.; Dorée, C.; Sau, J. Data assimilation for real-time estimation of hydraulic states and unmeasured perturbations in a 1D hydrodynamic model. *Math. Comput. Simul.* **2011**, *81*, 2201–2214. [CrossRef]

20. Jeroen, V.; Linden, V. Volumetric water control in a large-scale open canal irrigation system with many smallholders: The case of chancay-lambayeque in Peru. *Agric. Water Manag.* **2011**, *98*, 705–714.

21. Islam, A.; Raghuwanshi, N.S.; Singh, R. Development and application of hydraulic simulation model for irrigation canal network. *J. Irrig. Drain. Eng.* **2008**, *134*, 49–59. [CrossRef]

22. Renault, D. Aggregated hydraulic sensitivity indicators for irrigation system behavior. *Agric. Water Manag.* **2000**, *43*, 151–171. [CrossRef]

23. Cornish, G.; Bosworth, B.; Perry, C.; Burke, J. Water Charging in Irrigated Agriculture: An Analysis of International Experience. Available online: http://www.fao.org/docrep/008/y5690e/y5690e00.htm (accessed on 6 June 2018).

24. Laycock, A. *Irrigation Systems: Design, Planning and Construction*; CAB International: Wallingford, UK, 2007; pp. 1–285.

25. Molle, F. Water scarcity, prices and quotas: A review of evidence on irrigation volumetric pricing. *Irrig. Drain. Syst.* **2009**, *23*, 43–58. [CrossRef]

26. Food and Agriculture Organization of the United Nations (FAO). *The State of the World's Land and Water Resources for Food and Agriculture (SOLAW)—Managing Systems at Risk*; Food and Agriculture Organization of the United Nations: Rome, Italy, 2011; pp. 1–281.

27. Rault, P.K.; Jeffrey, P. On the appropriateness of public participation in integrated water resources management: Some grounded insights from the Levant. *Integr. Assess.* **2008**, *8*, 69–106.

28. Huang, Y.; Fipps, G. Developing a modeling tool for flow profiling in irrigation distribution networks. *Int. J. Agric. Biol. Eng.* **2009**, *2*, 17–26.

29. Tariq, J.A.; Latif, M. Improving operational performances of farmers managed distributary canal using SIC hydraulic model. *Water Resour. Manag.* **2010**, *24*, 3085–3099. [CrossRef]

30. Sau, J.; Malaterre, P.O.; Baume, J.P. Sequential Monte Carlo hydraulic state estimation of an irrigation canal. *Comptes Rendus Mécanique* **2010**, *338*, 212–219. [CrossRef]

31. Kumar, P.; Mishra, A.; Raghuwanshi, N.S.; Singh, R. Application of unsteady flow hydraulic-model to a large and complex irrigation system. *Agric. Water Manag.* **2002**, *54*, 49–66. [CrossRef]

32. Giupponi, C. Decision Support Systems for implementing the European Water Framework Directive: The MULINO approach. *Environ. Model. Softw.* **2007**, *22*, 248–258. [CrossRef]

33. Mateos, L.; Lopez-Cortijo, I.; Sagardoy, J. SIMIS the FAO decision support system for irrigation scheme management. *Agric. Water Manag.* **2002**, *56*, 193–206. [CrossRef]

34. Leenhardt, D.; Trouvat, J.L.; Gonzalès, G.; Pérarnaud, V.; Prats, S.; Bergez, J.E. Estimating irrigation demand for water management on a regional scale. *Agric. Water Manag.* **2004**, *68*, 207–232. [CrossRef]

35. Mailhol, J.-C. *Evaluation à l'échelle Régionale des Besoins en Eau et du Rendement des Cultures Selon la Disponibilité en eau. Application au bassin Adour-Garonne*; Agence de l'Eau Adour-Garonne: Toulouse, France, 1992; pp. 1–24.

36. Sousa, V.; Santos Pereira, L. Regional analysis of irrigation water requirements using kriging. Application to potato crop (*Solanum tuberosum* L.) at Tras-os-Montes. *Agric. Water Manag.* **1999**, *40*, 221–233. [CrossRef]

37. Heinemann, A.B.; Hoogenboom, G.; Faria de, R.T. Determination of spatial water requirements at county and regional levels using crop models and GIS: An example for the state of Parana, Brazil. *Agric. Water Manag.* **2002**, *52*, 177–196. [CrossRef]

38. Kinzli, K.-D.; Gensler, D.; Oad, R.; Shafike, N. Implementation of a decision support system for improving irrigation water delivery: Case study. *Irrig. Drain. Syst.* **2015**, *141*, 05015004. [CrossRef]

39. Miao, Q.; Shi, H.; Gonçalves, J.; Pereira, L. Basin irrigation design with multi-criteria analysis focusing on water saving and economic returns: Application to wheat in Hetao, Yellow River Basin. *Water* **2018**, *10*, 67. [CrossRef]

40. Yang, G.; Liu, L.; Guo, P.; Li, M. A flexible decision support system for irrigation scheduling in an irrigation district in China. *Agric. Water Manag.* **2017**, *179*, 378–389. [CrossRef]

41. Tanure, S.; Nabinger, C.; Becker, J.L. Bioeconomic model of decision support system for farm management: Proposal of a mathematical model: Bioeconomic model of decision support system for farm management. *Syst. Res. Behav. Sci.* **2015**, *32*, 658–671. [CrossRef]

42. Consortium of the Canale Emiliano Romagnolo (CER). Available online: http://www.consorziocer/ (accessed on 17 May 2018).

43. Istituto Nazionale di Statistica (ISTAT). Censimento Agricoltura 2010. Available online: http://www4.istat.it/it/censimento-agricoltura/agricoltura-2010 (accessed on 22 May 2018).

44. Munaretto, S.; Battilani, A. Irrigation water governance in practice: The case of the Canale Emiliano Romagnolo district, Italy. *Water Policy* **2014**, *16*, 578. [CrossRef]

45. Service IRRIFRAME. Available online: https://ssl.altavia.eu/Irriframe/ (accessed on 29 October 2017).

46. Delibera di Giunta Regionale (DGR). *Bollettino Ufficiale della Regione Emilia Romagna, Parte Seconda, n.9—Deliberazione della Giunta Regionale 21 Dicembre 2016, n. 2254*; Giunta Regionale: Bologne, Italy, 2016; pp. 1–10. (In Italian)

47. Mannini, P.; Genovesi, R.; Letterio, T. IRRINET: Large scale DSS application for on-farm irrigation scheduling. *Procedia Environ. Sci.* **2013**, *19*, 823–829. [CrossRef]

48. Singley, B.C.; Hotchkiss, R.H. Differences between Open-Channel and Culvert Hydraulics: Implications for Design. 2010. Available online: https://doi.org/10.1061/41114 (accessed on 27 July 2018).

49. Software SIC2. Available online: http://sic.g-eau.net/ (accessed on 30 October 2017).

50. Haque, S. Impact of irrigation on cropping intensity and potentiality of groundwater in murshidabad district of West Bengal, India. *Int. J. Ecosyst.* **2015**, *5*, 55–64.

51. Thenkabail, P.; Dheeravath, V.; Biradar, C.; Gangalakunta, O.R.; Noojipady, P.; Gurappa, C.; Velpuri, M.; Gumma, M.; Li, Y. Irrigated area maps and statistics of India using remote sensing and national statistics. *Remote Sens.* **2009**, *1*, 50–67. [CrossRef]

52. Tanriverdi, C.; Degirmenci, H.; Sesveren, S. Assessment of irrigation schemes in Turkey: Cropping intensity, irrigation intensity and water use. In Proceedings of the Tropentag 2015, Berlin, Germany, 16–18 September 2015.

53. Battilani, A. L'irrigazione del medicaio. *Agricoltura* **1994**, *3*, 31–33.

54. Battilani, A. Regulated deficit of irrigation (RDI) effects on growth and yield of plum tree. *Acta Hortic.* **2004**, *664*, 55–62. [CrossRef]

55. Battilani, A.; Anconelli, S.; Guidoboni, G. Water table level effect on the water balance and yield of two pear rootstock. *Acta Hortic.* **2004**, *664*, 47–54. [CrossRef]

56. Battilani, A.; Ventura, F. Influence of water table, irrigation and rootstock on transpiration rate and fruit growth of peach trees. *Acta Hortic.* **1996**, *449*, 521–528. [CrossRef]

57. Food and Agriculture Organization of the United Nations (FAO). Irrigation Water Management: Irrigation Scheduling, Training Manual n.4. 1989. Available online: http://www.fao.org/docrep/t7202e/t7202e08.htm (accessed on 21 June 2017).

58. Delibera di Giunta Regionale (DGR). *Bollettino Ufficiale della Regione Emilia Romagna, Parte Seconda, n.327—Deliberazione della Giunta Regionale 5 Settembre 2016, n. 1415*; Giunta Regionale: Bologne, Italy, 2016; pp. 1–11. (In Italian)

59. Artina, S.; Montanari, A. *Consulenza per lo Studio delle Perdite Idriche nelle Reti di Distribuzione dei Consorzi di Bonifica*; Università degli Studi di Bologna: Bologne, Italy, 2007; pp. 1–46. (In Italian)

60. Taglioli, G.; Cinti, P. Water Losses of Land Channels [Emilia-Romagna]. *Rivista di Ingegneria Agraria.* 2002. Available online: http://agris.fao.org/agris-search/search.do?recordID=IT2004060023 (accessed on 27 July 2018).

61. Consorzio della Bonifica Renana. Available online: https://www.bonificarenana.it/ (accessed on 11 July 2018).

62. Gejadze, I.; Malaterre, P.O. Discharge estimation under uncertainty using variational methods with application to the full Saint-Venant hydraulic network model: Discharge estimation under uncertainty using variational methods. *Int. J. Numer. Methods Fluids* **2017**, *83*, 405–430. [CrossRef]

63. Oubanas, H.; Gejadze, I.; Malaterre, P.O. River discharge estimation under uncertainty from synthetic SWOT-type observations using variational data assimilation. Contribution of the UMR G−eau investigation team, IRSTEA Montpellier, France to the SWOT mission. *La Houille Blanche* **2018**, *2*, 84–89. [CrossRef]

64. Baume, J.P.; Malaterre, P.O.; Benoit, G.B. *SIC: A 1D Hydrodynamic Model for River and Irrigation Canal Modeling and Regulation*; CEMAGREF, Agricultural and Environmental Engineering Research: Montpellier, France, 2005; pp. 1–81.

65. Sau, J.; Malaterre, P.-O.; Baume, J.P. Sequential Monte Carlo hydraulic state estimation of an irrigation canal. *Comptes Rendus Mécanique* **2010**, *338*, 212–219. [CrossRef]

66. Cunge, J.; Holly, F.M.; Verwey, A., Jr. *Practical Aspects of Computational Rive Hydraulics*; Pitman Advanced Pub. Program: Boston, MA, USA, 1980; pp. 1–420.

67. Chaudhry, H.M. *Open Channel Flow*, 2nd ed.; Springer: New York, NY, USA, 2007; pp. 1–517.

68. Baume, J.-P.; Belaud, G.; Vion, P.Y. *Hydraulique Pour le Génie Rural Notes de Cours*; Ecole Nationale Supérieure d'Agronomie, CEMAGREF: Montpellier, France, 2006; pp. 1–179.

69. Nielsen, K.D.; Weber, L.J. Submergence effects on discharge coefficients for rectangular. In Proceedings of the Joint Conference on Water Resource Engineering and Water Resources Planning and Management 2000, Minneapolis, MN, USA, 30 July–2 August 2000.

70. Nelder, J.A.; Mead, R. A simplex method for function minimization. *Comput. J.* **1965**, *7*, 308–313. [CrossRef]

71. United States Bureau of Reclamation (USBR). *Water Measurement Manual*; Water Resources: Englewood, CO, USA, 1997.

72. Wu, S.; Rajaratnam, N. Solutions to rectangular sluice gate flow problems. *J. Irrig. Drain. Eng.* **2015**, *141*, 06015003. [CrossRef]

73. Sauida, M.F. Calibration of submerged multi-sluice gates. *Alexandria Eng. J.* **2014**, *53*, 663–668. [CrossRef]

74. Chow, V.T. *Open-Channel Hydraulics*; McGraw-Hill: New York, NY, USA, 1959; pp. 1–680.

75. De Doncker, L.; Troch, P.; Verhoeven, R.; Bal, K.; Meire, P.; Quintelier, J. Determination of the manning roughness coefficient influenced by vegetation in the river Aa and Biebrza river. *Environ Fluid Mech.* **2009**, *9*, 549–567. [CrossRef]

76. Dyhouse, G.; Hatchett, J.; Benn, J. *Floodplain Modeling Using HEC-RAS*; Haestad Press: Aterbury, CT, USA, 2003; pp. 1–1261.

77. Tedeschi, L.O. Assessment of the adequacy of mathematical models. *Agric. Syst.* **2006**, *89*, 225–247. [CrossRef]

78. St-Pierre, N.R. Comparison of model predictions with measurements: A novel model-assessment method. *J. Dairy Sci.* **2016**, *99*, 4907–4927. [CrossRef] [PubMed]

79. Harrison, S.R. Validation of agricultural expert systems. *Agric. Syst.* **1991**, *35*, 265–285. [CrossRef]

80. Mayer, D.G.; Butler, D.G. Statistical validation. *Ecol. Model.* **1993**, *68*, 21–32. [CrossRef]

81. Mayer, D.G.; Stuart, M.A.; Swain, A.J. Regression of real-world data on model output: An appropriate overall test of validity. *Agric. Syst.* **1994**, *45*, 93–104. [CrossRef]

82. Dent, J.B.; Blackie, M.J. *Systems Simulation in Agriculture*; Applied Science: London, UK, 1979.

83. Mesplé, F.; Troussellier, M.; Casellas, C.; Legendre, P. Evaluation of simple statistical criteria to qualify a simulation. *Ecol. Model.* **1996**, *88*, 9–18. [CrossRef]

Article

Decision Support System Tool to Reduce the Energy Consumption of Water Abstraction from Aquifers for Irrigation

Juan Ignacio Córcoles [1], Rafael Gonzalez Perea [2,*], Argenis Izquiel [3] and Miguel Ángel Moreno [4]

[1] Renewable Energy Research Institute, Section of Solar and Energy Efficiency, C/ de la Investigación 1, 02071 Albacete, Spain; JuanIgnacio.Corcoles@uclm.es

[2] Department of Plant Production and Agricultural Technology, School of Advanced Agricultural Engineering, University of Castilla-La Mancha. Campus Universitario, s/n, 02071 Albacete, Spain; Rafael.GonzalezPerea@uclm.es

[3] Regional Centre of Water Research, Castilla-La Mancha University, Campus Universitario s/n, 02071 Albacete, Spain; aizquiel@yahoo.com

[4] Institute for Regional Development, University of Castilla-La Mancha. Campus Universitario, s/n, 02071 Albacete, Spain; MiguelAngel.Moreno@uclm.es

* Correspondence: Rafael.GonzalezPerea@uclm.es

Received: 20 December 2018; Accepted: 12 February 2019; Published: 14 February 2019

Abstract: In pressurized irrigation networks that use underground water resources, submersible pumps are one of the highest energy consumers. The objective of this paper was to develop a decision support system, implemented in MATLAB®, to reduce the energy consumption of the water abstraction process, from an aquifer to a reservoir in existing wells, by installing a frequency speed drive. An economic module with the aim to assess the economic profitability of the investment cost of the variable speed drive was also developed. This tool was used in three wells that were located in the Eastern Mancha Aquifer. Several scenarios and irrigation seasons were analyzed while considering the interannual and annual variation in ground water depth. In the three analyzed irrigation societies (named A, B, and C), energy savings were achieved using a variable speed frequency when compared with fixed speed. Considering the analyzed cases, when the dynamic water table level is higher, energy savings ranged from 4.4% and 24.4%, using a variable speed ratio of 0.9 and 0.82. The energy savings based on the variable speed frequency increased when the dynamic water table level was lower, with the average energy savings close to 23%, 22% and 6.8% for irrigation societies A, B, and C, respectively. The results also show that the investment costs of the variable speed drive in two of the three irrigation societies studied were highly profitable, with a payback that ranged from 4.5 to 10 years.

Keywords: irrigation network; energy consumption; variable speed; well; water depth

---

1. Introduction

In most countries of the world, the use of underground water resources has a relevant importance, mainly in arid and semiarid climates. Approximately one-third of the landmass in the world is irrigated by groundwater. This water source is widely used in agriculture, representing 45% of the irrigated land that uses groundwater in the United States of America, 58% in Iran, and 67% in Algeria [1]. Groundwater is important in regions of Spain, such as in the Castilla–La Mancha region, where that approximate source of water represents more than 65% of water use in irrigation and urban networks [2]. In these areas, it is necessary to improve the energy that is consumed, which guarantees the economic sustainability of irrigation societies, where energy costs have increased during the last years.

Regarding groundwater resources, one relevant point to highlight is water discharge and recharge phenomena. In this regard, several sources of recharge mainly come from precipitation and surface-water bodies, such as streams, ponds, or lakes. One of the most common recharge sources is the irrigation water, mainly when the amount of water applied to crops exceed the crop water requirements. Regarding the discharge, flow into streams or groundwater pumped from wells is also common. Water recharged to groundwater for years has a high variability, because it depends on the amount of precipitation or the local geology of each irrigable area.

Some tools have been developed to improve water and energy use in irrigation networks, considering the energy efficiency of the pumping systems or the irrigation scheduling management at the plot scale [3,4]. Most of those studies did not consider the energy efficiency from wells, which, in most of cases, they do not work with adequate efficiency.

In pressurized networks that use those resources, the energy consumed is a factor to be considered, which represents a high participation of the total management, operation, and maintenance costs [5]. In those areas, water is extracted from the aquifer using submersible pumps with high installed power depending on the water table level. The influence of the use of underground water resources can be highlighted when comparing the average values of the energy cost per unit of irrigation delivery in water users' associations (WUAs) [6]. For instance, in the Castilla-La Mancha Region (Spain), the energy costs per cubic meter delivery ranged from 0.022 € m^{-3} to 0.026 € m^{-3} in WUAs with sprinkler irrigation systems and with drip irrigation systems, respectively. These costs did not include the costs of water abstraction from wells. If these costs were considered, the energy costs calculated for sprinkler and drip irrigation systems would increase, reaching values close to 0.061 € m^{-3} and 0.071 € m^{-3}, respectively. These costs are very similar in both irrigation systems. It was explained because in all irrigation societies analyzed, the average head supply by the pumping systems were very similar between sprinkler irrigation systems (58 m) and drip irrigation systems (53 m), besides energy efficiency for each pump of the pumping stations.

Most of the studies carried out in pressurized irrigation networks are focused on the study of energy efficiency in pumping stations, which pump water from a reservoir. In some cases, these studies are based on determining the irrigation sectoring methodologies to improve energy management [7–11]. In other cases, a comparison of several types of regulation systems in pumping stations is evaluated, considering the efficiency of the pumping system for each combination of flow discharge and pressure head [9,12,13].

In some energy audits carried out in the Castilla–La Mancha region [14], it can be highlighted that the submersible pumps are one of the highest energy consumers in these types of associations, reaching up to 70% of the energy cost.

For that reason, it is necessary to analyze the performance and energy efficiency in wells. Some studies are focused on determining the optimal well discharge [15–18]. Other researchers [19] tried to optimize the characteristics and efficiency curves in the pumping wells that delivered to a pivot irrigation system. In that study, a methodology was carried out to determine the minimum total water application cost (investment plus operation costs), while optimizing the diameters of the pumping pipes, distribution pipes, and lateral pipes.

In wells, the most typical management approach is to work with a fixed speed frequency during the whole irrigation season; however, an analysis that can evaluate the energy savings of using a frequency speed drive to manage the water abstraction from wells has not been found.

The aim of this paper was to develop a decision support system tool, named DSSW (Decision Support System for Water abstraction), to determine the effect of installing frequency speed drives in wells on the energy savings and cost recovery in managing the underground water abstraction process. Several scenarios were analyzed in three wells with different characteristics that are located in the Eastern Mancha Aquifer (Spain). This work complements previous studies focused on the improvement of energy efficiency at irrigable areas, such as MOPECO [20] or GREDRIP model [4], used to determine irrigation schedules and to estimate the relationship between irrigated crop yield and net water applied. In this regard, the use of DSSW does not depend on the water applied at each irrigation network, because it establishes the energy savings and cost recovery independently of the total irrigation applied at each irrigable area. The variation in the dynamic water table level in different irrigation seasons was evaluated. For each scenario, the operating point was computed with several variable speed frequencies to determine the minimum energy consumed per cubic meter of water delivery, compared with the fixed speed frequency. An economic analysis of installing this type of device was also performed to evaluate the economic profitability of this action.

2. Methodology

2.1. Description of the DSSW Tool

The DSSW tool was developed in MATLAB®(MathWorks Inc., MA, USA) to simulate the performance of wells using fixed and variable speed pumping, as well as the energy consumption for each type of management approach. An energy analysis of different indicators and the main energy variables was accurately calculated. By applying the DSSW tool, it was possible to determine the viability of installing a frequency speed drive to minimize the energy consumption per cubic meter of water extracted from the aquifer. With this aim, the followings steps were taken:

1. Characterization of the analyzed wells by considering the actual water table levels during the irrigation season, along with the hydraulic characteristics of the pumping pipe diameter, material, length, location of the reservoir (distance from the well and difference in elevation), and the type and model of the pump (characteristic and efficiency curves).
2. Calculation of the system curve for the different analyzed scenarios throughout the irrigation season.
3. Calculation of the operating point of the pump and energy consumption, which determines the ratio of energy consumed per cubic meter (kWh m^{-3}).
4. Simulation of the well performance using a frequency speed drive with several variable speed ratios to calculate the most efficient pump speed.
5. Energy analysis to account for the energy savings by using a frequency speed drive.
6. Economic analysis.

The implemented tool has four modules: (1) Pumping system module, which simulates the variable speed pump performance; (2) hydraulic module, which determines the system curve accurately by considering the hydrogeologic parameter of the well and the characteristics of the pipes; (3) energy module, which determines the operating point of the system under steady state conditions and calculates the energy rates to determine the energy efficiency for each rotation speed of the pump; and (4) economic analysis module, which determines the economic profitability of the different scenarios studied in module 3.

2.1.1. Pumping System Module

The pumping system module considered the characteristics and efficiency curves of the analyzed pump, H-Q and η-Q, as indicated in Equations (1) and (2), respectively. These curves were obtained from the theoretical data provided by the pump manufacturer.

$$H = a + bQ + cQ^2 \tag{1}$$

$$\eta = eQ + fQ^2 \tag{2}$$

where H is the pumping head in m; Q is the flow discharge in l min^{-1}; η is the pump efficiency at fixed speed in percentage; and a, b, c, e, and f are coefficients of the adjusted model, which define the shape of the curves.

To define the performance of the pump at different rotation speeds, the characteristics and efficiency curves for each speed were obtained using the affinity laws. Therefore, for a variable speed drive, the characteristic curves can be defined as follows:

$$H_{vs} = a\alpha^2 + b\alpha Q + cQ^2 \tag{3}$$

$$\eta_{vs} = e\frac{Q}{\alpha} + f\frac{Q^2}{\alpha^2} \tag{4}$$

where H_{vs} is the pumping head at a variable speed in m; η_{vs} is the pump efficiency at fixed speed in %; and α is the ratio between the speed of the variable speed drive and the maximum speed as a fixed speed drive. In addition, the efficiency of the variable speed drive for different frequencies have been taken into account according to [9].

2.1.2. Hydraulic Model

This module allowed for the system curve to be determined. In this study, the system was composed of a pump, the pumping pipe from the submersible pump to the head of the well and the distribution pipe from the head of the well to the discharge (Figure 1). Accordingly, the system curve was computed to introduce the data related to the dynamic water table level (Z_o), the reference level of water discharged (Z_f) at the top of the reservoir, the pipe diameter (D), the pipe length (L), and the Hazen–William coefficient (C) of the pumping pipe and the distribution pipe [2].

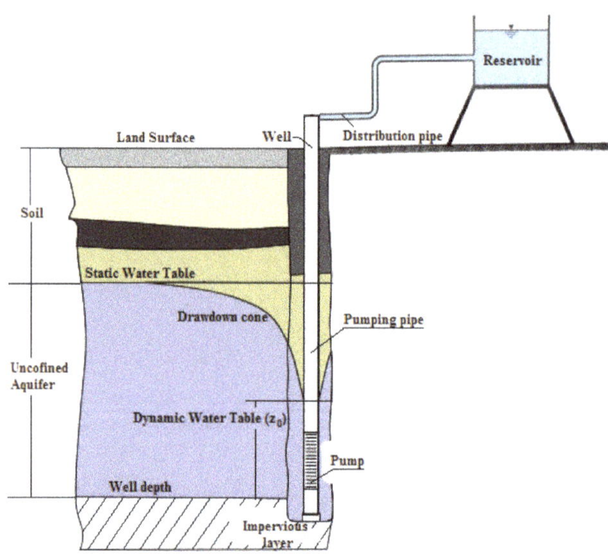

Figure 1. Infrastructure of the analyzed system.

2.1.3. Energy Module

Once the system curve and characteristic curves for the fixed pump were implemented, the operating point was computed, which can be defined as the intersection of the system curve and the pump characteristic curve [21,22]. Hence, according to the studied system curve, it was possible to determine the operating point at each variable speed. In the proposed tool, several variable speed ratios (α) were used, one while the pump was working at maximum speed ($\alpha = 1$) and the rest while the pump was activated with the frequency speed drive, which ranged from a minimum α to the maximum speed, at 0.02 intervals. Those parameters can be modified in the tool by the user.

According to the previous information, at each variable speed ratio, several parameters were automatically computed, such as the pumping head (H, m), the flow discharge (Q, l min^{-1}), efficiency (η, %), and absorbed power (Na, kW) by the pump. Moreover, at each variable speed ratio, the consumed energy was computed using the energy ratio indicator (ER, kW h m^{-3}), which is defined as the relationship between the energy consumed per volume of water delivered. Installed power describes the nominal power of the motor (mechanical power at the motor axis) while the absorbed power is the electrical power absorbed by the motor under each demand condition (low for low frequencies of the frequency speed drive, high for maximum frequency). The absorbed power also changes for the same frequency depending on the hydraulic power (i.e., changes in the water table level). Therefore, the variable speed drive that minimizes the ER was determined, which was useful for obtaining the energy savings and comparing the use of fixed and variable speed pumps.

2.1.4. Economic Analysis Module

With the aim to determine the economic profitability of the different scenarios analyzed by comparing the use of fixed and variable speed pumps, an economic analysis module was implemented in the DSSW tool. Considering the energy savings by comparing the use of fixed and variable speed pumps, this module analyzes the economic profitability of the variable speed drive investment. Thus, Equation (5) shows the relationship between the variable speed drive and its installed power.

$$C_{vs} = -0.0518 \times P_{vs}^2 + 82.46 \times P_{vs} + 47.62 \tag{5}$$

where C_{vs} is the variable speed drive cost in €; P_{vs} is the installed power in kW.

Equation (5) was obtained from the 13 main speed variable drive manufacturers in the Spanish market. According to these manufacturers, an annual maintenance cost of 5% of the variable speed drive cost and a lifespan of 15 years were considered in the economic analysis.

Based on Equation (5), which calculates the annual maintenance cost, lifespan, and the energy savings with variable speed pumps, this module computes the net present value (NPV$_{vs}$ in €), which determines the profitability of the investment, as well as the internal rate of return (IRR$_{vs}$) and the payback period of the variable speed drive investment. Thus, the DSSW tool can determine the economic profitability of each study case.

2.2. Case Studies

The DSSW tool was applied to three WUAs, named A, B, and C, from the Castilla–La Mancha region, which were representative of the irrigable areas of this region. All the WUAs had similar characteristics, with the use of underground water resources as the main source of water. At each area, groundwater was pumped to a storage reservoir from which it was then pumped into the irrigation network by a pumping station. These areas were managed under a rotational schedule with on-demand management, and the command area ranged from 267 to 863 ha (Table 1). In these areas, the dynamic water table (DWT) level in the 2007 ranged from 71 (WUA B) to 105 m (WUA A). There were annual variations of the DWT, but the interannual variation of this variable was much higher.

Table 1. Characteristics of the water user associations.

Irrigation Society	A	B	C
Command area (ha)	863	764	267
Wells (number)	4	6	1
Storage capacity (m^3)	130,000	5 of 5000 1 of 6000	20,000
Water distribution network management	Rotational schedule	Rotational schedule	On demand
Irrigation system	Sprinkler and drip irrigation	Sprinkler and drip irrigation	Drip irrigation

The main characteristics of the submersible pumps and the required data for the tool at each one of the analyzed WUAs are shown in Table 2.

Table 2. General information required by the developed tool.

Pump Model	WUA* A INDAR 423-3	WUA B INDAR BL-345-3	WUA C INDAR BL-385-3
Reference level of dynamic water table (Z_o, mm)	588.15	602.57	646.89
Reference level of water discharged (Z_f, mm)	697.00	680.10	751.03
Pumping pipe diameter (D, mm)	250	315	272
Hazen–Williams coefficient (C) at the pumping pipe	125	125	125
Pumping pipe length (L, m)	160	100	130
Distribution pipe diameter (D, mm)	588	315	272
Hazen–Williams coefficient (C) at the distribution pipe	125	125	125
Distribution pipe length (L, m)	2750	10	50

*WUA: Water User Association.

With the aim to compute the energy consumption by each WUA, the water volume applied in each irrigation season was recorded. The water volumes applied were similar in each irrigation season. Thus, average applied water volumes of 836,225 m^3, 601,136 m^3, and 170,535 m^3 for WUA A, WUA B, and WUA C, respectively, were considered. Finally, the unit energy cost of the study region was 0.10 € kWh^{-1}.

2.3. Data Acquisition

At each of the analyzed wells, the hydraulic, electrical and topographic parameters were measured during the peak period (July) of the 2007 irrigation season. This information was useful for determining the performance of each of the analyzed wells. Regarding the hydraulic data, the flow rate was measured using a portable ultrasound flow meter (2.5% accuracy) at the pump discharge pipe. In addition, the DWT level was measured with a portable electric contact meter. The electrical parameters, such as the current, voltage, power factor, and absorbed power, were obtained using an electrical network analyzer (1.5% accuracy). Regarding the topographic parameters, such as the top of the reservoir, where water was discharged, and the head of well, were measured using GNSS-RTK equipment (with an error of less than 1 cm). Thus, the operating point was measured (discharge, pressure, and efficiency) and compared with the theoretical characteristic and efficiency curves.

2.4. Water Table Level Analysis

In this study, an analysis of the influence of the water table level on the operating point at different pump rotation speeds was carried out. It was applied for several irrigation seasons, using the water table level information from the selected piezometer database.

In the whole Jucar Basin, in which the analyzed WUAs were located, there was a dense network of piezometers (www.chj.es). Monthly data from each piezometer since the early sixties were available. However, there were many gaps in the available data, and a proper selection of the most representative piezometers for each WUA was performed.

At each piezometer, the available information included monthly values of the static water table (SWL) level at each irrigation season. Table 3 shows the identification (Id) of each piezometer, as well as their UTM (Universal Transverse Mercator) coordinates and the data available period for each water users' association. In WUA A, information about two piezometers was available (08.29.102 and 08.29.313), with monthly data of the SWT level from 2007 to 2013. In WUA B, three piezometers were available (08.29.047, 08.29.051, and 08.29.060), with SWT information from 2007 to 2013 and 2015. In WUA C, four piezometers had available information, piezometers 08.29.033 and 08.29.058 (information from 2007 to 2013), 08.29.055 (information from 2007 to 2011), and 08.29.014 (from 2008 to 2013). The location for each piezometer is shown in Figure 2.

Table 3. Piezometer information for each irrigated area.

WUA	Id-Number	Coord X [1]	Coord Y [1]	Data Available (Years)
A	08.29.102	585981.8361	4304764.2977	From 2007 to 2013
	08.29.313	589075.8342	4303205.2979	
B	08.29.047	600345.1536	4334477.9268	From 2007 to 2013
	08.29.051	600729.1332	4332226.9447	2015
	08.29.060	599206.0866	4328434.9856	
C	08.29.033	577548.335	4358316.5237	From 2007 to 2013
	08.29.058	556391.9999	4344263.9132	From 2007 to 2013
	08.29.055	573112.1778	4347483.7795	From 2007 to 2011
	08.29.014	577430.1485	4346041.8566	From 2008 to 2013

[1] Coordinate system: UTM ETR89.

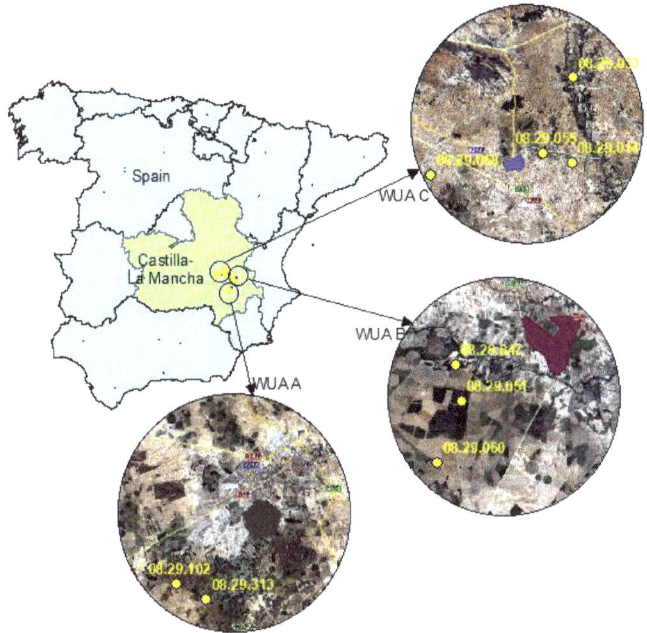

Figure 2. Location of the piezometers and analyses at Irrigation Society A (**a**), B (**b**), and C (**c**).

Once the piezometer database was available for each analyzed well, it was possible to estimate the actual DWT level for each month of the analyzed irrigation seasons. To compute the DWT, we had information about the DWT measured in July 2007, in all the analyzed cases. It was measured during energy audits carried out in that irrigation season [14]. It was measured using a portable SEBA-electric contact meter. Hence, using the DWT measured data as the main reference, the percentage of annual and interannual variation was calculated with the original data of the piezometers related to SWT. The same variation for the SWT was applied to the DWT measured in 2007 to obtain DWT values for the rest of analyzed irrigation seasons.

In addition, for each irrigation season, three cases (Case 1, Case 2, and Case 3) were proposed, with the aim to define an average DWT level for each case during the corresponding irrigation season. These cases were proposed to analyze the annual DWT variation. Therefore, Case 1 represented the average DWT level for a period with high crop water requirements (from June to August), Case 2 (from March to May), and Case 3 (from September to November) represented the average DWT level for a period with low crop water requirements. Although the developed tool is capable of evaluating as many scenarios as a user can define (monthly or even daily), we decided to simplify the case studies to demonstrate the results more clearly.

3. Results and Discussion

3.1. Operating Point

As a first step, the characteristic curves H-Q and efficiency-Q at each one of the analyzed wells are shown in Figure 3. In this case, the theoretical operating point and the measured operating point for the peak period (Case 1) of irrigation season 2007 are shown.

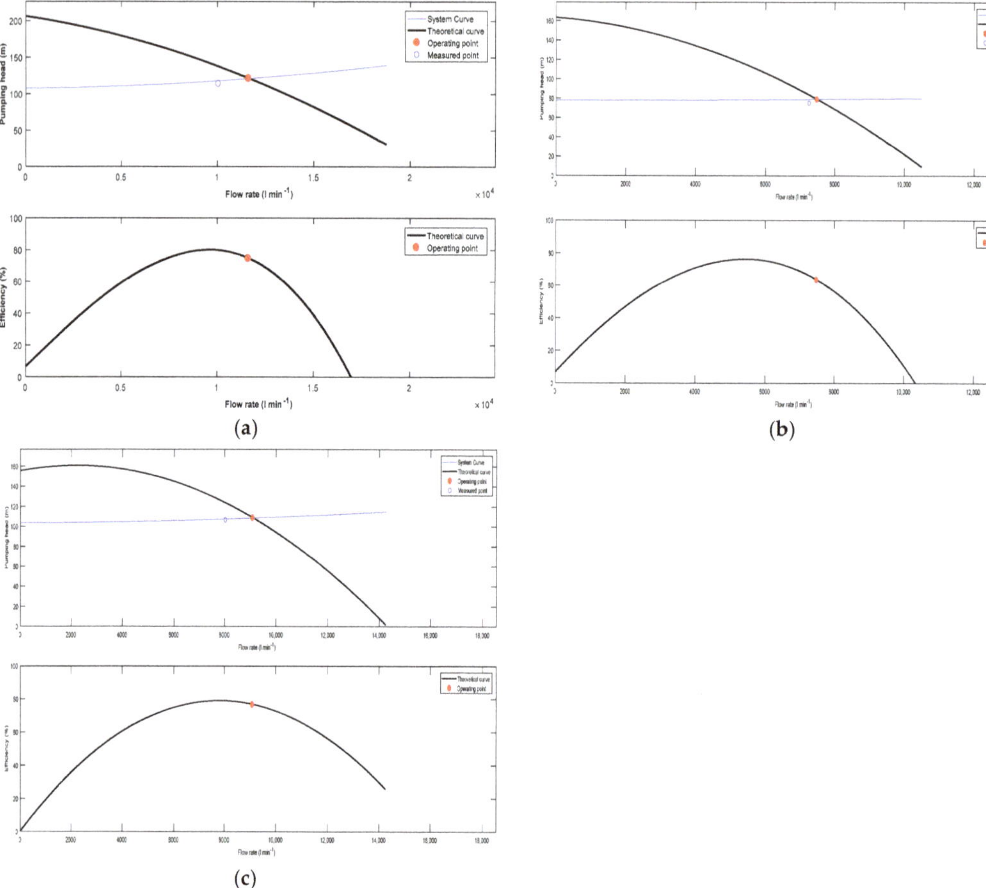

Figure 3. Characteristic curves H-Q, efficiency-Q, and operating point for Case 1 in WUA A (**a**), B (**b**), and C (**c**).

In all the analyzed pumps, the tendency was similar, and it can be highlighted that the operating point did not match with the best efficiency point. It is located instead in a region where the pump efficiency was decreasing. This finding was common in the analyzed WUAs, where the pumps were selected to work in a region where the operating point was located to the right of the maximum efficiency, and the pumps were installed in a way that it was expected to operate in normal conditions. This is related to the fact that the use of underground water resources can increase the water depth level, thereby increasing the dynamic water table (DWT), and increasing the pumping head can lead to the translation of its operating point to a region of maximum efficiency in the nearby future.

In these figures, the measured operating point was included to obtain similar values to the theoretical operating point. With regard to this, these figures are useful to highlight that all the pumps were properly selected and to obtain acceptable efficiency values, which ranged from 64% to 77% in WUA B and WUA C, respectively.

3.2. Piezometer Analysis

One representative piezometer was selected for each analyzed WUA. Two criteria were considered to select each one. First, a proximity criterion to the analyzed well was applied, and the closest piezometers to the analyzed WUA were selected. The second criterion was focused on the piezometer selection with the most available data from several irrigation seasons. In some of the available piezometers, some data were missing; thus, these piezometers were discarded. Therefore, the piezometers with the most available data were then determined.

In Figure 4, the evolution of the static water table (SWT) throughout all of the available irrigation seasons is represented at each piezometer, as well as the interannual variation of the SWT with respect to January of 2007. The SWT evolution of the two piezometers (08.29.102 and 08.29.313) of the WUA A is shown in Figure 4a. Both piezometers show a similar tendency, ranging in SWT values from 103 to 85 m. The interannual variation of the SWT values also showed a similar tendency, considering the range of values from 2007 to 2013 (Figure 5b). In this case, piezometers 08.29.313 and 08.29.102 were located at a distance of 9.6 and 12.3 km, respectively. According to the second selection criterion, piezometer 08.29.313 was chosen.

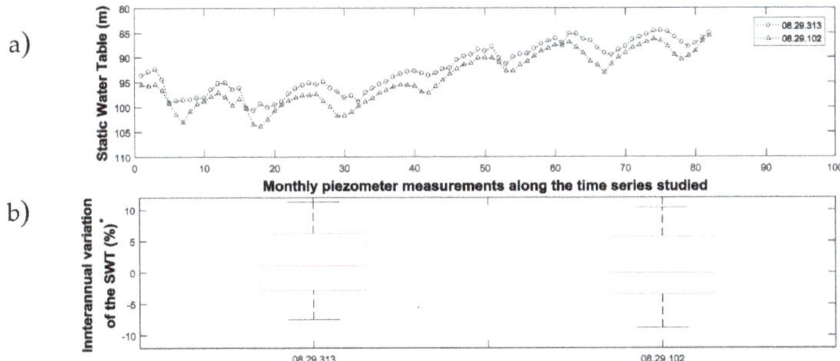

* Interannual variation of the SWT[1] with respect to January of 2007. [1]Static Water Table

Figure 4. Static water table (**a**) and monthly interannual variation (**b**) for each piezometer in WUA A.

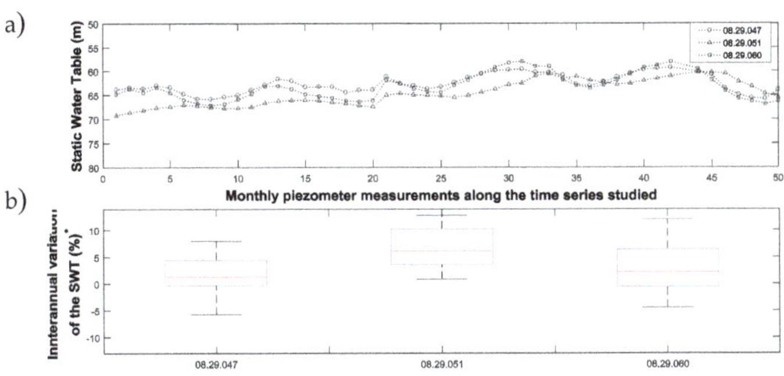

* Interannual variation of the SWT with respect to January of 2007

Figure 5. Static water table (**a**) and monthly interannual variation (**b**) for each piezometer in WUA B.

Concerning to WUA B (Figure 5a,b), the SWT evolution remains similar for the three piezometers ranging their SWT values from 62 to 70 m. In this irrigation society, although piezometer 08.29.047 was the closest to the analyzed well, it was located in a region close to a river; thus, the influence of this could affect to the data accuracy. Hence, in this area, piezometer 08.29.051 was selected, which was at a distance of 8.4 km.

For WUA C, the piezometer 08.29.058 (Figure 6a) had a different tendency in comparison to the other piezometers. In fact, it was too far (25 km) from the analyzed well and was; therefore, discarded. Piezometer 08.29.033 was discarded because it was close to a river. Considering the remaining piezometers with a similar tendency, piezometer 08.29.014 was selected because it is close to the well (distance of 7.5 km) and more information was available for it (Figure 6b).

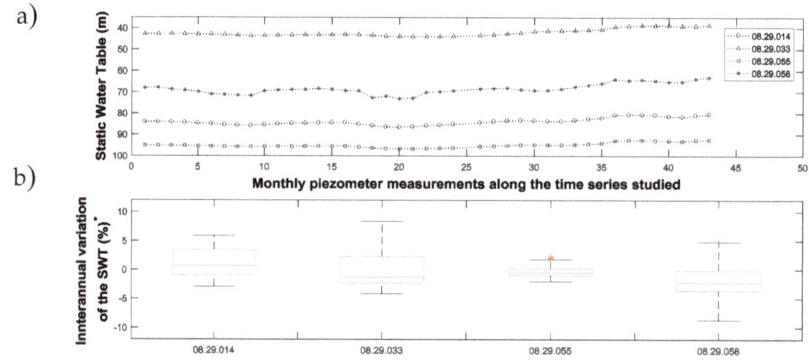

* Interannual variation of the SWT with respect to January of 2007

Figure 6. Static water table (**a**) and monthly interannual variation (**b**) for each piezometer in WUA C.

In Table 4, the DWT for each analyzed case and irrigation season is shown for each of the selected piezometers. In addition, during each irrigation season, the average values of the DWT for each case, along with the coefficient of variation (CV) for each one, is shown. Regarding the CV, it can highlight the low variability of the DWT, considering the selected months for each analyzed case.

Table 4. Average dynamic water table (DWT) and coefficient of variation (CV) for each irrigation season and analyzed case.

Piezometer	2007		2008		2009		2010		2011		2012		2013		2015	
08.29.313	DWT (m)	CV (%)	DWT (m)	CV (%)	DWT (m)	CV (%)	DWT (m)	CV (%)	DWT (m)	CV (%)	DWT (m)	CV (%)	DWT (m)	CV (%)	DWT (m)	CV (%)
Case 1	105.5	0.28	106.9	0.67	103.7	0.99	99.62	0.44	96.73	0.85	92.98	1.12	91.79	1.26	91.09	1.05
Case 2	99.70	1.20	102.5	0.76	101.7	0.26	99.80	0.54	94.43	0.50	91.75	1.28	90.68	0.50	89.44	2.64
Case 3	105.0	0.19	106.3	0.58	104.6	0.94	98.92	0.59	95.11	0.72	95.20	0.59	93.03	1.10	92.23	0.01
Case 1	71.03	0.38	70.54	0.53	71.01	1.31	68.69	0.21	64.98	0.69	65.44	2.21	64.96	0.71	67.62	1.39
Case 2	71.52	0.57	69.80	0.13	68.78	0.34	68.70	0.71	64.42	0.72	64.05	1.05	65.56	1.18	65.00	0.23
Case 3	71.42	0.20	71.17	0.00	71.87	0.31	68.60	0.83	66.08	0.35	68.83	0.66	66.05	0.62	69.15	0.05
Case 1	98.82	0.52	99.21	0.40	100.24	0.43	97.20	0.22	94.37	0.60	94.00	0.86	92.85	0.68	-	-
Case 2	98.12	0.16	97.96	0.35	98.35	0.35	97.59	0.71	93.91	0.53	92.24	0.23	92.35	0.37	-	-
Case 3	98.63	0.34	99.09	0.48	99.56	0.53	95.83	0.80	93.69	0.71	94.07	0.79	92.29	0.90	-	-

For each one of the selected piezometers, the high interannual variation of the DWT can be highlighted. This fact is important because it allows for a wide range of DWT values for each WUA, which is useful for validating the behavior and performance of the developed tool under different dynamic water table levels. Moreover, the selected piezometers show similar tendencies of decreasing the DWT level for the 2011 irrigation season. This fact is explained by the increasing energy tariffs during that period, which resulted in a reduction of the use of groundwater resources in these areas, thereby contributing to the recovery of the aquifers. Hence, several values for the DWT have been used in these WUAs, which are representative of the DWT levels in the Castilla–La Mancha region.

3.3. Energy Analysis

For each analyzed well of each WUA, the irrigation season with the highest DWT value and the last irrigation season whose DWT values are available were chosen for the energy analysis. The DWT levels in Case 3 for the three WUAs analyzed were very similar. Therefore, only Case 1 and Case 2 were considered for the energy analysis of the three WUAs studied. Consequently, for each WUA studied, two irrigation seasons and two cases (Case 1 and Case 2) were considered in this analysis. The irrigation seasons with highest DWT values and the last irrigation season with available data for WUA A, B, and C were 2008, 2007, and 2007; and 2015, 2015, and 2013, respectively.

Figures 7–15 show the energy analysis results for each WUA, each irrigation season and each studied case. In WUA A (Figure 7), the operating point that minimizes the energy ratio (ER) is shown for Case 1 and Case 2, for irrigation seasons 2008 and 2015. Both irrigation seasons were proposed because of the high difference between the DWT levels among them in each case (Table 4). The minimum ER values reached a value close to 0.399 (Case 1) and 0.373 kWh m^{-3} (Case 2), which represented variable speed ratios (α) of 0.90 and 0.86, respectively. With regard to the 2015 irrigation season, the minimum ER ranged from 0.342 and 0.336 kWh m^{-3} for Case 1 and Case 2, respectively, which were obtained for α = 0.82 in both of them.

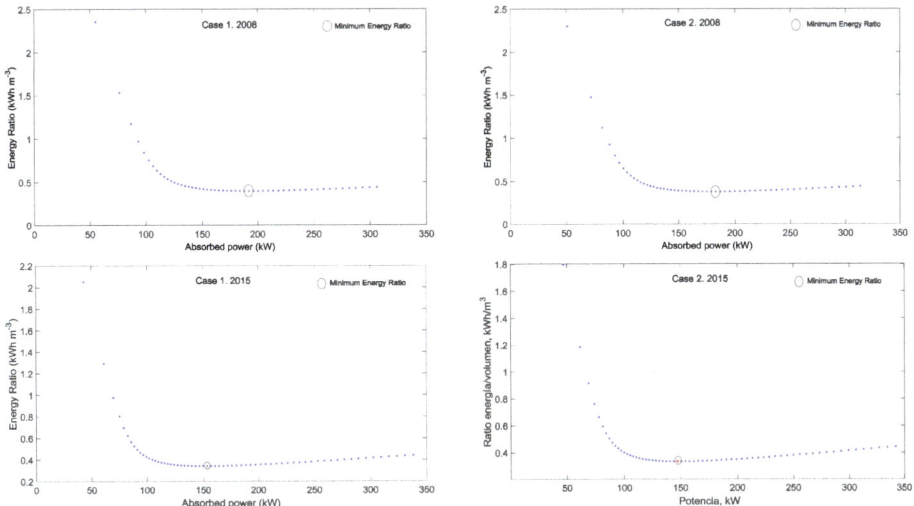

Figure 7. Energy ratio (kWh m^{-3}) at each variable speed ratio (α) in WUA A.

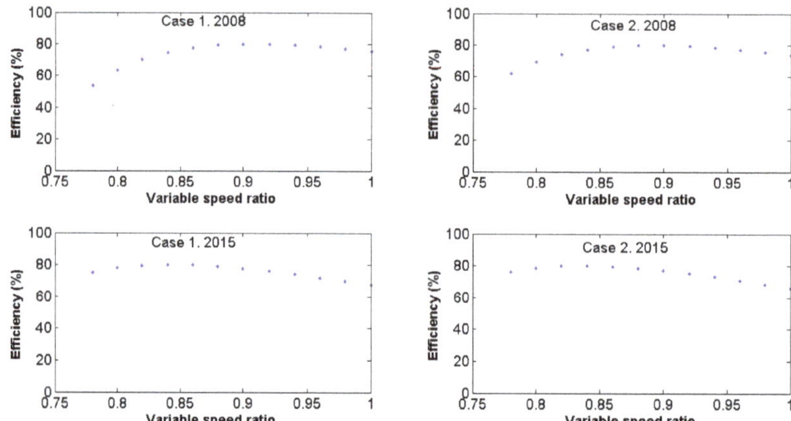

Figure 8. Pump efficiency (%) depending on the variable speed ratio in WUA A.

In 2008, when comparing the ER obtained with a fixed speed (at maximum α) and the variable speed pump that minimized the ER, the maximum energy saving obtained ranged from 9.9% (Case 1) to 13.1% (Case 2). Regarding the 2015 irrigation season, additional differences can be highlighted, with energy savings close to 22.8% (Case 1) and 24.4% (Case 2). In 2008, the difference between the efficiency at $\alpha = 1$ and α that minimizes the ER was not too high and was more important for the 2015 irrigation season. These differences can be explained when the pump efficiency at each variable speed ratio is represented (Figure 9), where the maximum pump efficiency could not be reached at a fixed speed. This finding is related to the fact that when the pump is working at a fixed speed, the operating point is located to the right of the maximum efficiency, as can be shown in Figure 9 for Case 2. Therefore, when the pump works at a variable speed, the operating point moves to the left, thus increasing the pump efficiency when a variable speed drive is used.

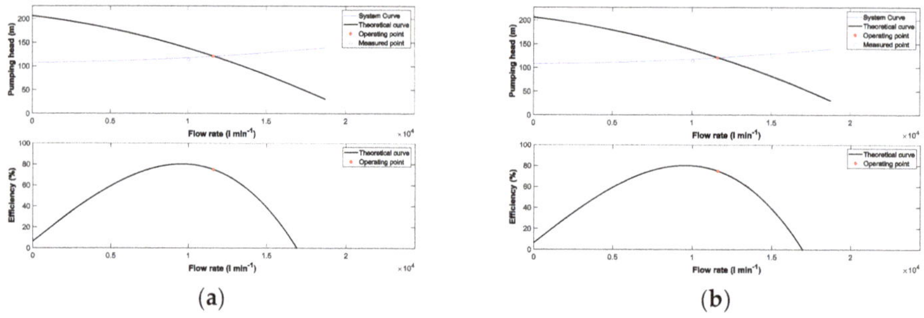

(a)   (b)

Figure 9. Operating point at fixed speed in WUA A for Case 2 in 2008 (a) and 2015 (b) irrigation seasons.

About WUA B, the results for the 2007 and 2015 irrigation seasons are shown. In 2007, no great differences between the ER at fixed speeds and α are shown for Case 1 and Case 2 (Figure 10), which is related to the fact that the DWT level was very similar during the 2007 irrigation season in both of cases. The minimum energy consumed was obtained when a variable speed of 0.82 was used, obtaining energy savings approximately 17.7% and 17.2% in Case 1 and Case 2, respectively, with respect to $\alpha = 1$. According to this, the minimum ER reached values close to 0.278 and 0.280 kWh m^{-3} for Case 1 and Case 2, respectively. In both cases, the α value that minimizes ER was close to 0.82. In 2015, the minimum ER was reached for α of 0.80 (Case 1) and 0.78 (Case 2), with ER values of 0.266 (Case 1) and 0.257 kWh m^{-3} (Case 2), respectively, obtaining energy savings between 20.7% and

23.0%, respectively. In both cases and irrigation seasons, the difference between the pump efficiency at maximum α and the pump efficiency for α that minimizes the ER was high (Figure 11), which is explained for WUA A, because when the pump was working at a fixed speed, the operating points for Case 1 and Case 2 were located to the right of the efficiency curve (Figure 12).

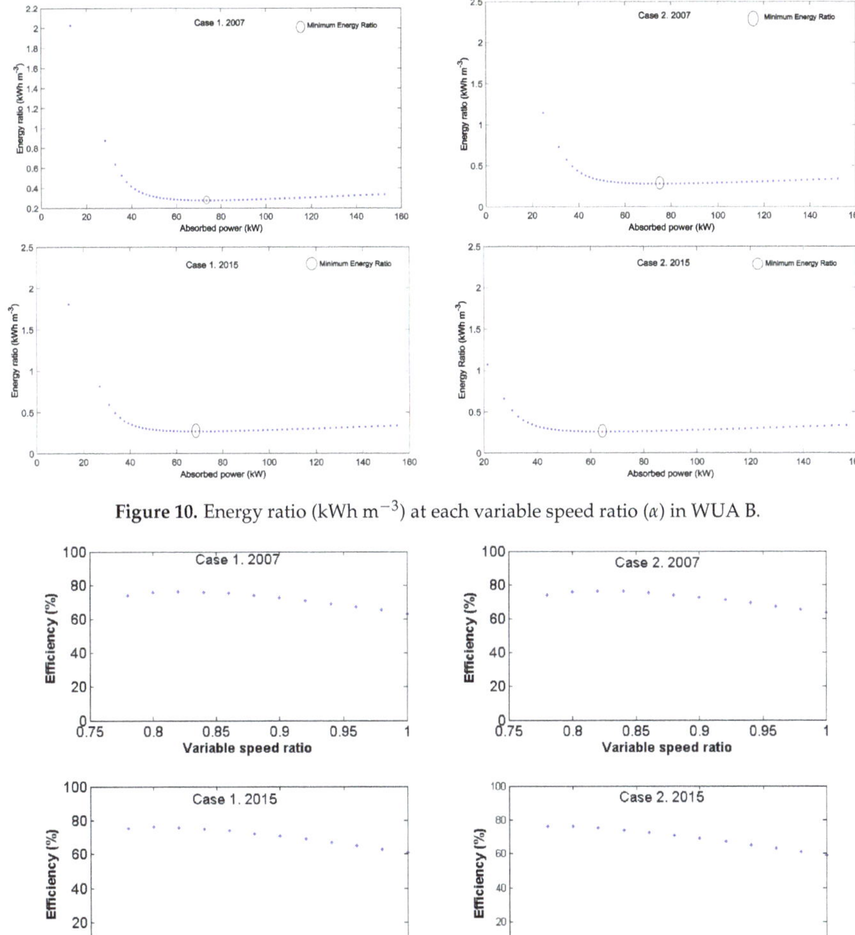

Figure 10. Energy ratio (kWh m^{-3}) at each variable speed ratio (α) in WUA B.

Figure 11. Pump efficiency (%) depending on the variable speed ratio in WUA B.

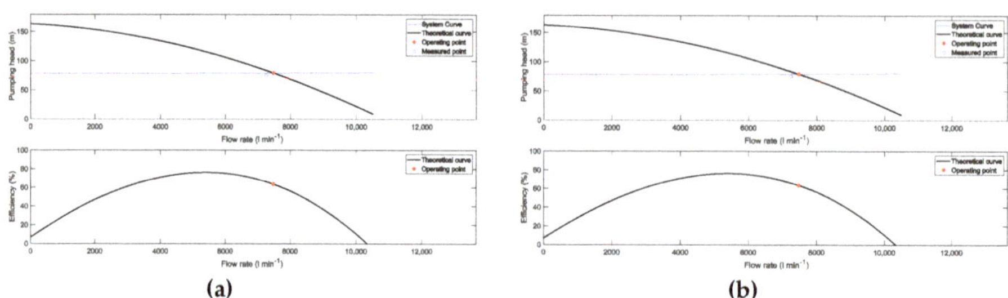

Figure 12. Operating point at fixed speed in WUA B for Case 2 in 2007 (**a**) and 2015 (**b**) irrigation seasons.

For WUA C, the tendency was very similar to the previously analyzed WUA. In 2007, the variable speed that minimizes the energy consumed per unit of water supplied was 0.9, for Case 1 and Case 2 (Figure 13), and the energy savings obtained ranged from 4.4% and 4.6%, respectively. In 2015, the energy savings slightly increased, ranging from 6.7% (Case 1) to 6.9% (Case 2), and were obtained when comparing the fixed speed with α of 0.88, which minimizes the ER. In both irrigation seasons and analyzed cases, the pump efficiency reached the highest values at a speed ratio of 0.9 (Figure 14), which contributes the energy savings obtained when using a variable speed. The operating point in Case 2 for both irrigation seasons (Figure 15) was to the right of the maximum efficiency, which was similar to that of WUAs A and B.

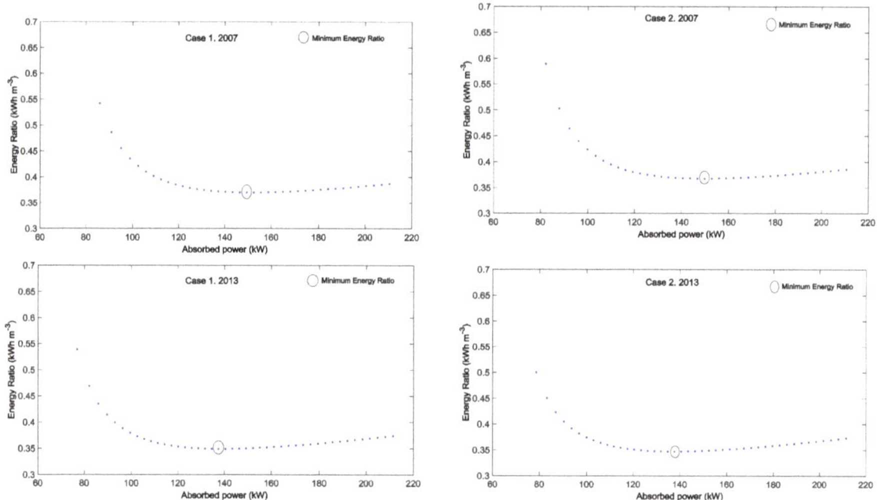

Figure 13. Energy ratio (kWh m^{-3}) at each variable speed ratio (α) in WUA C.

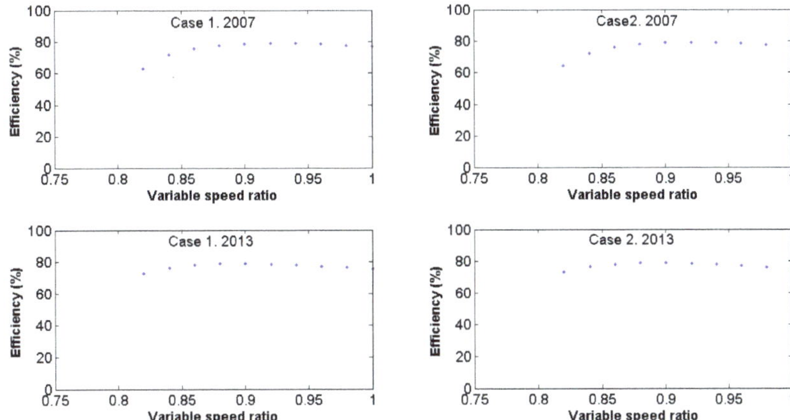

Figure 14. Pump efficiency (%) depending on the variable speed ratio in WUA C.

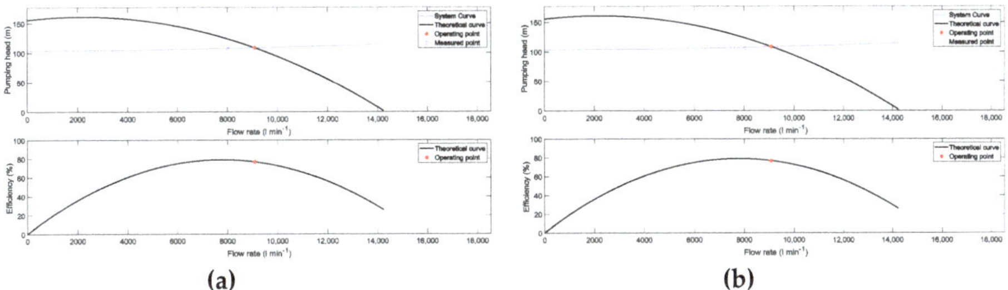

(a)   (b)

Figure 15. Operating point at fixed speed in WUA C for Case 2 in 2007 (**a**) and 2013 (**b**) irrigation seasons.

3.4. Economic Analysis

The economic analysis module assesses the profitability of installing a frequency speed drive vs. the option of a fixed speed. Thus, the speed variable drive investment was evaluated for each study case, irrigation season, and WUA. Table 5 shows a summary of the energy consumed by the fixed speed pump and the variable speed pump, its energy savings, and the absorbed power of the variable speed drive, which determines its investment costs according to equation 5 and the unit energy cost according to the study region. The energy savings obtained by the energetic module of the DSSW tool considering the use of variable speed pumps ranged from 9.9% to 24.4% for WUA A, from 17.2% to 23.0% for WUA B, and from 4.4% to 6.9% for WUA C. The energetic module of the DSSW tool also computes the absorbed power by each speed variable drive. Thus, the absorbed power and the investment cost of the variable speed drive according to equation 5 for WUA A were 371 kW and 23,510 €, respectively. Similarly, the absorbed power of the variable speed drive for WUA B and WUA C were 198 and 251 kW, respectively. The investment cost was 14,344 € for WUA B and 17482 € in the case of WUA C.

Based on the data shown in Table 5, the economic module of the DSSW tool computes the net present value, *NPV (in €)*, the internal rate of return, *IRR*, and the payback of each speed variable drive investment. Figure 16 shows the NPV values considering a discount rate which ranged from 1% to 100% (step 1%) for each study case, irrigation season, and WUA A (Figure 16a), WUA B (Figure 16b), and WUA C (Figure 16c). Concerning WUA A, the IRR values were 10% (Case 1) and 14% (Case 2) in 2008 and 26% (Case 1) and 27% (Case 2) in 2015. IRR is the discount rate that makes the NPV of all cash flows equal to zero. Furthermore, an NPV value equal to or higher than zero determines the profitability of potential investments. Consequently, higher NPV values with higher profitability is an investment, and a higher IRR value is the safer investment. The current discount rate is approximately 4.5%, which is far under the IRR value obtained for WUA A. Thus, the investment of the variable speed drive in the WUA A is very cost-effective and safely obtains NPV values for the current discount rate of 10,285, 18,218, 43,188, and 46,617 € for the two case studies and the two irrigation seasons analyzed, respectively, for the 15-year lifespan. The payback periods for the current discount rate were 10, 8, 5, and 4.5 years (Figure 17a).

Table 5. Energy consumption of fixed speed pump, variable speed pump, energy saving, absorbed power by the variable speed drive, and investment cost of the variable speed drive.

WUA	Cases	Energy Consumption (kWh m^{-3}) [1]	Energy Consumption (kWh m^{-3}) [2]	Energy Saving (%)	Absorbed Power by the Variable Speed Drive (kW)	Investment Cost of the Variable Speed Drive (€)
A	Case 1, 2008	0.439	0.399	9.92	371	23,510
	Case 2, 2008	0.422	0.373	13.10	371	23,510
	Case 1, 2015	0.420	0.342	22.84	371	23,510
	Case 2, 2015	0.418	0.336	24.44	371	23,510
B	Case 1, 2007	0.327	0.278	17.7	198	14,344
	Case 2, 2007	0.328	0.280	17.24	198	14,344
	Case 1, 2015	0.321	0.266	20.65	198	14,344
	Case 2, 2015	0.316	0.257	23.03	198	14,344
C	Case 1, 2007	0.387	0.371	4.35	251	17,482
	Case 2, 2007	0.385	0.368	4.64	251	17,482
	Case 1, 2015	0.372	0.349	6.70	251	17,482
	Case 2, 2015	0.372	0.348	6.90	251	17,482

[1] Energy consumption of fixed speed pump. [2] Energy consumption of variable speed pump.

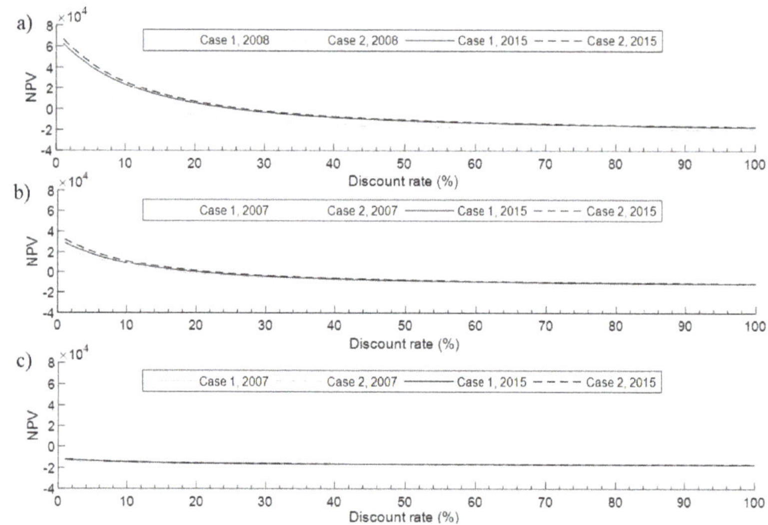

Figure 16. Net present value for several discount rates of the speed variable drive investment for WUA A (**a**), WUA B (**b**), and WUA C (**c**).

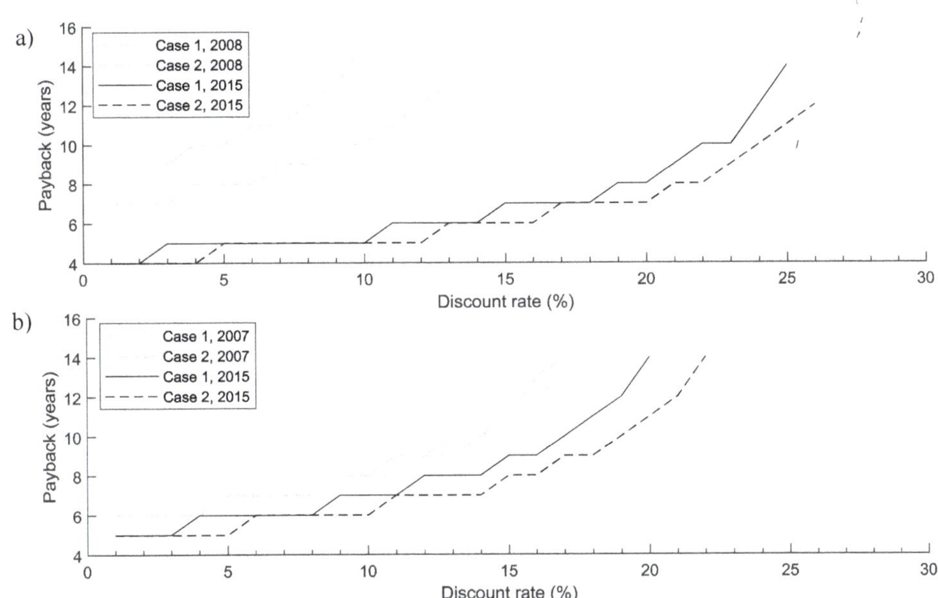

Figure 17. Payback period for several discount rates of the speed variable drive investment for WUA A (**a**) and WUA B (**b**).

Regarding the WUA B, in 2007, the IRR values were 18% for Case 1 and Case 2, whereas in 2015, they were 21% (Case 1) and 22% (Case 2). Consequently, although the percentages of energy savings between WUA A and WUA B were similar, the investment safety in WUA B was higher than that of WUA A, and the payback was lower (Figure 17b). These results are because the absorbed power by the variable speed drive in WUA B was much lower than that of WUA A, so the investment cost of the variable speed drive was also lower. Therefore, the NPV values for a discount rate of 4.5% were 15,861, 15,291, 19,376, and 21,988 € for the two case studies and the two irrigation seasons analyzed, respectively, for the 15-year lifespan. Because of the lower energy savings and the lower water volume applied in WUA B, the profitability was smaller. The payback periods for the current discount rate in WUA B were 6.5, 6.5, 6, and 5 years, respectively (Figure 17b).

However, because of the high absorbed power and the high cost of the variable speed drive in WUA C, and the low energy savings and the small amount of water applied, the NPV was negative for any discount rate considered (Figure 16c). Consequently, under the conditions of WUA C, the variable speed drive was unprofitable for the two case studies and the two irrigation seasons analyzed, respectively, for the 15-year lifespan.

4. Conclusions

A decision support system tool for water abstraction from aquifers for irrigation, the DSSW tool, was developed. The proposed tool can be useful for managers of irrigable areas once the crop distribution is known. The proposed methodology can be useful for reducing energy consumption during the water abstraction process from an aquifer to a reservoir in existing wells by installing a frequency speed drive. This tool might be combined with other tools focused on a centralized management, such as Irrigation Advisory Services, where managers can determine the optimum amount of water applied depending on the crop production costs and gross margin. It should be combined with tools that take into account the crop water demands and which give information about the influence of the water applied on crop yields. In this regard, according to the irrigation strategy followed at each area, and the total volume water supplied by wells, managers can decide whether a variable speed drive is or is not profitable.

In the three analyzed cases, energy savings, in comparison with the fixed speed, ranged from 4.4% and 24.4%, using a variable speed ratio of 0.9 and 0.82. The energy savings when using a variable speed frequency increased when the dynamic water table level was lower, and average energy savings close to 23%, 22%, and 6.8% were obtained for irrigation societies A, B, and C, respectively.

An economic analysis module was also developed in the DSSW tool. This module determines the economic profitability of the variable speed drive investment. The results show that the investment cost of the variable speed drive in most irrigation societies studied was very profitable, with a payback that ranged from 4.5 to 10 years. However, irrigation societies with very low energy savings and a high investment cost of the variable speed drive were linked to a low amount of applied water volume, and the investment was unprofitable. This decision support system tool has the potential to be useful for facilitating the transference of this methodology to engineers and managers of irrigable areas.

Author Contributions: Conceptualization, M.A.M. and J.I.C.; methodology, A.I. and R.G.P.; software, J.I.C.; validation, A.I., J.I.C. and R.G.P.; formal analysis, J.I.C., R.G.P., A.I. and, M.A.M.; investigation, J.I.C., R.G.P., A.I. and, M.A.M.; resources, J.I.C. and M.A.M.; data curation, J.I.C. and A.I.; writing—original draft preparation, J.I.C., R.G.P., A.I. and, M.A.M.; writing—review and editing, J.I.C. and R.G.P.; visualization, J.I.C., A.I. and R.G.P.; supervision, M.A.M.; project administration, M.A.M.; funding acquisition, M.A.M.

Funding: We would like to acknowledge the Spanish Ministry of Education and Science (MEC) for funding the AGL2007-66716-C03-03 project and the Consejería de Educación y Ciencia de Castilla-La Mancha for funding the project PCI08-0117. The authors wish also to express their gratitude to the Regional Agency of Energy in Castilla-La Mancha (AGECAM) for funding the project "Auditorías energéticas en Castilla-La Mancha", contributing funding and support that made the development of this study possible.

Conflicts of Interest: The authors declare no conflict of interest.

References

1. Carrión, F.; Sanchez-Vizcaino, J.; Corcoles, J.I.; Tarjuelo, J.M.; Moreno, M.A. Optimization of groundwater abstraction system and distribution pipe in pressurized irrigation systems for minimum cost. *Irrig. Sci.* **2016**, *34*, 145–159. [CrossRef]

2. Moreno, M.A.; Córcoles, J.I.; Moraleda, D.A.; Martinez, A.; Tarjuelo, J.M. Optimization of underground water pumping. *J. Irrig. Drain. Eng.* **2010**, *136*, 414–420. [CrossRef]

3. Khadra, R.; Moreno, M.A.; Awada, H.; Lamaddalena, N. Energy and Hydraulic Performance-Based Management of Large-Scale Pressurized Irrigation Systems. *Water Resour. Manag.* **2016**, *30*, 3493–3506. [CrossRef]

4. Lima, F.A.; Martínez-Romero, A.; Tarjuelo, J.M.; Córcoles, J.I. Model for management of an on-demand irrigation network based on irrigation scheduling of crops to minimize energy use Part I): model Development. *Agric. Water Manag.* **2018**, *210*, 49–58. [CrossRef]

5. Córcoles, J.I.; de Juan, J.A.; Ortega, J.F.; Tarjuelo, J.M.; Moreno, M.A. Management evaluation of Water Users Associations using benchmarking techniques. *Agric. Water Manag.* **2010**, *98*, 1–11. [CrossRef]

6. Córcoles, J.I.; De Juan, J.A.; Ortega, J.F.; Tarjuelo, J.M.; Moreno, M.A. Evaluation of Irrigation Systems by Using Benchmarking Techniques. *J. Irrig. Drain. Eng.* **2012**, *138*, 225–234. [CrossRef]

7. Rodríguez Díaz, J.A.; López Luque, R.; Carrillo Cobo, M.T.; Montesinos, P.; Camacho Poyato, E. Exploring energy saving scenarios for on-demand pressurised irrigation networks. *Biosyst. Eng.* **2009**, *104*, 552–561. [CrossRef]

8. Carrillo Cobo, M.T.; Rodríguez Díaz, J.A.; Montesinos, P.; López Luque, R.; Camacho Poyato, E. Low energy consumption seasonal calendar for sectoring operation in pressurized irrigation networks. *Irrig. Sci.* **2011**, *29*, 157–169. [CrossRef]

9. Fernández García, I.; Moreno, M.A.; Rodríguez Díaz, J.A. Optimum pumping station management for irrigation networks sectoring: Case of Bembezar MI (Spain). *Agric. Water Manag.* **2014**, *144*, 150–158. [CrossRef]

10. González Perea, R.; Camacho Poyato, E.; Montesinos, P.; Rodríguez Díaz, J.A. Critical points: Interactions between on-farm irrigation systems and water distribution network. *Irrig. Sci.* **2014**, *32*, 255–265. [CrossRef]

11. Rodríguez Díaz, J.A.; Montesinos, P.; Camacho Poyato, E. Detecting Critical Points in On-Demand Irrigation Pressurized Networks—A New Methodology. *Water Resour. Manag.* **2012**, *26*, 1693–1713. [CrossRef]

12. Jiménez-Bello, M.A.; Martínez-Alzamora, F.; Castel, J.R.; Intrigliolo, D. Validation of a methodology for grouping intakes of pressurized irrigation networks into sectors to minimize energy consumption. *Agric. Water Manag.* **2011**, *102*, 46–53. [CrossRef]

13. Córcoles, J.I.; Tarjuelo, J.M.; Moreno, M.A. Methodology to improve pumping station management of on-demand irrigation networks. *Biosyst. Eng.* **2016**, *144*, 94–104. [CrossRef]

14. Moreno, M.A.; Ortega, J.F.; Córcoles, J.I.; Martínez, A.; Tarjuelo, J.M. Energy analysis of irrigation delivery systems: Monitoring and evaluation of proposed measures for improving energy efficiency. *Irrig. Sci.* **2010**, *28*, 445–460. [CrossRef]

15. Helweg, O.J. DETERMINING OPTIMAL WELL DISCHARGE. *ASCE J Irrig Drain Div* **1975**, *101*, 201–208.

16. Scalmanini, J.C.; Scott, V.H.; Helweg, O.J. Energy and efficiency in wells and pumps. *Proc. 12th Bienn. Conf. Gr. Water, Sacramento, Calif. Calif. Water Resour. Cent. Rep.* **1979**, *45*, 1–210.

17. Helweg, O.J.; VonHofe, F.; Quek, P.T. Improving well hydraulics. In Proceedings of the Irrigation Drainage Special Conference, Honolulu, HI, USA, 22–24 July 1991.

18. Helweg, O.J.; Jacob, K.P. Selecting the optimum discharge rate for a water well. *Hydraul. Eng.* **1991**, *117*, 934–939. [CrossRef]

19. Moreno, M.A.; Medina, D.; Ortega, J.F.; Tarjuelo, J.M. Optimal design of center pivot systems with water supplied from wells. *Agric. Water Manag.* **2012**, *107*, 112–121. [CrossRef]

20. Domínguez, A.; Martínez, R.S.; Juan, J.A.; Martínez-Romero, A. Simulation of maize crop behaviour under deficit irrigation using MOPECO model in a semi-arid environment. *Agric. Water Manag.* **2012**, *107*, 42–53. [CrossRef]

21. Lamaddalena, N.; Sagardoy, J.A. *Performance Analysis of on-Demand Pressurized Irrigation Systems*; FAO: Rome, Italy, 2000; Volume 132. [CrossRef]

22. Calejo, M.J.; Lamaddalena, N.; Teixeira, J.L.; Pereira, L.S. Performance analysis of pressurized irrigation systems operating on-demand using flow-driven simulation models. *Agric. Water Manag.* **2008**, *95*, 154–162. [CrossRef]

 *water*

Article

Quantification of Daily Water Requirements of Container-Grown *Calathea* and *Stromanthe* Produced in a Shaded Greenhouse

Richard C. Beeson, Jr. and Jianjun Chen *

Department of Environmental Horticulture, Institute of Food and Agricultural Sciences, University of Florida, Apopka, FL 32703, USA; rcbeeson@ufl.edu
* Correspondence: jjchen@ufl.edu; Tel.: +1-407-814-6161

Received: 30 July 2018; Accepted: 3 September 2018; Published: 5 September 2018

Abstract: Irrigating plants based on their water requirements enhances water use efficiency and conservation; however, current irrigation practices for container-grown greenhouse plants largely relies on growers' experiences, resulting in leaching and/or runoff of a large amount of water. To address water requirements of greenhouse-grown plants, this study adapted a canopy closure model and investigated actual evapotranspiration (ET_A) of *Calathea* G. Mey. 'Silhouette' and *Stromanthe sanguinea* Sond. from transplanting to marketable sizes in a shaded greenhouse. The daily ET_A per *Calathea* plant ranged from 3.55 mL to 59.39 mL with a mean cumulative ET_A of 4.84 L during a 224 day growth period. The daily ET_A of *S. sanguinea* varied from 7.87 mL to 97.27 mL per plant with a mean cumulative ET_A of 6.81 L over a 231 day production period. The best fit models for predicting daily ET_A of *Calathea* and *Stromanthe* were developed, which had correlation coefficients (r^2) of 0.82 and 0.73, respectively. The success in modelling ET_A of the two species suggested that the canopy closure model was suitable for quantifying water use of container-grown greenhouse plants. Applying the research-based ET_A information in production could reduce water use and improve irrigation efficiency during *Calathea* and *Stromanthe* production.

Keywords: actual evapotranspiration (ET_A); *Calathea*; container-grown plants; daily water requirements; ornamental foliage plants; *Stromanthe*; water need index (WNI)

1. Introduction

Freshwater is one of our most precious natural resources. Agricultural use of freshwater has been under rigorous scrutiny since irrigation withdrawals represent over 70% of all freshwater use worldwide [1,2]. In the United States (U.S.), irrigation accounts for 68% of groundwater and 29% of surface water withdrawals, encompassing up to 62% of all freshwater use [3]. Container plant production is an important sector of agriculture, which refers to growing plants from seedlings, liners, rooted cuttings, or grafted plants in containers or pots filled with substrates to marketable sizes or harvestable stages [4]. Growing media or substrates consist of a mix of soil, peat, vermiculate, perlite, or other organic components in different proportions. Container production is a widely-utilized method for growing a variety of plants including fruit, vegetable, nursery, and floriculture crops. Floriculture and nursery crops comprise almost 30% of the specialty crops grown in the U.S. with a total of $11.7 billion in sales in 2009 [5]. Container production in the U.S. currently accounts for approximately 90% of greenhouse, nursery, and floriculture crops [6]. Since plants are grown in artificial substrates confined by limited volumes (containers), they have to be frequently irrigated ranging from daily to weekly to avoid drought stress. Current irrigation practices have been largely based on growers' intuition or experience, and as a result, plants are often overirrigated [7]. Overirrigation has been reported to result in 25% to 90% of irrigated water to be leached and/or runoff [8–11]. Overirrigation not only reduces substrate aeration but also results in irrigation leaching and/or runoff, which is accompanied with nutrient elements, primarily nitrogen (N) and phosphorus (P). The movement of N and P in waterways could potentially contaminate ground and/or surface water [8]. Thus, irrigation based on plant growth requirements is becoming increasingly important for sustainable production of container-grown plants.

Plant water use is mainly a function of transpiration through leaves. When evaporation from soil or other substrates is included, water use is termed actual evapotranspiration (ET_A, mL day^{-1}). Transpiration accounts for the majority of actual ET_A, but it can be affected by local weather conditions, cropping system, plant species, and growth stages [12]. As a result, ET_A is estimated by multiplying reference ET (ET_0, mL day^{-1}) with a corresponding crop coefficient (Kc, dimensionless) [12]. The Penman–Monteith equation has been recommended as the sole method of estimating ET_0 [12], which is calculated by multiplying weather-based estimates of ET from a reference crop, such as grass or alfalfa. There are several modifications to the Penman–Monteith model that provide the same values for ET_0. One is the grass reference option established by the American Society of Civil Engineering (ASCE) [13]. Another is the program provided by Campbell Scientific Inc. (Logan, UT, USA) for its weather stations [14] that uses the full ASCE Penman–Monteith equation [15]. Kc is a function of fraction of ground cover and crop height, and Kc values have been reported for a wide range of agronomic crops [12,16]. However, accurate Kc values are difficult and expensive to develop [17]. Using precision weighing lysimeters is the most accurate way of estimating crop water usage and developing Kc. Weighing lysimeters have generally been regarded as the standard measuring device for estimating ET; they do so by measuring changes in mass of a soil container with plants positioned on a scale or other weight device. Weighing lysimeters measure ET_A as volumes, rather than depths.

Quantification of daily water use of container-grown plants dates back to the 1980s. The Thorn thwaite equation [18] was the original method used [19,20] to calculate ET_0. Other researchers used the top diameter of container [21] for calculating the surface area to convert ET_A to a depth for 22 woody ornamental species, and Kc values ranging from 1.1 to 5.1 were reported. In agronomic crops, however, Kc values rarely exceed 1.3 [22]. This represents a major discrepancy between agronomic and container plants. Agronomic crops are produced in the field, and ET_A and Kc are estimated in reference to an irrigated area or a ground area, which includes ground cover and canopy. In container-grown plants, ET_A is mostly measured by weight loss from a container initially near 100% container capacity; as such, ET_A is a measure in volume. However, ET_0 is calculated as a depth (mL). In container plant production, projected canopy areas (PCA) generally exceed container surfaces by several times. Evapotranspiration, mainly transpiration, takes place throughout the canopy. Since the measure of ET_A is in volume, it has to be normalized by an area to be the same units as ET_0 for the calculation of Kc. An approach to normalized ET_A is to calculate it based on PCA. When ET_A was normalized by both ET_0 and PCA, Kc declined as PCA approached canopy closure and became relatively constant after canopy closure [23]. Thus, it was proposed that the calculation of Kc of container-grown plants was based on canopy closure [24]. When using this model, researchers determine average canopy area and calculates percent canopy closure (%CC) based on distance between adjacent containers and the canopies of similar plants. Based on the ET_A, ET_0, and PCA, a water needs index (WNI) is calculated, which is a function of canopy closure of a group of plants, relating individual plant actual evapotranspiration (ET_A) to plant size and canopy ventilation and radiation [25]. Daily ET_0 was calculated from a meteorological measurement on site, with ET_A determined by an autonomous weighing lysimeter system [10]. The model has been used to quantify daily ET_A of several woody ornamental plants to market size including *Ligustrum japonicum* Thunb. [24], *Viburnum odoratissimum* Ker Gawl. [26,27], *Rhaphiolepis indica* (L.) Lindl. ex Her Gawl. [28] as well as foliage plants of *Asplenium nidus* L. and *Chamaedorea elegans* Mart. [29]. The irrigation of container plants based on daily water use has been documented to reduce nursery runoff volume and nutrient load without reducing plant growth [30,31].

The objectives of the present study were to determine ET_A of two important container-grown ornamental foliage plants, *Calathea* G. Mey. 'Silhouette' and *Stromanthe sanguinea* Sond., in a shaded greenhouse from tissue-cultured liners grown in 15 cm containers to marketable sizes, and to develop models to predict daily ET_A rates using the PCA model developed for container-grown woody ornamental plants. Foliage plants are those produced in shaded greenhouses and used primarily for interior decoration [32]. Florida is the leading state in the production of foliage plants, accounting for 72% of wholesale value in the U.S. in 2015 [33]. Quantification of their daily water use could provide research-based information for improving irrigation efficiency during foliage plant production in greenhouse conditions.

2. Materials and Methods

2.1. Experimental Location

The experiment was conducted in a shaded greenhouse at the University of Florida's Mid-Florida Research and Education Center (MREC) in Apopka where a Florida Automated Weather Network (FAWN) station was 46 m east of the shaded greenhouse. The station provided readings of air temperatures at three elevations (0.6, 1.8, and 9.1 m), dew point, rainfall, soil temperature, relative humidity, wet bulb temperature, barometric pressure, and wind speed every 15 min daily as well as daily evapotranspiration (ET).

2.2. Experimental Setup

An automated Weatherhawk weather station (Campbell Scientific Inc., Logan, UT, USA) was installed inside the shaded greenhouse where a miniature weighing lysimeter system was built for this study [10]. In brief, the system consisted of a control/data collection apparatus connected to mini lysimeters where a SDM-AM16-32 multiplexer (Campbell Scientific Inc., Logan, UT, USA), a CR10X data logger (Campbell Scientific Inc., Logan, UT, USA), and SDM-CD16AC relay control module (Campbell Scientific Inc., Logan, UT, USA) were used for receiving and storing data from the mini-lysimeters. Each mini-lysimeter was installed with a load cell (SSM-50-AJ, Interface Inc., Scottsdale, AZ, USA) that was suspended from a miniature tripod with a plant support. All load cells were calibrated with a seven-point curve using known masses. Every half hour, the data logger program recorded the mass of each lysimeter and stored it for later retrieval. At midnight, actual evapotranspiration (ET_A) for each lysimeter was determined as the difference in mass between 0500 h and midnight. There was no transpiration between midnight and 0500 h.

2.3. Plant Materials and Their Growth

Tissue-cultured liners (plantlets grown in plugs of 72 cell trays) of *Calathea* 'Silhouette' and *S. sanguinea* were transplanted singly into a peat-based substrate composed of 60% Canadian peat, 20% vermiculite, and 20% perlite in 15 cm containers. *Stromanthe* plants were transplanted on 8 September 2008 and harvested on 23 April 2009; and *Calathea* plants were transplanted on 7 July 2009 and harvested on 18 February 2010. Plants were fertilized by top dressing 5 g of a controlled-released fertilizer (CRF) (Osmocote 19–5–9, 8–9 month, The Scotts Co., Marysville, OH, USA) per container three weeks after potting. The experiment was arranged as a completely randomized block design with four replications. Each block had 15 plants per species, they were spaced in three rows, five containers along the length of benches, 30 cm apart. The center plant was placed in a suspension-weighing lysimeter, and the four closest plants to the lysimeter plant were designated as the interior plants for repeated canopy measurements. Plants were produced in the aforementioned shaded greenhouse under a maximum photosynthetic active radiation (PAR) of 200 μmol m^{-2} s^{-1}.

Plants were irrigated between 0800 h and 0900 h and allowed to have about 10% leachate fraction. Irrigation was supplied through self-fashioned rings of pressure-compensated drip tubing (Netafilm, Fresno, CA, USA). Pressure-compensating emitters were made in 30.5 cm intervals along the length of the tubing. There were four emitters that were cut and rolled to form two loops joined by a T-barb. Each container had a joined loop which was connected to 19 mm polyethylene tubing using equal lengths of 6 mm tubing. Water application for 15 loops (15 plants) was calculated by Christensen's coefficients of uniformity on each bench. The mean coefficient was 0.94 and ranged from 0.93 to 0.96. Typical application rates per bench were 187 mL per minute.

2.4. Data Collection

Greenhouse air temperature, relative humidity, wind speed, and solar radiation required for calculation of reference evapotranspiration (ET_0) was collected from the automated Weatherhawk weather station. The algorithm used for calculating reference evapotranspiration (ET_0) was based on the Campbell Scientific program.

Plant growth data were collected a week after potting. The widest width, width perpendicular to the widest width, and average height of the canopy were measured every three weeks on the lysimeter plants and adjacent four interior plants of each replication. The two widths were multiplied to estimate the two-dimensional PCA. When PCA was multiplied by the average height, canopy volume or growth index (GI) was estimated [34,35], assuming the three-dimensional canopy resembled a rectangular form. Plant water-use efficiency was calculated as total dry matter produced (g) by actual amount of water used (ET_A) [36].

2.5. Modelling Plant Water Use

The %CC, ET_A, ET_0, and PCA [24,26,27] were utilized for modelling the daily water use of two plant species. The %CC at each measurement was calculated by adding half the PCA of each of the four border plants to the PCA of the lysimeter plant and dividing the sum by allocated bench space for each plant (929 cm^2). This was the squared distance, at the center, between each plant. Since the containers were not respaced, canopies could become overlapped as shoots per plant expanded outward and up. The overlapping could result in calculation of %CC greater than 100% since it was determined on a fixed allocated bed area.

Daily ET_A (cm^3) of each lysimeter plant was converted to a depth by dividing with its PCA (cm^2). ET_A depth (cm) was then divided by the corresponding ET_0 (cm) and averaged over the seven days to calculate WNI [WNI = (ET_A/PCA)/ET_0] [28] for each lysimeter plant at each measurement date. Obtained WNI values of the four lysimeter replicates for each date were plotted against their corresponding %CC values. The plot was fitted to a three-parameter exponential decay curve using SigmaPlot (Version10; SPSS Inc., Chicago, IL, USA). The plot was also fitted to a third-order inverse polynomial equation also using SigmaPlot. An equation for the nonlinear line was derived using a three-level inverse polynomial equation (Version10; SPSS Inc., Chicago, IL, USA).

3. Results

3.1. Reference Evapotranspiration

The ET derived from FAWN from September 2008 to April 2009 ranged from 1.02 to 4.83 mm with a mean of 2.36 mm. The ET from July 2009 to February 2010 varied from 1.02 to 5.58 mm with a mean of 2.75 mm. The ET_0 values in the shaded greenhouse were about 12% of that outside the shaded greenhouse as measured by the FAWN station during the mentioned time. Relative humidity was normally higher, more than 75% inside the greenhouse. Air movement happened during most afternoons from early-Spring until late Fall due to the operation of evaporative cooling fans, although there was no measurable wind movement. Temperatures in the shaded greenhouse were also more moderate than outside conditions, with minimums set at 18.3 °C for heating and 32.2 °C for evaporative cooling. ET_0 was highest in April and May (days 90 to 150), then declined the rest of the period with the onset of summer rains, then shorter days.

3.2. Plant Growth

Canopy height and widths of *Calathea* 'Silhouette' increased in a polynomial fashion (data not shown). Growth index also increased polynomially (Figure 1A). *Calathea* 'Silhouette' at the time of harvest produced 19.4 leaves with a total leaf area of 852 cm^2. Shoot and root fresh weights were 33.8 and 5.6 g; and shoot and root dry weights were 3.7 and 0.5 g, respectively (Table 1).

Canopy heights and widths of *S. sanguinea* increased linearly over the production time (data not shown), but canopy width and growth index increased polynomially (Figure 1B). At harvest, the mean number of leaves was 84 with a total leaf area of 2729.2 cm^2. Shoot and root fresh weights were 110.7 and 42.2 g, and corresponding dry weights were 14.4 and 3.0 g, respectively (Table 1).

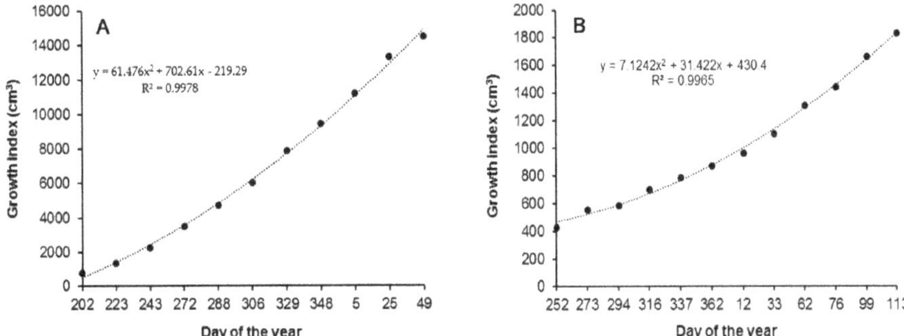

Figure 1. Growth indices (cm^3) of *Calathea* 'Silhouette' (**A**) and *Stromanthe sanguinea* (**B**) grown in 15 cm containers from tissue-cultured liners to marketable sizes. The equations are the best fit line (dash).

Table 1. Plant growth measurements at harvest by species. Plants were harvested when common commercial canopy sizes were attained [z].

Plant	Mean Leaf No	Leaf Area (cm^2)	Shoot Fresh Weight (g)	Root Fresh Weight (g)	Shoot Dry Weight (g)	Root Dry Weight (g)	Water Use Efficiency (g L^{-1}) [y]
Calathea	19.4 ± 0.59	852.0 ± 54.60	33.8 ± 2.40	5.6 ± 0.37	3.7 ± 0.33	0.5 ± 0.04	0.87
Stromanthe	84.0 ± 2.49	2729.2 ± 68.36	110.7 ± 3.10	42.2 ± 1.35	14.4 ± 0.17	3.0 ± 0.08	2.56

[z] Values represent the means ± standard errors of four replications. [y] Water use efficiency = the ratio of total dry weight (g) to total amount of water used (L).

3.3. Plant Actual Evapotranspiration (ET$_A$)

Daily ET$_A$ per *Calathea* 'Silhouette' plant ranged from 3.55 mL to 59.39 mL (Figure 2A) with an overall mean of 21.6 mL a day per plant. The mean cumulative ET$_A$ was 4.84 L over a 224 day production period that span mid-summer 2009 to mid-February 2010. Corresponding to the beginning of production in mid-summer and finishing in late winter, mean ET$_A$ was initially higher, up to 59.39 mL, then declined through the production period. By late October (day 120), mean ET$_A$ stopped declining, ranging generally from 11.83 to 23.66 mL per day until harvest.

Daily ET$_A$ for producing *S. sanguinea* ranged from 7.87 to 97.27 mL with a mean of 29.5 mL (Figure 2B). The mean cumulative ET$_A$ value was 6.81 L per plant during the entire production period. Mean daily ET$_A$ declined from transplanting through the end of the year. Mean ET$_A$ during this period decreased from 36.97 mL to 17.74 mL per day. Increases in mean ET$_A$ were slow to occur until early March 2009, when daily ET$_A$ increased from around 29.57 mL per day to a median of 73.93 mL per day over a 20 day period. The increase in mean daily ET$_A$ occurred due to large increases in daily variability. With the weeks before harvest, mean daily ET$_A$ ranged from 14.79 mL to nearly 97.27 mL per day.

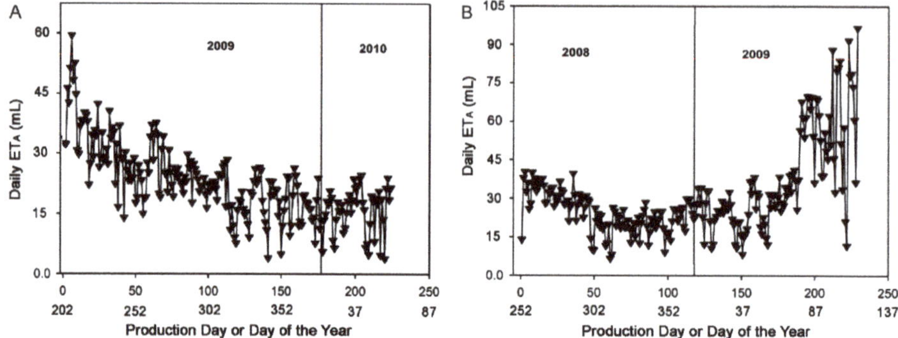

Figure 2. Mean daily actual evapotranspiration (ET$_A$) of *Calathea* 'Silhouette' (**A**) and *Stromanthe sanguinea* (**B**) grown in 15 cm containers during production from tissue-cultured liners to marketable sizes. Each triangle is the mean daily ET$_A$ of four plant replicates. The vertical line separates the year.

3.4. Data Analysis and Modeling

Application of the %CC model [26] was successful for both *Calathea* and *Stromanthe* with the best fit model presented in Table 2. The r^2 values were 0.82 and 0.73 for *Calathea* and *Stromanthe*, respectively, suggesting that strong correlations occurred between WNI and %CC for the two species. The equations predicted the water use of individual plants. If means could have been derived from these individual plant measurements, such that the relationships were derived for populations of plants, the r^2 would be even higher and predictions of overall crop water use would be more certain. The WNI coefficients declined as both plants' growth increased, which was illustrated by the increase in %CC (Figure 3A,B). The decline in WNI was due to the increase in canopy boundary layer resistance as plant foliage expanded and filled in the gaps between containers. When reaching near 100% CC, transpiration of all but the upper leaves become decoupled from the air above the canopy, resulting in 40% deceases in whole plant transpiration outdoors [26]. Conversely, random removal of approximately 33% plant canopy coverage has been shown to increase individual plant transpiration by 40% [26].

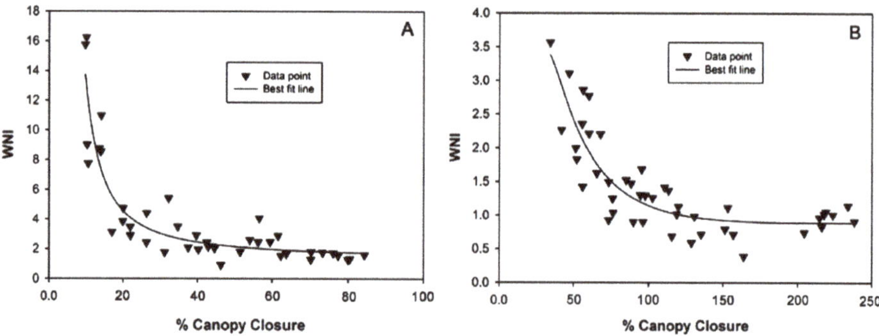

Figure 3. Inverse polynomial relationship between % Canopy Closure (CC) and the water need index (WNI) for *Calathea* 'Silhouette' (**A**) and *Stromanthe sanguinea* (**B**). Data points are four plant replicates, and the equation for the best fit line present in Table 2.

Table 2. Best fit models for predicting daily ET_A values of *Calathea* 'Silhouette' and *Stromanthe sanguinea* during production from tissue-cultured liners to marketable sizes in 15 cm containers.

Species	Model Equation	r^2
Calathea	WNI = $1.213 + 0.383/\%CC + 0.032/(\%CC)^2 + 0.005/(\%CC)^3$	0.82
Stromanthe	WNI = $1.136 - 1.131/\%CC + 1.395/(\%CC)^2 - 0.255/(\%CC)^3$	0.73

4. Discussion

The present study investigated daily ET_A of two important container-grown ornamental foliage plants, *Calathea* 'Silhouette' and *S. sanguinea*, from tissue-cultured liners to marketable sizes in a shaded greenhouse. Results showed that daily water requirements of *Calathea* varied from 3.55 mL to 59.39 mL over a 224 day production period, and the cumulative ET_A for producing this plant was 4.84 L. The daily water requirements of *S. sanguinea* ranged from 7.87 mL to 97.27 mL and its cumulative ET_A was 6.81 L. The cumulative ET_A values were comparable to two other container-grown foliage plants, *A. nidus* and *C. elegans* [29], in which both were grown in 15 cm containers with cumulative ET_A values of 7.95 L for *A. nidus* during a 294 day growth and 6.43 L for *C. elegans* during a 280 day production period. Although there has been little ET_A information available about container plants produced in greenhouse conditions, we believe that the cumulative ET_A data were valid. Researchers have produced foliage plants *Codiaeum variegaturm* 'Petra', *Dieffenbachia maculate* 'Camelle', and *Spathiphyllum* 'Petite' in 15 cm containers in an ebb-and-flow system from cuttings or tissue-cultured liners to marketable size [37]. Ebb-and-flow is a system where container plants on ebb-and-flow trays are subirrigated with recirculated nutrient solution. There is no irrigation water runoff, and the amount of water lost during the production can be quantified. The authors reported that the total amount of water used by the plants ranged from 6.8 L to 7.9 L depending on nitrogen rates. The same ebb-and-flow system has been used to produce 18 foliage plants across 15 genera in 15 cm containers in a shaded greenhouse and it has been found that the average amount of water required for producing these plants from cuttings or tissue culture liners to marketable sizes was 10.22 L [9]. The calculated amount of water loss on ebb-and-flow systems could be greater than those produced in the present study. This is because each ebb-and-flow tray has a surface area of 2.4 m^2, and water or nutrient solution are flooded in the trays to a depth of 2.5 cm for 10 min one to three times a week, during which a certain amount of water was lost due to evaporation.

Results from this study showed that the canopy closure model and WNI developed for woody ornamental plants are suitable for modelling daily water requirements of container-grown foliage plants in greenhouse conditions. In the present study, canopy widths and height were recorded every three weeks, and PCA and %CC were calculated based on container size and spacing. Using %CC, Kc values were calculated. The Kc values were multiplied by PCA and ET_0 to estimate the ET_A on daily base. WNI was then calculated by the formula $(ET_A/PCA)/ET_0$ [28]. The plots of WNI as a function of %CC for both *Calathea* and *Stromanthe* (Figure 3A,B) were fitted to the equations presented in Table 2 with the correlation coefficients (r^2) 0.82 and 0.73 for *Calathea* and *Stromanthe*, respectively. The modelling ET_A based on the aforementioned method has several advantages [25]. Kc calculation is based on canopy closure, it should be independent from container size. It also should avoid the need to use Fourier curve transformations to account for changes in ET_A or growth with season. Development of models based on canopy components should be easier to convert to predictive irrigation models than those that require complex transformation or provide Kc range only. On the other hand, if the %CC or PCA does not change, neither should the relationship of ET_A to ET_0, suggesting a period of canopy dormancy.

As far as is known, this is the first report on daily water use for *Calathea* and *Stromanthe*. This information along with ET_A values estimated in another two foliage plants, *Asplenium* and *Chamaedorea*, showed that foliage plants have rather lower cumulative ET_A values than other ornamental plants grown in greenhouses. For example, cumulative water use of two *Petunia hybrida* cultivars were 2.71 L and 4.08 L, respectively, when they were produced in 15 cm containers for 46 days [38]. The lower ET_A is likely related to the natural habits of their origin. Foliage plants were predominantly understory plants [39]; their leaves have thicker cuticles; the plants require lower light levels and lower nutrient input for growth. As a result, their net photosynthetic and transpiration rates are lower [40]. Commercial production of foliage plants, however, has not fully considered these characteristics. Traditional overhead irrigation resulted in the runoff of 90% of irrigation water. Subsequent improvement on irrigation with a capillary mat used 19 L to 23 L of irrigation water per plant, and drip irrigation used 10 L to 12 L of water [41]. Compared to the cumulative ET_A values ranging from 4.82 L to 7.95 L with those from either drip or capillary mat irrigation, irrigation of plants based ET_A could substantially save freshwater in container-grown foliage plant production. Plants from more than 1000 species across more than 100 genera are grown as ornamental foliage plants [32], and water requirements and water use efficiency could vary among species. For example, the water use efficiency of *Stromanthe* is three times greater than that of *Calathea* (Table 1). To irrigate container-grown foliage plants based on their requirements, the ET_A of each genus should be determined. The methodologies presented in this study provide an easy and affordable way to quantify daily water use, and the application of ET_A information into irrigation practices should significantly improve irrigation efficiency and conserve freshwater resources.

5. Conclusions

The daily ET_A established for *Calathea* and *Stromanthe* as well as those previously established for *A. nidus* and *C. elegans* suggest that the canopy closure model and WNI developed for woody ornamental plants are suitable for modelling daily water requirements of container-grown foliage plants produced in greenhouse conditions. The daily ET_A established for *Calathea* and *Stromanthe* could be used as reference guidelines for improving irrigation practices in commercial production of foliage plants. The application of ET_A should significantly reduce irrigation water runoff and leaching and conserve freshwater resources.

Author Contributions: Both authors conceptualized the present work, conducted in the experiments, analyzed the data, and drafted the manuscript.

Funding: This project was supported in part by the Southwest Florida Water Management District, Florida, USA.

Acknowledgments: The authors would like to thank Russell Caldwell for assistance in completion of the experiments and Caroline Roper for revising and editing the manuscript.

Conflicts of Interest: The authors declare no conflict of interest.

References

1. Doll, P.; Fiedler, K.; Zhang, J. Global-scale analysis of river flow alterations due to water withdrawals and reservoirs. *Hydrol. Earth Syst. Sci.* **2009**, *13*, 2413–2432. [CrossRef]
2. Food and Agricultural Organization of the United Nations. *AQUASTAT—FAO's Global Information System on Water and Agriculture*; FAO: Rome, Italy, 2010.
3. Kenny, J.F.; Barber, N.L.; Hutson, S.S.; Linsey, K.S.; Lovelace, J.K.; Maupin, M.A. *Estimated Water Use in the United States in 2005*; U.S. Geological Survey Circular: Reston, VA, USA, 2009.
4. Chen, J.; Wei, X. Controlled-released fertilizers as a means to reduce nitrogen leaching and runoff in container-grown plant production. In *Nitrogen in Agriculture-Updates*; Khan, A., Fahad, S., Eds.; InTech Open: Rijeka, Croatia, 2018.
5. Parrella, M.P.; Wagner, A.; Fujino, D.W. The floriculture and nursery industry's struggle with invasive species. *Am. Entomol.* **2015**, *61*, 39–50. [CrossRef]
6. CENSUS of Agriculture, United States Summary and State Data. 2007. Available online: http://www.agcensus.usda.gov/Publications/2007/Full_Report/usv1.pdf (accessed on 28 July 2018).
7. Belayneh, B.E.; Lea-Cox, J.D.; Lichtenberg, E. Costs and benefits of implementing sensor-controlled irrigation in a commercial pot-in-pot container nursery. *HortTechnology* **2013**, *23*, 760–769.

8. Chen, J.; Huang, Y.; Caldwell, R.D. Best management practices for minimizing nitrate leaching from container-grown nurseries. *Sci. World J.* **2001**, *1*, 96–102. [CrossRef] [PubMed]

9. Chen, J.; Beeson, R.C., Jr.; Yeager, T.H.; Stamps, R.H.; Felter, L.A. Evaluation of captured rainwater and irrigation runoff for greenhouse foliage and bedding plant production. *HortScience* **2003**, *38*, 228–233.

10. Beeson, R.C., Jr. Suspension lysimeter systems for quantifying water use and modulating water stress for crops grown in organic substrates. *Agric. Water Manag.* **2011**, *98*, 967–976. [CrossRef]

11. Majsztrik, J.C.; Ristvey, A.G.; Lea-Cox, J.D. Water and nutrient management in the production of container-grown ornamentals. *Hortic. Rev.* **2011**, *38*, 253–297.

12. Allen, R.G.; Pereira, L.S.; Raes, D.; Smith, M. *Crop Evapotranspiration: Guidelines for Computing Crop Water Requirements—FAO Irrigation and Drainage Paper 56*; Food and Agriculture Organization of the United Nations: Rome, Italy, 1998.

13. Howell, T.A.; Evett, S.R. The Penman-Monteith Method. 2006. Available online: http://www.cprl.ars.usda.gov/wmru/pdfs/PM%20COLO%20Bar%202004%20cor-rected%209apr04.pdf (accessed on 4 September 2018).

14. Campbell Scientific. *Application Note 4-D*; Campbell Scientific Ltd.: Logan, UT, USA, 1991.

15. Jensen, M.E.; Burman, R.D.; Allen, R.G. Evapotranspiration and Irrigation Water Requirements. 1990. Available online: http://agris.fao.org/agris-search/search.do?recordID=US19910106366 (accessed on 28 July 2018).

16. Allen, R.G.; Wright, J.L.; Pruitt, W.O.; Pereira, L.S. Water requirements. In *Design and Operation of Farm Irrigation Systems*; ASAE: St. Joseph, MN, USA, 2007.

17. Bryla, D.R.; Trout, T.J.; Ayars, J.E. Weighing lysimeters for developing crop coefficients and efficient irrigation practices for vegetable crops. *HortScience* **2010**, *45*, 1597–1604.

18. Thornthwaite, C.W. An approach toward a rational classification of climate. *Geogr. Rev.* **1948**, *38*, 55–94. [CrossRef]

19. Fitzpatrick, G. Water budget determinations for container-grown ornamental plants. *Proc. Fla. State Hortic. Soc.* **1980**, *93*, 166–168.

20. Fitzpatrick, G. Relative water demand in container-grown ornamental plants. *HortScience* **1983**, *18*, 760–762.

21. Burger, D.W.; Hartin, J.S.; Hodel, D.R.; Lukaszewski, T.A.; Tjosvoid, S.A.; Wagner, S.A. Water use in California's ornamental nurseries. *Calif. Agric.* **1987**, *41*, 7–8.

22. Doorenbos, J.; Pruitt, W.O. *Guidelines for Predicting Crop Water Requirements—FAO Irrigation and Drainage Paper 24*; Food and Agriculture Organization of the United Nations: Rome, Italy, 1977.

23. Beeson, R.C., Jr. Penman crop Coefficients for container growth landscape ornamentals. In *Evapotranspiration and Irrigation Scheduling*; Camp, C.R., Sadler, E.J., Yoder, R.E., Eds.; American Society of Agricultural Engineers: San Antonio, TX, USA, 1996.

24. Beeson, R.C., Jr. Modeling actual evapotranspiration of Ligustrum japonicum from rooted cuttings to commercially marketable plants in 12 liter black polyethylene containers. *Acta Hortic.* **2004**, *664*, 71–77. [CrossRef]

25. Beeson, R.C., Jr. Modeling irrigation requirements for landscape ornamentals. *HortTechnology* **2005**, *15*, 18–22.

26. Beeson, R.C., Jr. Response of evapotranspiration of Viburnum odoratissimum to canopy closure and the implications for water conservation during production and in landscapes. *HortScience* **2010**, *45*, 359–364.

27. Beeson, R.C., Jr. Modeling actual evapotranspiration of Viburnum odoratissimum during production from rooted cuttings to market size plants in 11.4-L containers. *HortScience* **2010**, *45*, 1260–1264.

28. Beeson, R.C., Jr. Development of a simple reference evapotranspiration model for irrigation of woody ornamentals. *HortScience* **2012**, *47*, 264–268.

29. Chen, J.; Beeson, R.C., Jr. Actual evapotranspiration of Asplenium nidus and Chamaedorea elegans during production from liners to marketable plants. *Acta Hortic.* **2013**, *990*, 339–344. [CrossRef]

30. Hagen, E.; Mambuthiri, S.; Fulcher, A.; Geneve, R. Comparing substrate moisture-based daily water use and on-demand irrigation regimes for oakleaf hydrangea grown in two container sizes. *Sci. Hortic.* **2014**, *179*, 132–139. [CrossRef]

31. Pershey, N.A.; Cregg, B.N.; Andresen, J.A.; Fernandez, R.T. Irrigating based on daily water use reduces nursery runoff volume and nutrient load without reducing growth of four conifers. *HortScience* **2015**, *50*, 1553–1561.

32. Chen, J.; McConnell, D.B.; Norman, D.L.; Henny, R.J. The foliage plant industry. *Hortic. Rev.* **2004**, *31*, 47–112.

33. USDA, USDA National Agricultural Statistics Service. *Floriculture Crops 2015 Summary*; USDA: Washington, DC, USA, 2016.

34. Beeson, R.C., Jr. Relationship of plant growth and actual evapotranspiration to irrigation frequency based on managed allowable deficits for container nursery stock. *J. Am. Soc. Hortic. Sci.* **2006**, *131*, 140–148.

35. Henny, R.J.; Holm, J.R.; Chen, J.; Scheiber, M. In vitro induction of tetraploids in Dieffenbachia x 'Star Bright M-1' by colchicine. *HortScience* **2009**, *44*, 646–650.

36. Stanhill, G. Water use efficiency. *Adv. Agron.* **1987**, *39*, 53–85.

37. Poole, R.T.; Conover, C.A. Fertilizers levels and medium affect foliage plant growth in an ebb and flow irrigation system. *J. Environ. Hortic.* **1992**, *10*, 81–86.

38. Kim, J.; van Iersel, M.W.; Burnett, S.E. Estimating daily water use of two Petunia cultivars based on plant and environmental factors. *HortScience* **2011**, *46*, 1287–1293.

39. Henny, R.J.; Chen, J. Cultivar development of ornamental foliage plants. *Plant Breed. Rev.* **2003**, *23*, 245–290.

40. Wang, Q.; Chen, J. Variation in photosynthetic characteristics and leaf area contributes to Spathiphyllum cultivar differences in biomass production. *Photosynthetica* **2003**, *41*, 443–447. [CrossRef]

41. Neal, C.A.; Henley, R.W. Water use and runoff comparisons of greenhouse irrigation systems. *Proc. Fla. State Hortic. Soc.* **1992**, *105*, 191–194.

Article

Characterization and Simulation of a Low-Pressure Rotator Spray Plate Sprinkler Used in Center Pivot Irrigation Systems

Cruz Octavio Robles Rovelo [1,*], Nery Zapata Ruiz [1], Javier Burguete Tolosa [1], Jesús Ramiro Félix Félix [2] and Borja Latorre [1]

[1] Department of Soil and Water, Estación Experimental de Aula Dei, CSIC, 50059 Zaragoza, Spain; v.zapata@csic.es (N.Z.R.); jburguete@eead.csic.es (J.B.T.); borja.latorre@csic.es (B.L.)

[2] Inter-American Institute of Water Technology and Sciences, Autonomous University of the State of Mexico, San Cayetano de Morelos, 50200 Toluca, Mexico; ramiroflfl@gmail.com

* Correspondence: coctaviorobles@eead.csic.es; Tel.: +34-976-716-080

Received: 15 July 2019; Accepted: 8 August 2019; Published: 14 August 2019

Abstract: Spray sprinklers enable to operate at low pressures (<103 kPa) in self-propelled irrigation machines. A number of experiments were performed to characterize the water distribution pattern of an isolated rotator spray plate sprinkler operating at very low pressure under different experimental conditions. The experiments were performed under two pressures (69 kPa and 103 kPa) and in calm and windy conditions. The energy losses due to the impact of the out-going jet with the sprinkler plate were measured using an optical technique. The adequacy to reproduce the measured water distribution pattern under calm conditions of two drop size distribution models was evaluated. A ballistic model was used to simulate the water distribution pattern under wind conditions evaluating three different drag models: (1) considering solid spherical drops; (2) a conventional model based on wind velocity and direction distortion pattern, and (3) a new drag coefficient model independent of wind speed. The energy losses measured with the optical method range from 20% to 60% from higher to lower nozzle sizes, respectively, for both evaluated working pressures analyzing over 16,500 droplets. For the drop size distribution selected, Weibull accurately reproduced the water application with a maximum root mean square error (RMSE) of 19%. Up to 28% of the RMSE could be decreased using the wind-independent drag coefficient model with respect to the conventional model; the difference with respect to the spherical model was 4%.

Keywords: rotator spray sprinkler; low-pressure; ballistic simulation; modified drag model; energy losses

---

1. Introduction

Self-propelled sprinkler irrigation systems and lateral move and center pivot systems have become an alternative for irrigation modernization, particularly for large scale land holding [1–3]. According to the last update of the census of agriculture in Spain, the irrigated land with self-propelled systems has increased by approximately 7% over the last five years [4]. In the same way, in the USA the increment reached 9% between 2008 and 2013 [5].

At the end of the twentieth century, fixed spray plate sprinklers (FSPS) and rotating spray plate sprinklers (RSPS) contributed to an important improvement on irrigation performance and a reduction of pressure requirements [3,6] compared with the previous impact sprinklers that equipped the irrigation machines. Currently, FSPS and RSPS are the most common sprinklers used in self-propelled irrigation machines. In Spain, the escalating electricity cost and consumption in the last ten years [7,8] are motivating farmers to look for more energy-efficient alternatives. Low-pressure devices (with pressure requirements lower than 103 kPa) have been commercialized in the last 40 years as an alternative to reduce energy bills.

No information is provided by the manufacturer regarding the water application patterns of this low-pressure sprinkler; moreover, it is well known that reducing the pressure in the sprinklers could modify the radial application pattern and reduce the irrigation quality in terms of uniformity. Therefore, it is important to provide technical data to assess the viability of this low-pressure sprinkler, particularly under overlapping scenarios on self-propelled irrigation machines.

A number of simulation models have been developed based on experimental data in order to improve the sprinkler irrigation designs. The characterization and simulation of the spatial water distributions patterns, the energy losses, the drop size distribution, and the wind drift and evaporation losses for FSPS and RSPS have been subject of numerous studies [9–14].

Mugele and Evans [9] and Solomon et al. [10] proposed the upper limit log normal (ULLN) model to characterize drop size distributions for impact sprinklers used in sprinkler irrigation machines as pivot or linear-move. Li et al. [15] and Kincaid et al. [16] proposed the Weibull model to describe the drop size distribution for impact sprinkler of solid-set systems. Both models are based on the measurement of the drop sizes emitted by the sprinklers. The first methodologies to characterize the drop sizes have been replaced by non-intrusive methods such as the disdrometer [17], the photographic method [18], the particle image velocimetry (PIV) [14], or particle tracking velocimetry (PTV) [19].

Recently, Zhang et al. [14] used the PIV technique to characterize the initial drop velocity of a FSPS and simulated the velocities with Computational Fluid Dynamic (CFD). One of their results indicates an important energy loss of the FSPS jets when impacting the deflecting plate for operating pressures lower than 100 kPa (between 28% and 51% of energy losses). Higher energy losses were presented by Ouazaa et al. [13], measuring initial drop velocities with the photographic method from Salvador et al. [18]. They found energy losses ranging from 35% to 75%. The differences between the results of Zhang et al. [14] and Ouazaa et al. [13] could be attributed to the different methodologies used in each research work.

A semi-empirical model to simulate the spatial distribution of water application pattern in sprinkler irrigation machines was presented by Molle and Le Gat [11]. The authors used a combination of the beta function for adjusting the radial water application of a two nozzle sprinkler used in center pivot system. Their analysis was based on their previous theoretical work [12]. They proposed two models: one for indoor and a second model for windy conditions (up to 6 m s^{-1}). In order to obtain the distributions curves, they divided the drops population into three groups: the ones generated by the deflecting plate of impact sprinkler (to break the main jet) and the two of the jet nozzles. Moreover, they performed a large number of calibrations for the three drop populations through experiments. Their results show almost negligible differences between measured and simulated values in the validation and calibration processes and their statistical indexes also indicated a satisfactory predictive ability.

In addition, the ballistic theory has been commonly used to describe drop dynamics in solid-set sprinkler irrigation models [20–26]. To simulate the center pivot sprinkler droplets, first it is necessary to know where the drops are formed, its initial velocity, and the volumetric drop size distribution [13].

In the ballistic model proposed by Fukui et al. [20], the drops trajectories are subjected to a drag coefficient (C) that for spherical drops depends on the Reynolds number (*Re*). This proposal was later modified by Seginer et al. [27] for analyzing the wind effect and was later adapted by Tarjuelo et al. [21] for introducing two factors, K_1 and K_2, affecting the drag coefficient (C′), to analyze the effect of the wind velocity in leeward and windward directions on water application patterns:

$$C' = C(1 + K_1 \times \sin \alpha - K_2 \times \cos \beta), \tag{1}$$

with C the drag coefficient of a spherical drop proposed by Fukui et al. [20] based on a previous theoretical work:

$$
\begin{array}{ll}
100 \leq Re & C = -0.0033\,Re + 1.2 + \frac{33.3}{Re}, \\
100 \leq Re \leq 1000 & C = -0.0000556\,Re + 0.48 + \frac{72.2}{Re}, \\
Re \geq 1000 & C = 0.45,
\end{array}
\tag{2}
$$

where α is the angle formed by the drop velocity vector with respect to the air (**V**) and drop velocity respect to the ground, and β is the angle formed by the vectors **V** and the wind velocity [21].

Ouazaa [28] in his efforts to reproduce the water application pattern of RSPS and FSPS used the ballistic model of Fukui et al. [20] considering the drag coefficient of Tarjuelo et al. [21]. Ouazaa [28] did not find a significant relationship of Tarjuelo's drag coefficients and the wind velocity. The authors adjusted the K parameters to force their relationships with the wind velocity by introducing the Rayleigh distribution functions using two K' coefficients for each sprinkler type RSPS and FSPS. They found that both models, K and K', accurately reproduced the RSPS, but not for the FSPS.

For low wind conditions, the model C' (Equation (1)) should not introduce considerable variations of the irrigation simulated in comparison with the measurements. The K_1 parameter does not affect the simulations because the factor "sin·α" tends to zero at low wind velocities; nevertheless, the K_2 parameter has an effect because of the values of "cos β", producing important changes in the water distribution.

Moreover, in the simulation model of Ouazaa [28] to estimate the drops trajectories of RSPS and FSPS, the K parameters respond to an average wind velocity losing the wind variability (different intensities and directions during an irrigation event) and its representativeness on drop dynamics. Therefore, it is necessary to establish alternative drag models that allow for generalizing its application to solve these problems.

The objectives of this research are: (1) to characterize the water distribution of a very low-pressure RSPS under different combination of nozzle sizes, working pressures and meteorological conditions; (2) to measure the drops velocities emitted by the sprinkler to estimate the energy losses of the out-going jet with the sprinkler plate using an optical technique; (3) to calibrate and validate the ballistic model with the existing drag models and with alternative models.

2. Materials and Methods

2.1. Sprinkler Features

The RSPS analyzed in this research is the Nutator N3000 (Figure 1c), equipped with the green deflector plate. The irrigation performance was measured for 6 different nozzle sizes from the 42 listed in the catalog 3000 Series 3TN Nozzle System. The sprinkler and the nozzles were manufactured by Nelson Irrigation Co. (Walla Walla, WA, USA, mention of trade marks does not imply the endorsement). The sprinkler, that combines spinning action with a continuously offset plate axis, can operate at two low-pressures (69 kPa and 103 kPa), both tested in this research. The green deflecting plate has a total of nine grooves (three different grooves repeated three times). The grooves are formed from the center of the plate with a depth and a curvature. The jet impacts with the sprinkler plate, dividing it into smaller jets (by the grooves) and in individual drops. The energy of this impact is used for the plate rotation and to do precession-nutation movements, throwing the nine jet drops with different horizontal angles up to 21° (no information about the lower horizontal angle is given by the manufacturer). The sprinkler plate is 11.7 cm in circumference and could approximately rotate from 50 RPM–160 RPM, considering both low pressures of 69 kPa and 103 kPa and all nozzle sizes. Then, the rotation speed of the sprinkler plate ranged between 0.12 m s^{-1} and 0.29 m s^{-1}.

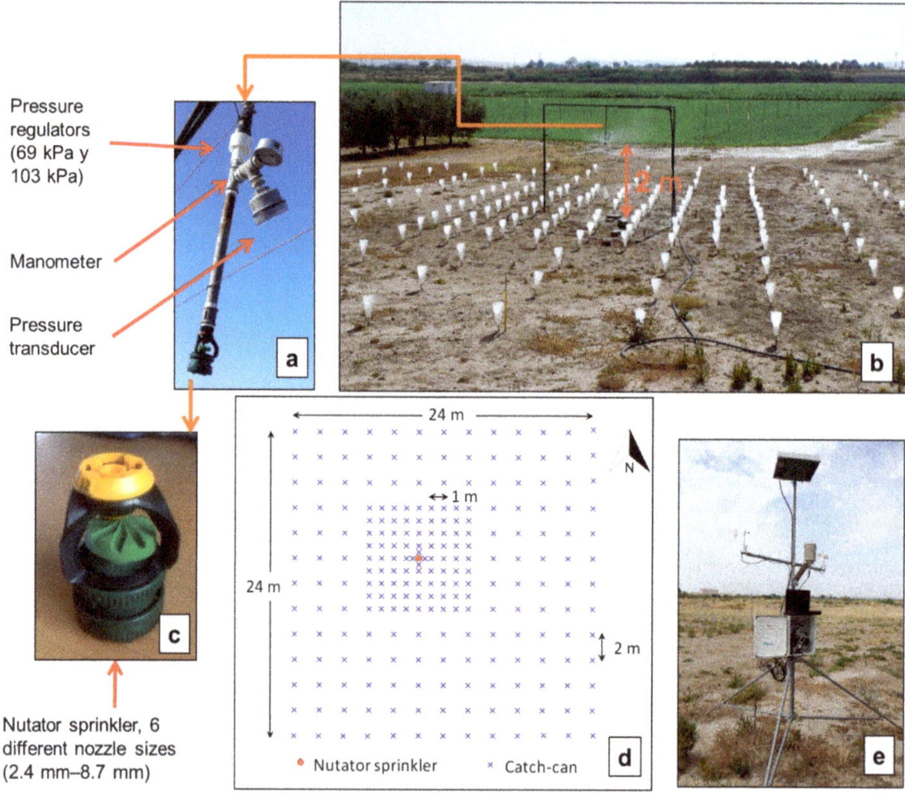

Figure 1. Experimental set-up for the characterization of the water application patterns of the Nutator sprinkler: (**a**) isolated Nutator sprinkler and its components (pressure regulators, manometer and pressure transducer); (**b**) field experimental layout; (**c**) Nutator sprinkler; (**d**) experimental catch-can set-up; (**e**) meteorological weather station.

2.2. Experimental Set-Up to Characterize the Water Application Patterns

The field experiments were carried out at the facilities of the Agrifood Research and Technology Centre of Aragón located in Montañana (Zaragoza), Spain. The Nutator N3000 sprinkler was independently evaluated using a support structure to locate its deflecting plate at 2.0 m a.g.l (Figure 1a,b). The experiments were performed for the two low working pressures of 69 kPa and 103 kPa and for the nozzle sizes: 2.4, 3.8, 5.2, 6.7, 7.9, and 8.7 mm (N12, N19, N26, N34, N40 and N44, respectively). The nozzle sizes selected for this research work are commonly used in the design of the center pivot irrigation systems.

A number of devices were installed in the pipeline before the sprinkler in order to maintain the irrigation performance: (1) a sediment filter to avoid blocking the out-going jet particularly for the smaller nozzle sizes; (2) a conventional flow meter with 1 L accuracy for verifying the discharge; (3) a pressure regulator to guarantee the operating pressure of 69 kPa and 103 kPa, respectively; (4) a glycerin manometer to visually verify the pressure during each irrigation event and (5) a pressure transducer (®Dickson) with a data logger to register the working pressure every minute (Figure 1a,b).

Flow meter readings were visually taken at the beginning and at the end of the irrigation events in order to compute the total amount of applied water. Pressure transducer data was downloaded after each irrigation event. Irrigation time was registered for each event using a conventional chronometer. The experiments were carried out while preventing the catch-cans overflow by checking the water level at different times during each event. A number of experiments were performed for each nozzle size and working pressure to evaluate the water application patterns under a range of wind velocities (calm, medium and windy) from 0.4 m s^{-1} to 9.5 m s^{-1}.

The catch-can configuration was a square network with a 1 m side for the 4 m closest to the nozzle and a 2 m side from 4 m to 20 m from the nozzle location (Figure 1d). In order to accurately characterize the water distribution near the sprinkler, a finer catch–can network was installed close to the sprinkler where the maximum application rate was expected for this kind of sprinklers. This configuration was similar to that presented in Faci et al. [6] for the RSPS evaluations, but with reinforcement in the area surrounding the sprinkler (Figure 1d). The range of the network was increased from 20 m to 24 m in the Southeast direction due to the predominant Northwest wind direction in the area (Figure 1d). The installed network had a total of 189 catch-cans covering an area of 576 m^2 (Figure 1d). The catch-cans had a conical shape, with a height of 0.40 m, a total capacity of 45 mm, and they were placed at 0.05 m on the soil surface (Figure 1b). Irrigation water collected in the catch-cans was measured as soon as the irrigation was completed for each experiment. A meteorological station was installed in situ to measure the air temperature and humidity and the wind velocity and direction (Figure 1e). All meteorological variables were measured every ten seconds, and averages were recorded with a data logger every sixty seconds.

The data of each irrigation event (total amount of water applied, pressure from the data logger, irrigation collected within every catch-can and the irrigation time) were sorted in individual spreadsheets classifying the experiments according to its wind velocity per nozzle size and per operating pressure. Spatial variability in water distribution patterns was assessed using contour line maps made with the ®SURFER software (Golden Software Inc., Golden, CO, USA) for each irrigation event.

In this research, comparisons of the irrigation performance were done with experiments under calm and windy conditions, choosing the smaller and the larger nozzle size in order to describe the radial application pattern of the Nutator sprinkler. Moreover, two radial application patterns of the Nutator sprinkler were compared with the experiments of FSPS from Ouazaa et al. [13] using the same nozzle sizes and working pressure (103 kPa).

The field data of these isolated experiments were used to calibrate the optimal parameters of the ballistic model. The simulated water distribution patterns were compared with the measured ones.

2.3. Experimental Set-Up for Drops Characterization

The Particle Tracking Velocimetry (PTV) technique of Bautista-Capetillo et al. [19] modified by Félix-Félix et al. [29] was used to characterize the sprinkler drops. The objective was to quantify the energy losses due to the impact of the jet with the sprinkler's plate. The experiments were carried under indoor conditions at the laboratory of flow visualization of the Inter-American Institute of Technology and Water Sciences of the Autonomous University of Mexico State in Toluca, Mexico.

The experimental set-up is shown in Figure 2. The installation was composed by a pressurized irrigation system and the optical PTV system. The pressurized system was integrated by: a 0.80 m^3 capacity water tank, a 0.37 kW power hydropneumatic pump and a pressure regulating tank, two 400 kPa glycerin manometers, two pressure regulators (69 kPa and 103 kPa), 22 mm PVC pipeline, and a Nutator N3000 sprinkler located at an elevation of 1 m above the soil using eight different nozzle sizes (2.0, 2.4, 3.2, 3.8, 4.4, 5.2, 6.7 and 7.9 mm corresponding to N10, N12, N16, N19, N22, N26, N34 and N40, respectively). The N40 nozzle size (7.9 mm) was not characterized with the PTV optical technique at 103 kPa due to experimental limitations affecting the pressure.

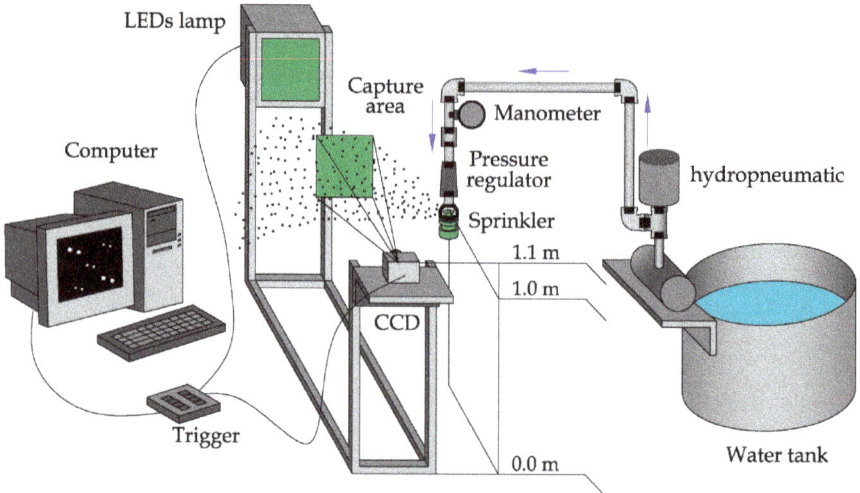

Figure 2. Experimental set-up for drops characterization. Spacings and components of the hydraulic and optical systems are shown.

Note that the nozzle sizes used to characterize the energy losses were not the same used to characterize the application patterns. The sprinkler was surrounded with a plastic sheet for watering a minimum area, leaving a zone enough for the outgoing jet. Drop characteristics were obtained fixing the sprinkler in a position so the jets and drops came out between two frames without modifying the initial conditions of the throwing.

The PTV system with in-line volumetric illumination was integrated by an illumination system, a high-speed charge-coupled device (CCD) camera and a computer. The illumination consisted in a high-power LEDs system. The CCD camera had a temporal resolution of 250 frames per second and a spatial resolution of 1024 by 1040 pixels (the CCD pixel size was 7.4 μm) equipped with a 50 mm lens. Moreover, a synchronizer (trigger) was programed with the computer in order to control the image acquisition sequence (camera) and the light (LEDs). The particle tracking vvelocimetry sediment (PTV-SED) v2.1 algorithm from Salinas-Tapia et al. [30] was used for the image processing. Tests were performed at night, illuminating the capture zone with the LED system. Every time the camera shutter opens, the lamp is activated in two pulses originating a photograph with two groups of drops: the ones captured in the first pulse and its respective pairs in the second pulse. PTV-SED was used to obtain drops diameters, drops velocities (both horizontal and vertical components), and drops angles, taking into account the coordinates of the drop centroids and their frequency in time. With the irrigation system working, pictures were taken at a distance of 0.30 m from the sprinkler at a height of 1.10 m for characterizing the drops (Figure 2). The capture zone at each height was 0.14 m × 0.11 m.

The PTV technique could eventually identify incorrect drops pairs because of the high density of particles in an image and the high velocity of the drops (close to the initial velocity of the drops ~14 m s^{-1} at 103 kPa). A post-processing analysis was performed in order to exclude invalid drops, so drop sizes smaller than 0.2 mm were neglected. In the first stage, the drops that do not follow a consistent behavior in their velocity and angle were rejected. Then a second filter was applied computing the inverse trajectory of the drops following the methodology of Sánchez-Burillo et al. [31]. The drops were returned to the sprinkler position using the ballistic theory with a negative time step. Drops differing 0.05 m of the sprinkler position were neglected.

Finally, the energy losses were computed for each nozzle as the difference between the theoretical velocity (Torricelli's equation) and the volumetric averaged drop velocity after the post-processing analysis. Power regressions were obtained to predict the energy losses for a nozzle size up to 9.9 mm (# 50 of the catalog) for both operating pressures, 69 kPa and 103 kPa.

The distance of 0.30 m between the sprinkler and the capture zone was enough accurate in order to obtain the energy losses due to the jet impact with the sprinkler plate minimizing the computing time for the inverse trajectory of the drops and its error. Due to the experimental set-up, it was not possible to measure drop features at closer distances.

Further information about PTV technique and image processing could be found in Bautista-Capetillo et al. [19] and Félix-Félix et al. [29] and the research work of Sánchez-Burillo et al. [31] for the inverse simulation.

2.4. Simulation of the Water Application Patterns

2.4.1. Ballistic Simulation

Considering weight, buoyancy and drag forces; the trajectory of a spherical drop is given by:

$$\mathbf{A} = \left(1 - \frac{\rho_a}{\rho_w}\right)\mathbf{g} - \frac{3\,\rho_a\,C'|\mathbf{U} - \mathbf{W}|}{4\,\rho_w\,d}\,(\mathbf{U} - \mathbf{W}), \tag{3}$$

where $\mathbf{A}$ is the drop acceleration vector, $\mathbf{g}$ is the gravity vector, ρ_a and ρ_w are the air and water density, respectively, $\mathbf{U}$ is the drop velocity vector with respect to the ground and $\mathbf{W}$ is the wind velocity vector.

Among the numerical methods to solve droplet dynamics, the fourth order Runge-Kutta (RK4) with fixed time step is the most extensively used method [20,22–24,32]. Recently, Robles et al. [26] used the third order Runge-Kutta (RK3) with variable time step for solving the Equation (3). Establishing a maximum error of 10 cm in drops trajectories, a decrease of 8.5% in the calculation time was obtained with respect to the RK4 method. This methodology was used in this research. The following procedures regarding drops generation, drop size distributions and the calibration processes were done based on the results of Robles et al. [26].

2.4.2. Drop Size Distribution

The experiments of the isolated sprinkler under the lowest wind velocity (<1.5 m s^{-1}) for each nozzle size and working pressures were used to calibrate two volumetric drop size distributions, ULLN (Equation (4)) and Weibull (Equation (5)):

$$f(d) = \frac{\alpha\,\exp\left[-\frac{1}{2}\left(\frac{\beta - \mu}{\sigma}\right)^2\right]}{\sigma\,d\,(\alpha - d)\,\sqrt{2\pi}}, \tag{4}$$

where α is the maximum drop diameter and $\beta(d) = \ln\left[d/(\alpha - d)\right]$

$$f(d) = 0.693\,n\frac{d^{n-1}}{d_{50}^n}\,\exp\left[-0.693\left(\frac{d}{d_{50}}\right)^n\right], \tag{5}$$

where $f(d)$ is the volumetric probability density function of the total discharge from the sprinkler, d is the drop diameter, μ and σ are the mean and the standard deviation of β, respectively, and d_{50} is the volume mean drop diameter. The ULLN distribution has three parameters α, μ and σ. The Weibull distribution has two parameters, d_{50} and n.

The parameters of both distributions were calibrated using the free software MPCOTool [33]. This module implements a number of optimization algorithms: regular systematic sampling, Monte-Carlo, orthogonal sampling, hill climbing, and genetic algorithms. The explanation of the optimization algorithms, their delimitation, cell size and drops simulated, was described in Robles et al. [26].

After the calibration process for both distributions, the water distribution pattern was simulated with the ballistic model. The Root Mean Square Error (RMSE) and the coefficient of correlation (r) of the 189 measured and simulated pluviometries were obtained for ULLN and Weibull models. Both models were compared and one of them was selected based on the accuracy of the water distribution simulations. Relationships between the model parameters and the nozzle size were also established.

$$\text{RMSE} = \frac{1}{N} \sqrt{\sum (S_i - O_i)^2} \tag{6}$$

$$r = \frac{\sum (S_i - \overline{S})(O_i - \overline{O})}{(N-1)\, Sd_s\, Sd_O} \tag{7}$$

where N is the total pluviometries number, S_i and O_i are the simulated and observed pluviometries, respectively, $\overline{S}$ and $\overline{O}$ are the simulated and observed pluviometries averages, respectively, Sd_s and Sd_o are the standard deviations of simulated and observed pluviometries values, respectively.

2.4.3. Drag Model, Calibration and Validation

In this research, a modification of the Tarjuelo's drag equation has been implemented. In the proposed model (L model), the K_1 and K_2 parameters of the Equation (1), are replaced with $L_1 \cdot W$ and $L_2 \cdot W$, respectively:

$$C' = C \times G, \tag{8}$$

where C is the Fukui's drag coefficient; $G = \max(1 + L_1 \cdot W \cdot \sin\ \alpha - L_2 \cdot W \cdot \cos\ \beta,\ 0.1)$; L_1 and L_2 are dimensionless parameters and W is the wind velocity module (m s^{-1}), and the 0.1 constant was used in order to avoid non-physical results (negative drag resistance values).

L_1 and L_2 are now independent of the wind velocity so they can be calibrated for each group of experiments of the same nozzle size and pressure. In comparison with the Equation (1), where the K parameters are constant values during the whole trajectory of drops in one experiment, with the L model (Equation (6)), C' can be computed with the same frequency as the measured meteorological conditions. Since the meteorology of each experiment was recorded every minute with a weather station located in situ, the C' value changes in accordance with the measured wind conditions.

A simplification on the definition of the K parameters has been commonly practiced obtaining the drag coefficient for impact sprinklers, RSPS and FSPS [13,21–24,28]. This simplification considers the K parameters as constant values for the experiment. In this research, the meteorological variability of the experiments could be analyzed, since the L model incorporates this possibility.

For assessing the performance of the isolated very low-pressure RSPS, three different drag models were evaluated and compared based on the calibration and validation phases. The drag models assessed were (1) the proposed by Fukui et al. [20]; (2) the K coefficients according to Tarjuelo et al. [21]; and (3) the L coefficients, the model proposed on this research work. The adequacy of the simulations with the drag models 2 and 3 was compared with respect to that of model 1.

In the Fukui et al. [20] drag model, no parameters need to be calibrated since drop drag coefficient is determined by its size and its velocity. The RMSE of this model was computed in every experiment for both pressures between the 189 measured and simulated pluviometries, and compared with the RMSE of the calibration and the validation processes of the other drag models.

The parameters of the Tarjuelo's drag model were calibrated for each individual experiment of the isolated sprinkler for both pressures. In contrast, the parameters of the new model L were calibrated by nozzle size groups and working pressure. Calibrating each individual experiment could introduce an overfitting problem and the failure to predict experiments with different conditions. On the other hand, calibrating the parameters by groups represents an advantage for generalizing the model and minimizing the errors with less probability to overfit [34].

The validation phase was based on the leaving-one-out-cross validation (LOOCV) method. In LOOCV only one experiment is selected as the validation set and the rest of the experiments are used to calibrate. The process is repeated to validate every experiment. In the Tarjuelo's model, a linear interpolation between the closest points is used to estimate the coefficients of the validation case. In the L model, the constant parameters of the group calibration are introduced in the validation case. The predicted ability of both models was assessed in terms of RMSE (between measured and simulated irrigations) and r in the validation phase.

3. Results and Discussion

3.1. Water Application Patterns

Table 1 shows the features of the water application pattern experiments performed on the isolated sprinkler plot. The wind velocity range of the experiments was between 0.6 m s^{-1} and 9.7 m s^{-1}. A total of 33 tests for both operating pressures (69 kPa and 103 kPa) and 7447 pluviometer readings were performed during two evaluation seasons. The experiments under the working pressure of 69 kPa were conducted in 2016 (between May–September) while the ones for 103 kPa were carried out between October–December 2017 and January-February 2018. An average of 5 experiments per nozzle size and operating pressures were evaluated, covering calm, medium, and high wind conditions.

Table 1. Experimental features of the measurements of the isolated Nutator sprinkler.

Nozzle Size (mm)	69 kPa				103 kPa			
	Exp [+]	Wind Velocity (m s^{-1})	Irrigation Time (h)	Pressure SD (kPa) ψ	Exp [+]	Wind Velocity (m s^{-1})	Irrigation Time (h)	Pressure SD (kPa) ψ
2.4	4	1.1–6.9	1.8–3.0	0.4	6	0.7–8.3	2.8–3.4	0.7
3.8	5	1.6–6.1	1.8–3.0	0.7	6	0.9–6.1	2.1–3.1	1.0
5.2	5	1.3–7.7	1.0–2.8	0.4	6	0.4–9.4	1.5–2.1	0.7
6.7	4	0.9–7.4	1.0–1.9	0.5	5	0.8–8.5	1.0–1.1	0.8
7.9	7	0.9–5.7	1.0–1.3	0.5	6	0.6–8.2	1.0–1.1	0.9
8.7	8	1.2–7.6	1.0–1.4	0.5	4	0.6–9.7	1.0	0.9

ψ mean standard deviation-SD-of the pressure measured with the pressure data logger per nozzle size experiment.
[+] Number of experiments per nozzle size for both operating pressures.

The average wind velocity of the experiments at 103 kPa was higher than those for 69 kPa (Table 1). Maximum wind velocities of 7.7 m s^{-1} and 9.7 m s^{-1} were registered for both operating pressures and minimum wind velocities of 0.9 m s^{-1} and 0.6 m s^{-1} were observed for 69 kPa and 103 kPa, respectively. Depending on the nozzle size and working pressure, the experiments lasted from a minimum of 1 h to a maximum of 3.4 h. The operating pressure along the irrigation events registered with the Dickson datalogger, suggests small variations in the out-going flow in each nozzle size, which indicates a correct operation of the pressure regulators. The maximum standard deviation-SD-observed was 0.7 kPa and 0.9 kPa for 69 kPa and 103 kPa, respectively. That barely represents 1% for each operating pressure.

The Figure 3 shows illustrative examples of the measured individual water distributions patterns at 103 kPa with the nozzle N12 (Figure 3a,c) and nozzle N44 (Figure 3b,d) under calm and windy conditions. The water application rate (mm h^{-1}) is shown in gray scale for each experiment (note that the scale color is different between nozzle sizes). The maximum application rate measured on the N12 nozzle was 6.4 mm h^{-1} and 3.3 mm h^{-1} for calm and windy conditions, respectively. While the maximum irrigation measured on the N44 was 41 mm h^{-1} and 25 mm h^{-1} for calm and windy conditions, respectively. These differences on irrigation between the two wind velocity intensities could be due to the variations in the wind velocity and in the wind direction that distort the pluviometry pattern. Moreover, the wind and drift evaporation losses (not evaluated in this work) and to the sampling variance because of low pluviometers density at measurements with high wind velocities modify the maximum application rates. Under low wind velocity, the wetted diameter by the N12 and N44 nozzle sizes reaches up to 14 m and 18 m, respectively. The experiments with low wind velocity

(Figure 3a,b) were performed at 1 m s⁻¹, the experiment at high wind velocity for N12 nozzle was 6 m s⁻¹ (Figure 3c) and the scenario of N44 nozzle (Figure 3d) was performed at 9 m s⁻¹.

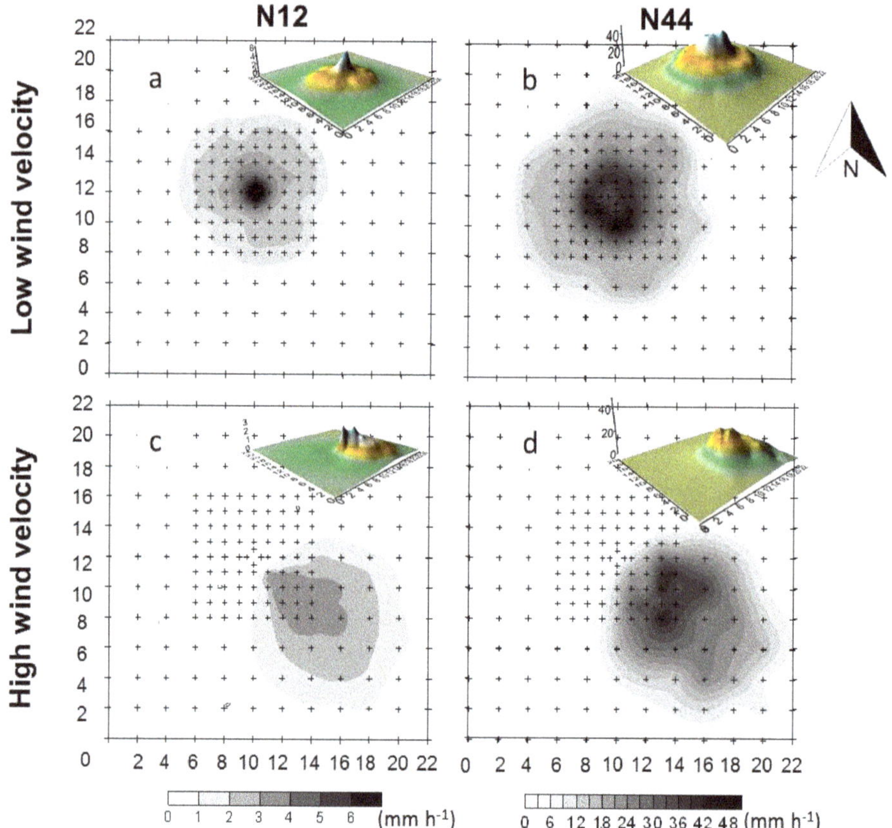

Figure 3. Measured water distribution at 103 kPa for the nozzle N12 (2.4 mm) and nozzle N44 (8.7 mm) for calm conditions (**a,b**) and windy conditions (**c,d**).

Table 2 shows the experimental features of the isolated Nutator sprinkler at the two pressures under low wind conditions. The irrigation volume collected from the pluviometers and the wetted radii (obtained from the boundary pluviometers wetted in each event) are also shown in the Table 2. The maximum application rate for all the experiments occurred between 0.0 m and 1.5 m from the sprinkler location. Under low-wind conditions, the radial application pattern (cross section of the spatial distribution pattern) for all nozzle sizes and both pressures were triangular in shape. The correlation coefficients (r) shown in the Table 2 represent the relationship between the application rate and the distance from the sprinkler. The r values suggest that the radial application becomes more triangular as the nozzle size and the operating pressure increases.

The triangular shape of the Nutator sprinkler is similar to the impact sprinklers commonly used in solid-set systems, but different from the fixed spray plate sprinklers such as the FSPS (with a doughnut/ring-shape). Figure 4 shows a comparison of the radial application patterns performed on two different nozzle sizes with a Nutator sprinkler (used in this research) and a FSPS at the same operating pressure of 103 kPa under no wind conditions. The data of the FSPS was obtained from Ouazaa et al. [13]. The nozzles compared were N26 (5.2 mm) (Figure 4a) and N44 (8.7 mm) (Figure 4b). According to Christiansen [35], theoretically, the triangular shape of the radial application, as the one of the Nutator sprinkler, could produce the most uniform distribution at overlapping the water distribution patterns of the emitters integrating the irrigation machine. There is a difference of 1.5 m in the wetted radii between the two sprinklers for the N26 nozzle and a difference of 1.0 m for the N44 nozzle, with the largest irrigated area for the Nutator sprinkler. This could be explained by a number of factors of each sprinkler FSPS and Nutator at 103 kPa, such as: drop sizes, deflecting plate geometry, jet break-up and therefore, the energy losses.

Table 2. Experimental features of isolated sprinklers under low wind conditions.

Pressure (kPa)	Nozzle Size (mm)	Measured Volume (L h^{-1})	Wetted Radii (m)	Maximum Application Rate (mm h^{-1})	r
	2.4	130.5	6.5	2.2	0.93
	3.8	418.0	8.7	5.9	0.93
69	5.2	816.0	8.6	10.0	0.98
	6.7	1366.7	8.4	17.3	0.97
	7.9	1927.9	8.4	25.3	0.97
	8.7	2295.7	9.0	30.0	0.97
	2.4	209.2	7.6	6.4	0.87
	3.8	560.4	8.6	10.2	0.94
103	5.2	1053.0	8.6	16.2	0.96
	6.7	1701.3	9.2	24.0	0.98
	7.9	2445.6	9.2	32.2	0.98
	8.7	2901.0	9.9	40.3	0.97

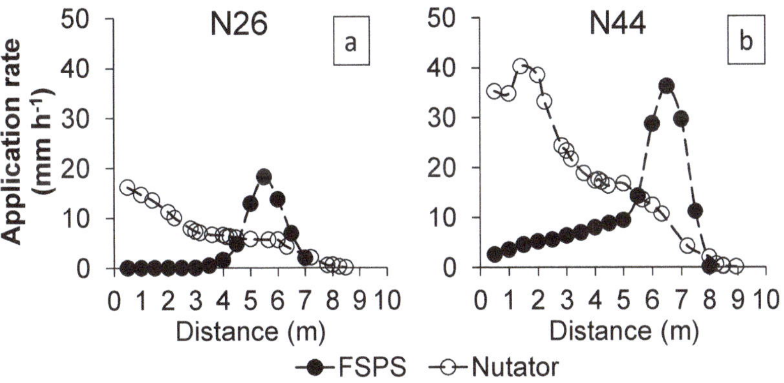

Figure 4. Measured radial water application at 103 kPa for (**a**) the nozzle N26 (5.2 mm) and (**b**) nozzle N44 (8.7 mm) of the Nutator sprinkler and FSPS (from Ouazaa et al. [13]).

On the other hand, the radial application pattern of the Nutator sprinkler is similar to the one of Ouazaa et al. [13] using a RSPS model R3000 ®Nelson (red plate with 6 grooves). The main difference between R3000 sprinkler and the Nutator is that the first requires pressures from 138 kPa, while the Nutator can operate from 69 kPa.

3.2. Energy Losses

Table 3 shows the drops features obtained with the PTV technique for each nozzle size and for both pressures. For the pressure of 69 kPa, a total of 10,227 droplets were analyzed from which about 53% (5514 drops) were neglected after the post-processing, leaving 4713 valid drops. For this pressure, the arithmetic average of drop diameter, the minimum and the maximum drop diameters reached values of 1.0 mm, 0.2 mm, and 4.0 mm, respectively. The arithmetic average velocity of the drops at 69 kPa was 8.0 m s^{-1} with a standard deviation of 2.4 m s^{-1}. The drops flew with an arithmetic average angle of 18° with respect to the horizontal plane.

Table 3. Drop features as arithmetic average values defined with the PTV technique after the post-processing for 69 kPa and 103 kPa, and for each of the nozzle sizes analyzed.

Pressure (kPa)	Nozzle Size (mm)	Number of Drops Analyzed	Drop Diameter (mm)	Drop Velocity (m s^{-1})	Drop Angle (°)
69	2.0	220	1.1	7.3	17.8
	2.4	187	1.1	7.6	17.6
	3.2	470	1.1	7.5	18.0
	3.8	345	1.2	7.7	17.5
	4.4	1058	1.0	7.1	18.2
	5.2	751	0.9	8.2	18.3
	6.7	629	1.0	9.0	17.7
	7.9	1053	1.0	8.9	18.2
103	2.0	180	0.9	7.8	17.1
	2.4	542	1.0	8.8	17.0
	3.2	326	1.1	9.3	17.2
	3.8	636	1.1	9.7	17.1
	4.4	538	1.0	10.0	17.5
	5.2	710	1.1	10.5	17.4
	6.7	1909	0.9	10.8	17.4

For a pressure of 103 kPa, a total of 6475 droplets were analyzed and about 25% (1634) of them were removed after the filtering, remaining 4841 valid drops (see Table 3). Values of the average drops diameter, the minimum and the maximum drop diameters of 1.0 mm, 0.2 mm and 4.8 mm were obtained, respectively. At 103 kPa, the drops flew to an average of 2.0 m s^{-1} faster than at 69 kPa (average drops velocity 10 m s^{-1}) with a standard deviation of 3.0 m s^{-1}. The average outgoing angle of the drops was 1° lower than for the pressure of 69 kPa (average drops angle of 17° with respect to the horizontal plane).

A considerably large drops number (~17,000) was analyzed with the semiautomatic PTV technique [19,29] compared with the labor-intensive low-speed photographic technique [13,18] where hundreds of drops are commonly characterized.

The radial velocity component of the drops ranged between 7.1 m s^{-1} and 10.8 m s^{-1}. Then, the tangential velocity component is between 1.8% and 2.7% of the total velocity. These values have been considered as negligible for PTV drop measurements and also for drop ballistic simulation. There are another uncertainty sources as sprinkler vibrations, pressure fluctuations, advection streams, etc. that, as the tangential velocity, has been also neglected. The Figure 5 shows the relationships between drop diameter and drop velocity for the pressures of 69 kPa (Figure 5a,b) and 103 kPa (Figure 5c,d) for two nozzle sizes N10 (2.0 mm) and N34 (6.7 mm). In general, for the Nutator sprinkler some particularities are noticeable: the drops velocity increases with nozzle size and the operating pressure, the drop diameters increases with the nozzle size (up to 4.8 mm) and finally, a large number of small drops are generated for the larger nozzle sizes (from nozzle N26-5.2 mm onwards) with a wide velocities range for both pressures. The variability on drop velocity for the small drops in large nozzle sizes (from N26) are due to the high uncertainty of the inverse trajectory resolution (small drops subjected to high drag).

The energy losses of the RSPS presented in Figure 6 were calculated with the volumetric averaged drop velocity for each nozzle size and operating pressure shown in Figure 5. The energy losses decrease with the nozzle size. Power regressions between nozzle size and energy losses relationships were stablished for both operating pressures. For the pressure tests of 69 kPa, minimum losses of 29% and maximum of 50% were observed and for the experiments of 103 kPa, the energy losses ranged between 19% and 60%.

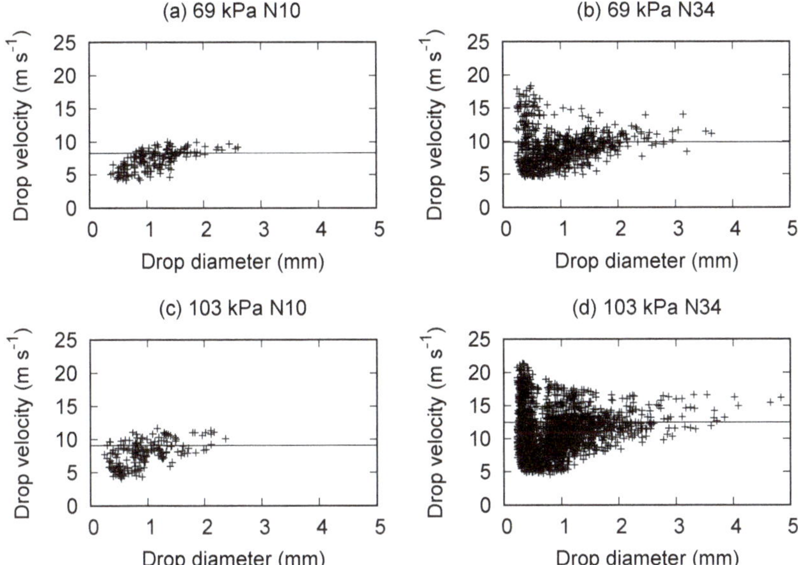

Figure 5. Measured drop diameter vs. measured drop velocity obtained with the PTV optical technique for the pressures of 69 kPa (**a,b**) and 103 kPa (**c,d**) (nozzle sizes N10-2.0 mm and N34-6.7 mm are shown). The continuous horizontal line represents the volumetric average drops velocity for each nozzle size and pressure.

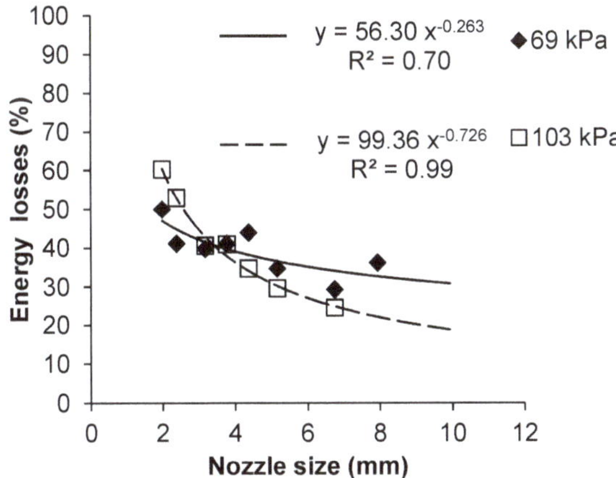

Figure 6. Estimated energy losses as function of the nozzle size. Energy loss regressions are shown for both pressures 103 kPa and 69 kPa with different symbols.

The energy losses observed of the Nutator sprinkler of this research work can be compared with those obtained by other authors using fixed spray plate sprinklers (FSPS). Ouazaa et al. [13] found higher energy losses for FSPS working at 69 kPa with respect to the Nutator sprinkler (from 40% to 70%) using the low-speed photographic technique to characterize the drops of five nozzle sizes between 2.4 mm and 6.7 mm. Zhang et al. [14], using an optical technique (PTV) to characterize the drops of a FSPS, obtained energy losses ranging between 28% and 50% for a pressure of 100 kPa using a 4.8 mm nozzle size. Sánchez-Burillo et al. [31] used the photographic technique of Salvador et al. [18]

and, after returning the drops to its initial position found energy losses between 33% and 74% using a FSPS with an operating pressure of 138 kPa for the nozzles 3.75 mm, 6.75 mm and 7.97 mm.

In agreement with Ouazaa et al. [13] and Zhang et al. [14] and the results of this research, sprinkler design should be reviewed for FSPS and for the RSPS included the Nutator sprinklers to minimize the energy losses, mainly for the smaller nozzle sizes.

3.3. Ballistic Model

3.3.1. Drop Size Distribution

A total of six experiments (one per nozzle size) of the isolated Nutator sprinkler under low wind conditions (<1.5 m s^{-1}) were selected to calibrate the parameters of both drop size distributions (Weibull and ULLN) for both operating pressures.

A comparison of both drop size distributions is shown in Figure 7, assessing the RMSE between the measured and simulated pluviometries of the six experiments. Each relationship is presented with a different symbol for each pressure. The 1:1 line is also represented in the figure. Although a slightly higher RMSE was observed using the Weibull model for both pressures, the statistical difference between ULLN and Weibull RMSE's is not significant at the 95.0% confidence level. Based on previous comparisons and considering the computational costs, a Weibull drop size distribution was selected to simulate the water application patterns of the experiments.

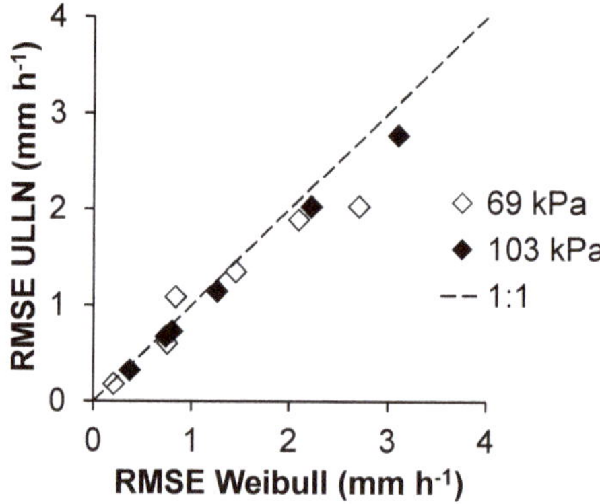

Figure 7. Comparisons of both drop size distributions, ULLN and Weibull, for 103 kPa and 69 kPa. Both operating pressures are represented with a different symbol. The dashed line represents the 1:1 relationship.

Table 4 shows the optimal parameters of the Weibull distribution (d_{50} and n) for both operating pressures and for each nozzle size evaluated. The RMSE increases with the nozzle size. The maximum RMSE was found for 8.7 mm nozzle size, with 2.7 mm h^{-1} and 3.1 mm h^{-1} for pressures of 69 kPa and 103 kPa, respectively. According to our measurements, the maximum RMSE reaches 24% and 27% of the total amount of water applied in one hour for 69 kPa and 103 kPa, respectively, RMSE increases with the nozzle flow. The Weibull model presents good adjustments, with correlation coefficients between 0.93 and 0.99.

Table 4. Optimal parameters of Weibull drop size distribution model for both operating pressures and nozzle sizes.

Pressure (kPa)	Nozzle Size (mm)	d_{50} (mm)	n	Application Rate (mm h^{-1})	RMSE (mm h^{-1})	R
	2.4	1.35	1.33	0.99	0.21	0.96
	3.8	1.34	1.51	1.75	0.75	0.93
69	5.2	1.47	1.47	3.51	0.84	0.97
	6.7	1.47	1.41	6.21	1.45	0.96
	7.9	1.45	1.45	8.76	2.10	0.97
	8.7	1.45	1.21	11.3	2.70	0.98
	2.4	0.92	1.34	1.14	0.37	0.95
	3.8	1.09	1.53	2.41	0.74	0.96
103	5.2	1.15	1.63	4.52	0.80	0.99
	6.7	1.12	1.58	6.43	1.26	0.99
	7.9	1.14	1.76	9.25	2.22	0.98
	8.7	1.14	1.76	11.5	3.10	0.97

The relationship between the nozzle size and the Weibull model parameters was presented in Figure 8. A linear regression was adjusted between nozzle sizes and the parameters d_{50} and n, for the pressure of 69 kPa (Figure 8a,b) and for the pressure of 103 kPa (Figure 8c,d). As expected, the tendency of d_{50} is to increase with the nozzle size, because larger nozzle generates larger drops, and to decrease with the working pressure. The average values of d_{50} were 1.42 mm and 1.10 mm for 69 kPa and 103 kPa, respectively. There is a significant statistical difference between the values of n for the two working pressures at the 95.0% confidence level after a Fisher's least significant difference (LSD) procedure. Average values of 1.4 and 1.6 were found for n at 69 kPa and 103 kPa, respectively. Kincaid et al. [16] and Ouazaa [13] working with RSPS, obtained a similar relationship between the model parameters, d_{50} and n, and the nozzle size and working pressure.

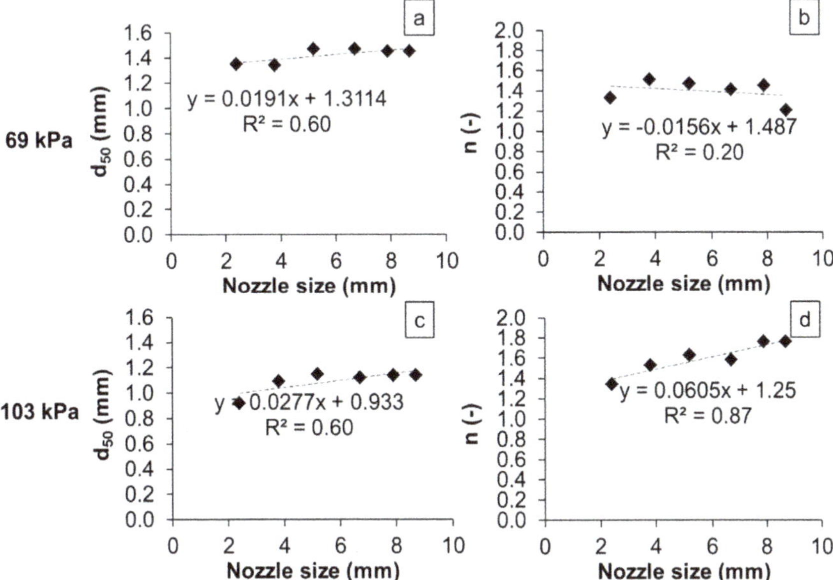

Figure 8. Optimal parameters of Weibull drop size distribution model (d_{50} and n) for both operating pressures 69 kPa (**a**,**b**) and 103 kPa (**c**,**d**). The dashed line represents the linear regression for each case.

3.3.2. Drag Coefficient, Calibration and Validation

In the calibration phase, the RMSE using the model of Fukui et al. [20] was related with the models of Tarjuelo et al. [21] and L for both pressures (Figure 9a,b). The three models accurately simulated the experimental measurements. The values of RMSE for the three models ranged between 0.18 mm h^{-1} and 5.10 mm h^{-1} for the pressure of 69 kPa, while for the pressure of 103 kPa the range was 0.3 mm h^{-1} to 6.2 mm h^{-1}. There was not statistically significant difference between the RMSE of the three models for both pressures. For the pressure of 69 kPa the RMSE of the L model were slightly lower compared with the Fukui et al. [20] model. Figure 9b (103 kPa) shows more variability of the RMSE's between the three models compared with the lowest pressure of 69 kPa (Figure 9a). For both pressures, the largest RMSE were obtained for the larger nozzle sizes (8.7 mm) using any model and the highest errors are associated with the high wind velocities (>5 m s^{-1}) and with the higher water application rate of the large nozzles. For the pressure of 69 kPa, the maximums RMSE were 4.5 mm h^{-1} (K model), 4.6 mm h^{-1} (L model) and 5.1 mm h^{-1} (Fukui's model), representing 28%, 30% and 32%, respectively, of the water application rate for the 8.7 mm nozzle size. For a pressure of 103 kPa, the maximums RMSE were 5.1 mm h^{-1} (K model), 5.2 mm h^{-1} (L model) and 6.2 mm h^{-1} (Fukui's model), representing 22%, 27% and 30%, respectively, of the water application rate for the 8.7 mm nozzle size. The computational time for the calibrations and the validations were similar for both models K and L, 5 days and 5 h for the L model, and 5 days and 7 h for the K model.

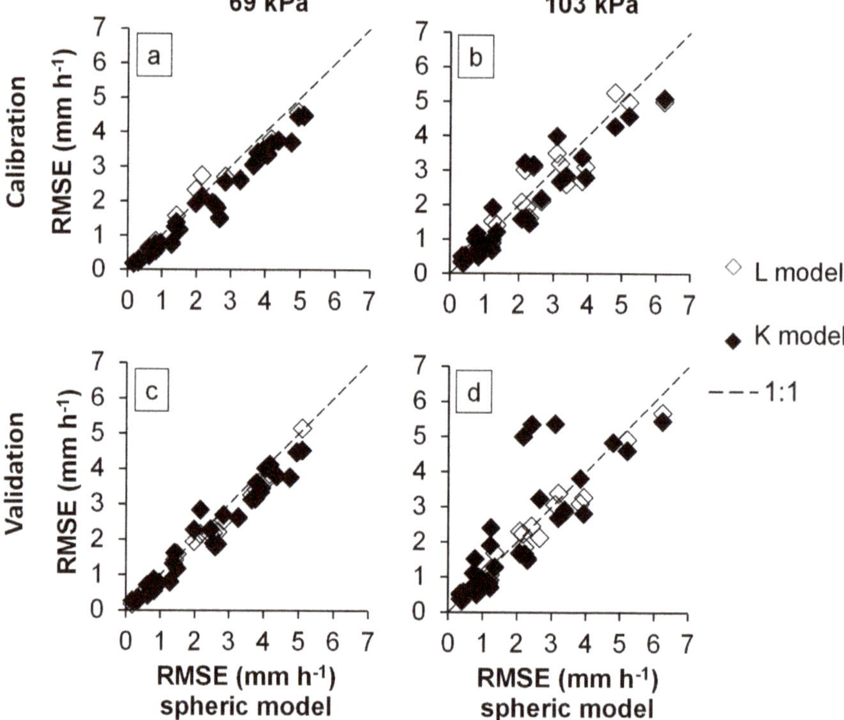

Figure 9. RMSE comparisons of the calibration (**a,b**) and validation (**c,d**) phases for the operating pressures of 69 kPa and 103 kPa. RMSE of the Fukui et al. [20] drag model vs. RMSE of the L model and Tarjuelo et al. [21] drag model for both pressures are shown in each figure. Each model is shown with a different symbol. The 1:1 relationship is represented with a dashed line.

Although no significant differences were observed between the three models in the calibration phase, slightly larger differences were found in validation. The LOOCV methodology of the validation phase used the 33 experiments performed for both pressures 69 kPa (Figure 9c) and 103 kPa (Figure 9d). The assessment in validation phase is more relevant, since the blind simulation represents the performance of a model in a real scenario under unknown conditions. There is not significant statistical difference on the RMSE of the Fukui et al. [20] model and the K model for both pressures at 95.0% confidence level. However, for the pressure of 103 kPa, three simulations had a higher RMSE using the K model. The largest RMSE of the validation phase corresponds to the 8.7 mm nozzle size experiments, with values of 5.4 mm h^{-1} for the K model and 3.1 mm h^{-1} for the L and Fukui's models. These differences represent 24% (K model) and 13% (with L model and Fukui's model) of the total amount of water applied in one hour.

Moreover, there is no statistically significant difference between both RMSE of the Fukui et al. [20] model and the L model for both pressures at 95.0% confidence level in the validation phase. However, slightly higher RMSE's were observed using the Fukui et al. [20] model with respect to the L model for 31 cases in the pressure of 69 kPa and for 24 cases for the pressure of 103 kPa. The maximum RMSE, 6.2 mm h^{-1}, was found for the 8.7 mm nozzle size working at 103 kPa using Fukui's model which represents 29% of the total amount of water applied. For the same pressure of 103 kPa, the maximum RMSE value with the L model is 5.7 mm h^{-1}, which represents 25% of the total amount of water applied in one hour.

The worst cases reproduced in the validation phase with the K model (Figure 9d), were under wind velocities lower than 1 m s^{-1}, which indicate a failure in the model of Tarjuelo et al. [21] at null wind velocities. It is common to establish the K parameters as zero but the threshold is not clear. On the other hand, the L model has a natural transition to Fukui's model.

In summary, at the validation phase and under the worst scenario, the L model improves the irrigation performance on 11% with respect to the K model and 4% respect to the Fukui et al. [20] model.

Figure 10 shows the relationship between nozzle size and the parameters L_1 and L_2 for a working pressure of 69 kPa (Figure 10a,b) and for a working pressure of 103 kPa (Figure 10c,d). The resulting linear regression models showed that there is not a clear effect of the operating pressure on the values of the L_1 parameters. Moreover, the optimal values of the parameters L_1 for both pressures can be considered as zero (Figure 10a,c). As the L_2 values increase, the drag coefficient of the drops decreases. The values of the parameter L_2 decrease as the nozzle size increases for the low pressure (Figure 10b). For the 103 kPa working pressure, the relationship between nozzle size and L_2 parameter is very weak; therefore it can be considered as a constant with an average value of 0.06. The previous recommendations given for the Nutator sprinkler used in this research could differ in values and/or tendencies for other sprinklers as RSPS or FSPS.

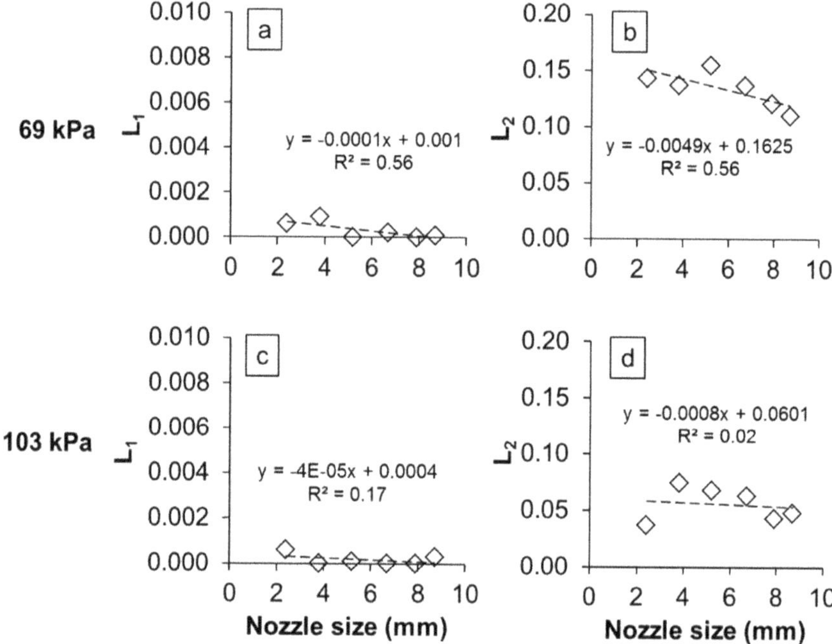

Figure 10. Optimal values of the L model related to the nozzle size for both operating pressures 69 kPa (**a,b**) and 103 kPa (**c,d**). Linear regressions are shown for each parameter.

4. Conclusions

The experimental characterization of a low-pressure rotator spray plate sprinkler was performed covering a wide range of meteorological conditions, nozzle sizes, and working pressures.

The radial application pattern of the Nutator sprinkler was triangular shape ensuring and adequate distribution uniformity of the overlapped sprinklers. The reported technical data of the isolated sprinkler under low wind conditions could be useful for designing center pivot irrigations systems.

The PTV technique allowed us to measure the features of 16,702 droplets originated from the sprinkler that were post-processed in order to obtain reliable data for energy losses characterization. This semiautomatic technique represents an advantage in the number of drops analyzed vs. calculation time with respect to the low-speed photographic technique. Moreover, the PTV technique was adequate to estimate the energy losses of the Nutator sprinklers, which ranged from 19% to 60%.

ULLN drop size distribution reproduced accurately the water application pattern of the RSPS at low pressure as much as the Weibull model, while the latter model was selected for simplicity and to save computational cost.

The properties of the L model proposed in this work represent an advantage over the previous models by improving the drops physics, and generalize its use for unknown conditions and new irrigation material. The new proposed model improves the calibration/validation processes by improving the nozzle size and pressure groups (avoiding overfitting), the drag force definition that is computed within a variable meteorology, and the continuous transition of the L model to the Fukui's model for low wind velocities.

At the validation phase, the L model in the Nutator sprinkler resulted in lower errors than the K model (28%). The K model was not accurate enough for this kind of sprinklers, even compared with the simple drag ballistic model of Fukui without corrections by the wind. The L model also resulted in slightly lower errors than the Fukui's drag model (4%).

Five regressions per operating pressure were proposed (energy losses, d_{50}, n, L_1 and L_2) in order to obtain the parameters model for the non-evaluated nozzle sizes in between the minimum (2.4 mm) and the maximum (8.7 mm) evaluated.

The L drag model proposed in this research represents an opportunity to reproduce the irrigation of other spray plate sprinklers, RSPS and FSPS. Further work is necessary in order to assess the improved model for impact sprinklers.

Author Contributions: C.O.R.R., N.Z.R., J.B.T., J.R.F.F. and B.L. contributed to all efforts leading to the production of this paper.

Funding: This research was funded by MICINN of the Government of Spain through the grants AGL2013-48728-C2-1-R and AGL2017-89407-R. Octavio Robles received a scholarship granted by the Minister of Economy and Competitiveness of the Spanish Government.

Acknowledgments: The authors would like to thank to the Inter-American Institute of Technology and Water Sciences in Toluca Mexico for providing the facilities of the flow visualization laboratory and to Humberto Salinas-Tapia and Boris López for their support.

Conflicts of Interest: The authors declare no conflict of interest.

References

1. Keller, J.; Bliesner, R.D. *Sprinkle and Trickle Irrigation*; Van Nostrand Reinhold: New York, NY, USA, 1991.
2. Tarjuelo, J.M. *El Riego por Aspersión y su Tecnología*, 3rd ed.; Ediciones Mundi-Prensa: Madrid, España, 1999.
3. Playán, E.; Garrido, S.; Faci, J.M.; Galán, A. Characterizing pivot sprinklers using an experimental irrigation machine. *Agric. Water Manag.* **2004**, *70*, 177–193. [CrossRef]
4. ESYRCE. *Encuesta de Superficies y Rendimientos de Cultivos en España*; Ministerio de Agricultura Pesca y Alimentación: Madrid, España, 2017; p. 101.
5. National Agricultural Statistics Service (USDA). *2014 Census of Agriculture. Farm and Ranch Irrigation Survey*; USDA: Washington, DC, USA, 2013; Volume 3, pp. 100–103.
6. Faci, J.M.; Salvador, R.; Playán, E.; Sourell, H. A comparison of fixed and rotating spray plate sprinklers. *J. Irrig. Drain. Eng. ASCE* **2001**, *127*, 224–233. [CrossRef]
7. Moreno, M.A.; Ortega, J.F.; Córcoles, J.I.; Martínez, A.; Tarjuelo, J.M. Energy analysis of irrigation delivery systems: Monitoring and evaluation of proposed measures for improving energy efficiency. *Irrig. Sci.* **2010**, *28*, 445–460. [CrossRef]
8. Tarjuelo, J.M.; Rodríguez-Díaz, J.A.; Abadía, R.; Camacho, E.; Rocamora, C.; Moreno, M.A. Efficient water and energy use in irrigation modernization: Lessons from Spanish case studies. *Agric. Water Manag.* **2015**, *162*, 67–77. [CrossRef]
9. Mugele, R.A.; Evans, H.D. Droplet size distribution in sprays. *Ind. Eng. Chem.* **1951**, *43*, 1317–1324. [CrossRef]
10. Solomon, K.H.; Kincaid, D.C.; Bezdek, J.C. Drop size distribution for irrigation spray nozzles. *Trans. ASAE* **1985**, *28*, 1966–1974. [CrossRef]
11. Molle, B.; Le Gat, Y. Model of water application under pivot sprinkler: II. Calibration and results. *J. Irrig. Drain. Eng.* **2000**, *126*, 348–354. [CrossRef]
12. Legat, Y.; Molle, B. Model of water application under pivot sprinkler. 1: Theoretical grounds. *J. Irrig. Drain. Eng.* **2000**, *126*, 343–347.
13. Ouazaa, S.; Burguete, J.; Paniagua, P.; Salvador, R.; Zapata, N. Simulating water distribution patterns for fixed spray plate sprinkler using the ballistic theory. *Span. J. Agric. Res.* **2014**, *12*, 850–863. [CrossRef]
14. Zhang, Y.; Sun, B.; Fang, H.; Zhu, D.; Yang, L.; Li, Z. Experimental and simulation investigation on the kinetic energy dissipation rate of a fixed Spray-plate sprinkler. *Water* **2018**, *10*, 1365. [CrossRef]
15. Li, J.; Kawano, H.; Yu, K. Droplet size distributions from different shaped sprinkler nozzles. *Trans. ASAE* **1994**, *37*, 1871–1878. [CrossRef]
16. Kincaid, D.C.; Solomon, K.H.; Oliphant, J.C. Drop size distributions for irrigation sprinklers. *Trans. ASAE* **1996**, *39*, 839–845. [CrossRef]
17. Montero, J.; Tarjuelo, J.M.; Carrión, P. Sprinkler droplet size distribution measured with an optical spectropluviometer. *Irrig. Sci.* **2003**, *22*, 47–56. [CrossRef]

18. Salvador, R.; Bautista-Capetillo, C.; Burguete, J.; Zapata, N.; Playán, E. A photographic method for drop characterization in agricultural sprinklers. *Irrig. Sci.* **2009**, *27*, 307–317. [CrossRef]

19. Bautista-Capetillo, C.; Robles, O.; Salinas-Tapia, H.; Playán, E. A particle tracking velocimetry technique for drop characterization in agricultural sprinklers. *Irrig. Sci.* **2014**, *32*, 437–447. [CrossRef]

20. Fukui, Y.; Nakanishi, K.; Okamura, S. Computer evaluation of sprinkler irrigation uniformity. *Irrig. Sci.* **1980**, *2*, 23–32. [CrossRef]

21. Tarjuelo, J.M.; Carrión, P.; Valiente, M. Simulación de la distribución del riego por aspersión en condiciones de viento. *Investig. Agrar. Prod. Prot. Veg.* **1994**, *9*, 255–272.

22. Carrión, P.; Tarjuelo, J.M.; Montero, J. SIRIAS: A simulation model for sprinkler irrigation: I. Description of the model. *Irrig. Sci.* **2001**, *20*, 73–84. [CrossRef]

23. Dechmi, F.; Playán, E.; Cavero, J.; Faci, J.M.; Martínez-Cob, A. Wind effects on solid set sprinkler irrigation depth and yield of maize (Zea mays). *Irrig. Sci.* **2003**, *22*, 67–77. [CrossRef]

24. Playán, E.; Zapata, N.; Faci, J.M.; Tolosa, D.; Lacueva, J.L.; Pelegrín, J.; Salvador, R.; Sánchez, I.; Lafita, A. Assessing sprinkler irrigation uniformity using a ballistic simulation model. *Agric. Water Manag.* **2006**, *84*, 89–100. [CrossRef]

25. Zerihun, D.; Sanchez, C.A.; Warrick, A.W. Sprinkler irrigation droplet dynamics. I: Review and theoretical development. *J. Irrig. Drain. Eng.* **2016**, *142*, 04016007. [CrossRef]

26. Robles, O.; Latorre, B.; Zapata, N.; Burguete, J. Self-calibrated ballistic model for sprinkler irrigation with a field experiments data base. *Agric. Water Manag.* **2019**, in press. [CrossRef]

27. Seginer, I.; Nir, D.; von Bernuth, D. Simulation of wind distorted sprinkler patterns. *J. Irrig. Drain. Eng. ASCE* **1991**, *117*, 285–306. [CrossRef]

28. Ouazaa, S. Development of Models for Solid-Set and Center-Pivot Sprinkler Irrigation Systems. Ph.D. Thesis, University of Zaragoza, Zaragoza, Spain, July 2015. Available online: http://digital.csic.es/handle/10261/117636 (accessed on 10 February 2019).

29. Félix-Félix, J.R.; Salinas-Tapia, H.; Bautista-Capetillo, C.; García-Aragón, J.; Burguete, J.; Playán, E. A modified particle tracking velocimetry technique to characterize sprinkler irrigation drops. *Irrig. Sci.* **2017**, *35*, 515–531. [CrossRef]

30. Salinas-Tapia, H.; García, A.J.; Moreno, H.D.; Barrientos, G.B. Particle tracking velocimetry (PTV) algorithm for non-uniform and nonspherical particles. In Proceedings of the Electronics, Robotics and Automotive Mechanics Conference CERMA, Cuernavaca, México, 26–29 September 2006; Volume 2, pp. 322–327.

31. Sánchez-Burillo, G.; Delirhasannia, R.; Playán, E.; Paniagua, P.; Latorre, B.; Burguete, J. Initial drop velocity in a fixed spray plate sprinkler. *J. Irrig. Drain. Eng.* **2013**, *139*, 521–531. [CrossRef]

32. Vories, E.D.; Von Bernuth, R.D.; Mickelson, R.H. Simulating sprinkler performance in wind. *J. Irrig. Drain. Eng.* **1987**, *113*, 119–130. [CrossRef]

33. GitHub. MPCOTool: The Multi-Purposes Calibration and Optimization Tool. By Burguete, J. and Latorre, B. Available online: http://github.com/jburguete/mpcotool (accessed on 9 January 2019).

34. Burnham, K.P.; Anderson, D.R. *Model Selection and Multimodel Inference*, 2nd ed.; Springer: New York, NY, USA, 2002; p. 488.

35. Christiansen, J.E. *Irrigation by Sprinkling*; Bulletin 670; University of California, College of Agriculture, Agricultural Experiment Station: Berkeley, CA, USA, 1942; p. 124.

Article

Feasibility of the Use of Variable Speed Drives in Center Pivot Systems Installed in Plots with Variable Topography

Victor Buono da Silva Baptista [1], Juan Ignacio Córcoles [2], Alberto Colombo [3] and Miguel Ángel Moreno [4,*]

[1] Engineering Dept., University of Lavras. Campus Universitario, s/n, Lavras 3037, Brazil; victor.buonosb@ufla.br
[2] Renewable Energy Research Institute, Section of Solar and Energy Efficiency, C/ de la Investigación 1, 02071 Albacete, Spain; JuanIgnacio.Corcoles@uclm.es
[3] Water Resources and Sanitation Dept., University of Lavras. Campus Universitario, s/n, Lavras 3037, Brazil; acolombo@ufla.br
[4] Institute for Regional Development, University of Castilla-La Mancha. Campus Universitario, s/n, 02071 Albacete, Spain
*   Correspondence: MiguelAngel.Moreno@uclm.es

Received: 11 September 2019; Accepted: 17 October 2019; Published: 21 October 2019

Abstract: Pumping systems are the largest energy consumers in center pivot irrigation systems. One action to reduce energy consumption is to adjust the pumping pressure to that which is strictly needed by using variable speed drives (VSDs). The objective of this study was to determine the feasibility of including VSDs in pumping systems that feed center pivot systems operating in an area with variable topography. The VSPM (Variable Speed Pivot Model) was developed to perform hydraulic and energy analyses of center pivot systems using the EPANET hydraulics engine. This tool is able to determine the elevation of each tower for each position of the center pivot using any type of digital elevation model. It is also capable of simulating, in an accurate manner, the performance of the center pivot controlled with a VSD. The tool was applied to a real case study, located in Albacete, Spain. The results show a reduction in energy consumption of 12.2%, with specific energy consumptions of 0.214 and 0.244 kWh m^{-3} of distributed water obtained for the variable speed and fixed speed of the pumping station, respectively. The results also show that for an irrigation season, to meet the water requirements of the maize crop in the region of the study (627 mm), an average annual savings of 14,107.35 kWh was obtained, which resulted in an economic savings of 2821.47€.

Keywords: hydraulic model; variable topography; energy consumption; variable speed; center pivot system

---

1. Introduction

The quantity and quality of food needed to satisfy all the demands of the population will become a major concern worldwide in the following years. It is expected that by the year 2050, there will be a world population of 9.15 billion people [1]. The increase of food production to satisfy demand is a challenge for agricultural professionals, who require the use of techniques focused on increasing production efficiency. Increasing use efficiency in food production requires the use of improved technology in all production processes, as well as improving the efficiency of irrigation systems. Irrigated agriculture accounts for 16% of the world's cultivated area and is expected to produce 44% of world food by 2050 [1–3].

Sustainable use of water resources could be accomplished increasing the efficiency of irrigation systems, thereby reducing the amount of water and energy to satisfy crop water requirements. In this regard, technological developments in the infrastructures of irrigation systems contribute to increasing water use efficiency. However, this efficiency increase is related to a significant increase in energy consumption in recent years [4].

According to [5], sprinkler irrigation systems represent around 11% (35 million hectares) of the total irrigated areas in the world. Within this method, the most outstanding systems are conventional sprinklers, traveling guns, center pivots, and linear-moving laterals. With eight million hectares of irrigated area, the center pivot system represents 23% of the area irrigated by sprinkler irrigation systems.

Center pivot irrigation consists of the application of water through a moving lateral line with several water outlets supported on moving towers, which revolve around a fixed pivot point (center tower) and irrigate a circular area [6,7]. This equipment is considered a highly efficient system compared to other irrigation systems. It is flexible and easily operable, which reduces labor and maintenance costs. It can also be operated on surfaces with variable topography, resulting in conservation of water, energy, and time [8]. However, the initial cost of the equipment and the required energy demands at the pumping station are some of its main limitations.

The reduction of the energy consumption of pumping stations that feed irrigation systems has been studied by several researchers. Pumping stations are the largest consumers of energy in pressurized irrigation systems, especially in situations where underground water resources are used.

Gilley et al. [9] suggested several changes in center pivot systems: switching high and medium pressure sprinklers to low pressure emitters, changing nozzle size and spacing, or changing irrigation intervals and maintenance to increase pumping station efficiency. Moreno et al. [10] developed a new methodology to obtain the characteristic curves of the pump, thereby minimizing the costs of the pumping station. In a later study, Moreno et al. [11] stated that the energy consumption of the center pivot, which irrigates an area of 75 ha, can be minimized by adopting measures like increasing the lateral line's diameter, reducing the equipment's operating time, and increasing the flow per unit of area. Barbosa et al. [12] concluded that constant center pivot monitoring is essential to maintaining adequate energy efficiency levels when assessing the behavior of different energy efficiency indicators for a center pivot operating in variable topography.

In center pivot equipment, the pumping station is designed to meet the most critical situation, i.e., the position of the lateral line where there is the highest elevation point and the greatest need for higher pumping pressure [13]. However, this situation is variable along with the lateral line rotation in the irrigated area due to topographic differences. This means that the pumping station is oversized during most of the lateral line rotation, and the energy is wasted. Pressure regulating valves are installed before the emitters, so pressure fluctuations that are due to topographical differences and oversizing of the pumping station do not influence the flow rate of the emitters [14].

An option to adjust the pumping pressure of this equipment is the use of variable speed drives (VSDs) to control the speed of these pumps [15,16]. This control provides a substantial reduction in power in relation to the reduction of flow and pressure in pressurized irrigation systems [13,17].

Several researchers have used variable speed drives to control pumping stations to reduce energy consumption. Hanson et al. [18], in a center pivot system with a well pumping unit, concluded that variable speed pumps save about 32% of the energy. Lamaddalena et al. [19] showed that 27% to 35% of energy savings can be achieved using VSD in two Italian irrigation districts operating with three parallel horizontal axis pumps. Brar et al. [20] concluded that a 9.6% energy reduction is possible for a 13.6 m difference in the irrigated area for a study containing 100 center pivots in Nebraska (USA), with each pivot containing a pumping station. In this study, digital elevation models (DEMs) with a spatial resolution of 10 m × 10 m were used to obtain the topographic characteristics of the irrigated areas.

King et al. [13] stated that the optimum efficiency in terms of energy and water use can be achieved when the pumping station is able to maintain the required minimum pressure regardless of the operating conditions. Scaloppi et al. [21] stated that the point of minimum pressure is constantly moving along the lateral line because it is influenced by the topography of the irrigated area. They also presented theoretical bases for the calculation of this movement.

However, working with the minimum pressure does not always guarantee the lowest power consumption due to variations in the operating point of the pump and variable performance when working at low frequencies [17,22]. In addition, most of these studies did not take into account the effect of the VSD efficiency on the final result. Therefore, it is important to consider VSD efficiency in energy savings accounting and not assume that this efficiency will always be constant and high [17].

In a study on VSD efficiency, King et al. [13] reported that a pumping station with a VSD can save 15.8% and 20.2% of its energy compared to fixed speed pumping (for uniform and variable rate irrigation, respectively) without considering the inefficiency of the VSD. When the efficiency of the VSD is accounted for, there are 7.5% and 12.4% energy savings, respectively.

The main novelty of the present study is developing a simulation model of hydraulic behavior by considering the hydraulic elements of the irrigation system in detail and integrating the energy characteristics of the pumping station and topographic characteristics through digital elevation models (DEMs). In this way, energy efficient technologies and management strategies can be developed to reduce energy use to ensure sustainable irrigation without reductions in the efficiency of water application.

The objective of the present study was to determine the feasibility of including variable speed drives in pumping systems that feed center pivot irrigation systems operating in areas with variable topography. Considering this objective, the VSPM (Variable Speed Pivot Model) tool was developed, in which the hydraulic simulation model of the pivot was integrated into a simulation model of the pumping station so that, with data related to topography, flow rate, and pressure, the power supply could be adjusted to the actual demand of the system. In addition, it is useful to determine the potential of reduction from the perspective of the energy, economics, and sustainability of the installation of the VSDs in these irrigation systems.

2. Material and Methods

2.1. Proposed Procedure

The proposed methodology has the following steps, as presented in 1. For the development of this model and data acquisition, a real center pivot was used. However, this model can be used for pivots of different sizes and with different numbers of towers.

The characterization of the irrigated area, along with the characteristics of the pumping station and irrigation system from the manufacturers' technical data, were inserted into the VSPM (Variable Speed Pivot Model) tool to process and edit the input data. Subsequently, the hydraulic simulation was calculated using the EPANET hydraulic engine. The hydraulic simulation results were then used to determine the energy consumption of the irrigation system under study.

2.2. Topography of the Irrigated Area

As shown in Figure 1, the elevation values of the moving towers were obtained through Digital Elevation Models (DEMs) (with a spatial resolution of 5 m × 5 m), which were freely obtained from the PNOA (Spanish National Program of aerial photogrammetry). These products are freely available for any location in Spain. The developed tool is able to use any type of DEM.

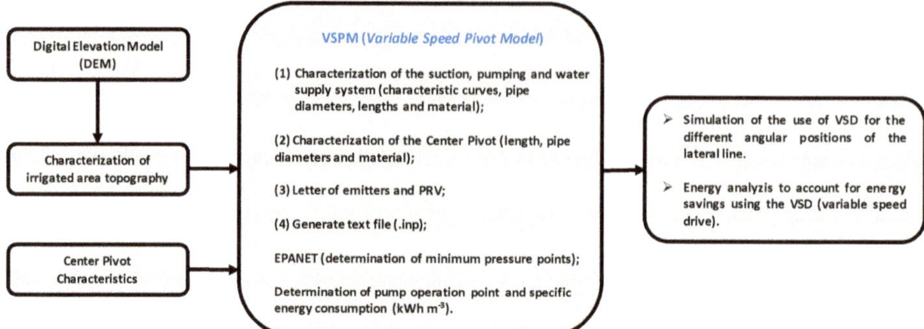

Figure 1. Diagram of the proposed procedure.

The utilized DEMs (.tif formatted images) were loaded into the QGIS® Desktop 3.2.1 software (QGIS Development Team, Open Source Geospatial Foundation). In this software, the geographical coordinates (X, Y) of the point referring to the center of the pivot (X_0, Y_0) were defined.

To obtain the X and Y coordinates of all moving towers in different angular positions relative to the center pivot's lateral line, a computational routine was developed in the MATLAB® 2018b software. Automatically, with the values of the geographic coordinates and the distance between towers, the elevation values were determined for 36 angular positions of the lateral line, equally spaced by 10°, with the North position of the lateral line as the position of angle 0°, as in Figure 2.

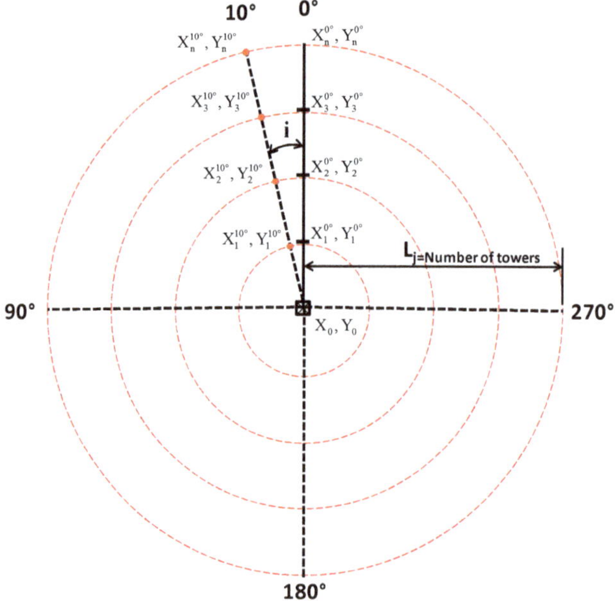

Figure 2. The geographic coordinates during the rotation of the lateral line at the center pivot.

The calculation of the geographical coordinates X, Y of each moving tower (j) was carried out as follows:

$$X_j^i = X_0 + L_j \cdot \cos(i), \text{ with } 1 \leq j \leq \text{Number of towers} \tag{1}$$

$$Y_j^i = Y_0 + L_j \cdot \sin(i), \text{ with } 1 \leq j \leq \text{Number of towers} \tag{2}$$

where i is the angular position of the lateral line $(0°, 10°, \ldots, 350°)$, j is the number of moving towers, in this case, $1 \leq j < \leq 9$, and L_j is the distance from the center tower to the index tower j, m.

Thus, for the center pivot under study, the dimensions of the nine moving towers were obtained in 36 different angular positions of the lateral line, resulting in 324 elevation values.

2.3. Hydraulic Model Description

The simulation of the operation of the center pivot irrigation system, for different angular positions of the lateral line of the center pivot, was carried out using EPANET software [23]. With this aim, the VSPM tool was developed. The function of this tool is to edit the input data required by the EPANET software in addition to energy analysis, as shown in Figure 1.

Center pivot irrigation system simulations, for different lateral line angular positions, was carried out using EPANET software [23]. With this aim, the VSPM tool was developed. The function of this tool is to edit the input data required by the EPANET software in addition to energy analysis, as shown in Figure 1. Figure 3a shows the dimensions of the lateral line span and Figure 3b shows the EPANET network map of the first span of the center pivot lateral line.

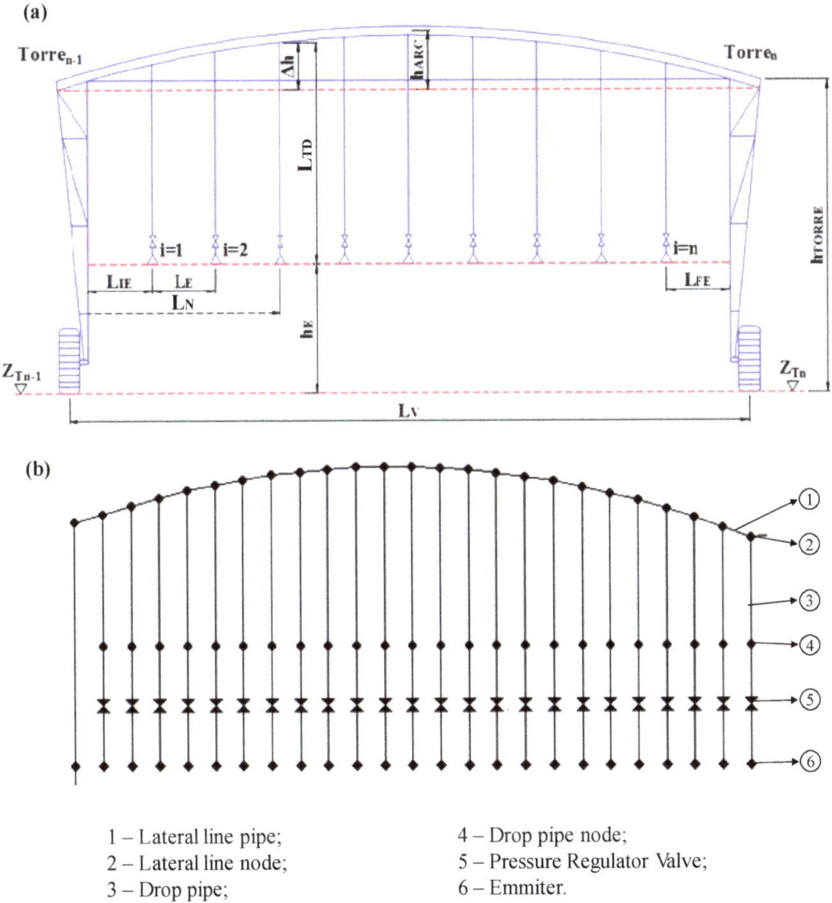

1 – Lateral line pipe;
2 – Lateral line node;
3 – Drop pipe;
4 – Drop pipe node;
5 – Pressure Regulator Valve;
6 – Emmiter.

Figure 3. (**a**) Center pivot lateral line span dimensions, (**b**) EPANET network map of the first span of the center pivot lateral line.

The VSPM tool has four steps: (1) characterization of the suction, pumping, and water supply systems, (2) characterization of the center pivot, (3) emitters and pressure regulating valves (PRV), (4) generation of the text file in the EPANET format.

The system that includes suction, pumping, and water supply was pre-determined containing the following hydraulic elements: a fixed level reservoir (water source), a pumping station, and a supply pipe composed of three different links. The required information for these components are: elevation of the water level in the reservoir, elevation of the ground where the pumping station is located, length, pipe diameter and roughness coefficient (depending on the selected head loss equation) of all links, and the pairs of values (Q-H) constituting the pump characteristic curve, as taken from the manufacturer's technical data.

In the center pivot characterization module, the center pivot's equipment information is inserted: the elevation of the pivot point and the elevation of the moving towers, as well as pipe lengths, pipe diameters, and the roughness coefficient of the pivot point, lateral line, and drop pipes.

In the module related to emitters and pressure regulating valves (PRVs), the emission coefficients of the emitters and the working pressure of the PRVs are inserted. If information about the emitter's distribution on the pivot is not available, it remains possible to generate and simulate a commonly used commercial emitter distribution with this tool.

Finally, the VSPM tool generates a text file (in .inp format) with all the hydraulic model information entered by the user. This file consists of input data, which is required by EPANET to perform the hydraulic simulation. In the VSPM modules, the flow rate and discharge coefficient of each emitter, the node elevations, and the pipe lengths were determined.

2.3.1. Flow Rate of the Emitters

The total flow of the irrigation system is the sum of the flow rate of each emitter along the lateral line. The flow rate of the emitters (in $m^3\ h^{-1}$) was calculated according to the area corresponding to each water outlet, and the gross depth to be applied to the rotation time was specified by the manufacturer, as proposed by Valiantzas et al. [24]. The flow rate for the first outlet of each span (Equation (3)) and the remainder of each outlet (Equation (4)) was calculated. The lengths used are shown in Figure 3a:

$$q_{x=1} = \frac{2\pi L_{IE}^2}{1000}\frac{Lb}{Tg}, \text{ for } x = 1 \tag{3}$$

$$q_x = \frac{2\pi R_{inst}L_E}{1000}\frac{Lb}{Tg}, \text{ for } 2 \leq x \leq n \tag{4}$$

where q_x is the flow rate of the outlet with order number x ($1 \leq x \leq n$), $m^3\ h^{-1}$, L_{IE} is spacing between the tower and the first emitter (x = 1), m, L_E is the spacing between the emitters with order number x ($2 \leq x \leq n$), m, R_{inst} is the radius of the installation of the emitter, relative to the center tower, m, Lb is the gross irrigation depth (9 mm), and Tg is the rotation time (21 h).

After the acquisition of the flow rate values at each water outlet (q_x), the discharge coefficient (k_e, Equation (5)) of each emitter was determined. The value of the pressure in the emitter was the same as the value of the PRV nominal pressure (H_{prv}), assuming an ideal valve:

$$k_e = \frac{q_i}{H_{prv}^\beta} \tag{5}$$

where k_e is emitter discharge coefficient, $m^{2.5}\ h^{-1}$, H_{prv} is the PRV nominal pressure, m, and β is the pressure exponent, in this case with a value of 0.5.

2.3.2. Determination of Elevations and Lengths

The spans lengths (Equation (6)) and the water outlet relative distance to the previous moving tower (Equation (7)) were calculated:

$$L_S = L_{IE} + L_{FE} + [(N_O - 1)L_E] \tag{6}$$

$$L_N = \begin{cases} N = 1 \Rightarrow R_T + L_{IE} \\ 2 \leq N < n \Rightarrow L_{N=1} + L_E \\ N = n \Rightarrow L_{2 \leq N < n} + L_{FE} \end{cases} \tag{7}$$

where L_S is the length of span (m), L_{FE} is the distance between the last emitter and next tower of the span (m), N_O is the number of water outlets in the span, L_N is the distance of the node referring to the water outlet in relation to the previous tower (m), N is the water outlet in the span, and R_T is the radius of rotation of the tower relative to the centre tower (m).

The irrigation system is composed of nodes, which are the connections between the links. Each water outlet was determined as a set of six hydraulic components, as shown in Figure 3b. Elevation is the main feature required for the nodes (Equation (8)). The slope between the towers (Equation (9)), the variable height between the lateral line and the height of the tower (Equation (10)), and the length of the drop pipe (Equation (11)) were also calculated:

$$Z_{n=x} = Z_{Tn-1} + (S_T L_N) + \Delta h \tag{8}$$

$$S_T = \frac{Z_{Tn} - Z_{Tn-1}}{L_S} \tag{9}$$

$$\Delta h = \frac{4 h_{arc} L_N^2}{L_S^2} \left(\frac{L_S}{L_N} - 1 \right) \tag{10}$$

$$L_{DP} = h_T - h_E + \Delta h \tag{11}$$

where $Z_{n=x}$ is the node elevation x, m. Z_{Tn-1} is the tower elevation previous to node n (m). Z_{Tn} is the tower elevation posterior to node n (m). S_T is the slope between towers n e n-1. Δh is the length of the drop pipe between the lateral line and the tower are variable for each lateral line water outlet (m). h_{arc} is the maximum height of the lateral line arc (m). L_{DP} is the total length of the drop pipe, which is variable for each lateral line water outlet (m). h_T is the height of the moving towers (m). h_E is the height of the emitter relative to the ground (m).

For the case of the overhanging nodes, the same slope (S_T) of the last span of the lateral line was assumed in the calculation of the dimensions ($Z_{n=x}$).

2.4. Calculation of the Pumping Operation Point and Energy Consumption

Hydraulic Model

Using EPANET, the hydraulic simulation was performed at each angular position of the center pivot lateral line with the pumping station operating at a maximum fixed speed. The head loss was computed based on the Hazen–Williams roughness coefficient (C_{HW}) values of 100 (for the lateral line and water supply pipe) and 140 (for drop pipes).

Head and efficiency curves of the pump are considered. Affinity laws were implemented to the regulation of variable frequency drive. For an accurate analysis of the efficiency, all the components of the pumping station were considered, as proposed by Fernández García et al. [17]. Also, energy losses in cables were computed.

$$\eta_t = \eta_p \cdot \eta_m \cdot \eta_v \cdot \eta_c \tag{12}$$

where η_t is the total efficiency of the pumping station, η_p is the pump efficiency, η_m is the motor efficiency, η_v is the VSD efficiency, and η_c is the cable efficiency.

The motor efficiency for each angular position i was determined through the exponential model proposed by Bernier et al. [25].

$$\eta_{m_i} = 0.94187 \cdot \left(1 - e^{-0.0904 \cdot \left(\frac{P_{Abs(i)}}{P_{Nom}} \right)} \right) \tag{13}$$

where P_{Nom} is the nominal power in kW and $P_{Abs(i)}$ is the absorbed power in kW.

To compute the VSD efficiency values at each angular position i of the lateral line, Equation (14) was used. This value should be supplied by the VSD manufacturer but is commonly not supplied. In this case study, we utilized the value obtained by Moreno et al. [22] for a pump with a similar power:

$$\eta_{v_i} = 70.126 - 232.47\alpha + 582.032\alpha^2 - 323.134\alpha^3 \tag{14}$$

where α is the ratio between the speed of the variable speed drive and the maximum speed as a fixed speed drive (n_v/n_f)

2.5. Determination of Specific Energy Consumption (CEE)

At each angular position of the lateral line (36 angular positions spaced 10°) the minimum pressure required to pressurize the irrigation system was determined. Therefore, at each position, the adequate pumping pressure head (H_i) was estimated:

$$H_i = H_p - \left(H_{min(i)} + H^*_{prv} \right) \tag{15}$$

where H_p is the pressure head at the fixed pumping speed (m), $H_{min(i)}$ is the minimum pressure head along the lateral line at each angular position i, m (obtained in EPANET), and H^*_{prv} is the PRV pressure head, including the pressure regulator loss (69 kPa + 34 kPa).

The specific energy consumption using the VSD at each angular position i of the lateral line (CEE^v_i, em kWh m^{-3}, Equation (16)) for $0° \leq i \leq 350°$ was calculated by the values of the pumping pressure head (H_i, m), the water specific weight (γ, N m^{-3}), and the efficiencies (η) of the pump, motor, VSD, and cable:

$$CEE^v_i = \frac{H_i \, \gamma}{3600 \, \left(\eta_{b_i} \, \eta_{m_i} \, \eta_{v_i} \, \eta_{c_i} \right)} \tag{16}$$

The specific energy consumption of the equipment, considering a pumping station with a fixed speed (CEE^f, em kWh m^{-3}), was determined through Equation (17):

$$CEE^f = \frac{H_p \, \gamma}{3600 \, \left(\eta_b \, \eta_m \, \eta_c \right)} \tag{17}$$

The energy reduction (ER, %) was determined by Equation (18). The energy consumption (EC, kWh) in the pump station for each position (fixed speed and variable speed) was determined from Equations (19) and (20), respectively.

$$ER = \left(\frac{CEE^f - \left(CEE^v_i \right)_{av}}{CEE^f} \right) \times 100 \tag{18}$$

$$EC_f = Q \times T_o \times CEE^f \tag{19}$$

$$EC_v = Q \times T_o \times \left(CEE^v_i \right)_{av} \tag{20}$$

where $(CEE_i^v)_{av}$ is the average of the specific energy consumption in the different angular positions of the lateral line, and T_o is the operating time of the irrigation system.

2.6. Case Study

This study was developed in a center pivot irrigation system located in the "La Felipa" district, which belongs to Chinchilla de Monte-Aragón (Albacete) in the Castilla La-Mancha region (Spain). The geographical coordinates of the latitude and longitude of the center point of the pivot are 39° 4′44.43″ N and 1°39′35.27″ W. The elevations of the water level in the reservoir, the pumping station, and the center pivot point are, 675.91 m, 676.91 m, and 661.91 m, respectively.

With regard to the center pivot, the lateral line of this equipment operates without an end gun, irrigating an area with a total radius of 488.6 m, equivalent to a surface area of 76 ha. The lateral line is composed of nine spans, comprising four spans with a length of 57 m, five spans with a length of 51 m, and an overhang of 5.6 m. The lateral line has an internal pipe diameter of 162.27 mm to the last tower, and the overhang has an internal pipe diameter of 108.74 mm. Along the lateral line, there are 164 emitters (type SP4 + PL / R), which are all equipped with pressure regulator valves (69 kPa) that have a pressure regulator loss of 34 kPa, with 3 m spacing. The first outlet is located 2.10 m from the pivot point, mounted at the end of flexible drop pipes, with an internal pipe diameter of 19.05 mm at a height of 1.80 m from the ground surface. The towers have a fixed height of 3.54 m, and the largest arc of the spans has a value of 0.7 m in height. The highest elevation of the irrigated area is 667.60 m, and the lowest value is 652.29 m.

According to the specifications of the pivot model, the model's total flow rate is 326.61 m^3 h^{-1} and has a supply pipe with a length of 920 m made of PVC with a nominal pipe diameter of 300 mm, which leads to the pivot point's water from a fixed level reservoir with a capacity of approximately 7000 m^3. The pumping station is composed of a pump (from the brand KSB, model WKL 150/1), with rotor 360 mm in size, which is driven by a three phase electric motor whose nominal voltage, power, and rotation values are 400V, 90 kW, and 1750 rpm, respectively. The electrical installation, with a power factor of 0.85, has electric copper cables with a length of 50 m and a cross-section of 16 mm^2.

The time of operation of the center pivot was determined to supply the water requirements of a maize crop in the study region (Domínguez et al., 2012). The gross irrigation water requirement (GIWR) applied in this period was 627 mm, with an operating time (T_o) of 1440 h.

In order to perform an economic analysis of the use of VSD in center pivot irrigation system pumping stations, the average value of 0.2 € kWh^{-1} was adopted as the reference price of the electric energy in the study region.

3. Results and Discussion

3.1. Hydraulic Model

The use of the VSPM tool made it possible to accurately characterize the irrigation system in EPANET (Figure 4). Hence, the hydraulic simulation was performed at each angular position of the center pivot lateral line, thereby obtaining the pressure distribution.

In Figure 4, two angular positions of the lateral line, 250° and 40°, are represented. These positions are those with the highest level of differences in relation to the pivot point, representing 5.68 m uphill and 9.62 m downhill, respectively.

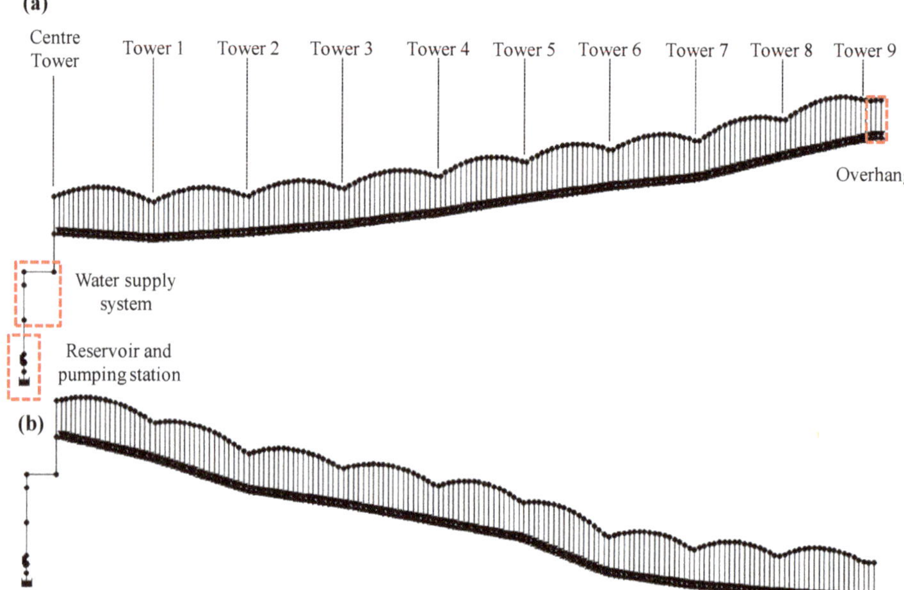

Figure 4. Network maps of the center pivot system from EPANET inp files generated by the VSPM (Variable Speed Pivot Model) tool considering two different lateral line angular positions (**a**) 250° uphill and (**b**) 40° downhill.

It can be seen that the hydraulic simulations were not performed with constant topographic slopes, as done in other works [13,20,21,24,26]. This fact allowed a greater precision in the determination of the speed of the pumping station to guarantee the lowest energy consumption with the appropriate pressurization of the lateral line in different angular positions.

The EPANET nodes and links corresponding to the fixed-level reservoir, pumping station, suction pipe, and supply pipe are highlighted in Figure 4. It should be noted that these hydraulic elements with their characteristics are constant in all angular positions of the lateral line of the center pivot. The pipe in the overhang is also highlighted in Figure 4, showing that the slope at the end of the lateral line is maintained with the same slope value of the last span. In addition, the correct characterization of the topography of the irrigated area allowed precise characterization of the energy consumption of the irrigation system operating with a maximum fixed and variable speed for the pumping station.

3.2. Operating Point

The characteristic curves of the pumping station (Q-H_p, Q-η_p, and Q-P_{Abs}), adjusted through data taken from the manufacturer, are shown in Figure 5. The operating point of the index characteristic curve "f", relative to the configuration of the pumping unit with a fixed speed is shown. In addition, the operating point of the index characteristic curve "v" refers to the configuration with a variable speed. In both cases, the angular position of the lateral line of 40° was represented with the fixed speed (n_f = 1750 rpm and α = 1.00) and variable speed pump (n_v = 1523 rpm and α = 0.85). This position was chosen because it has a minor pressure head value.

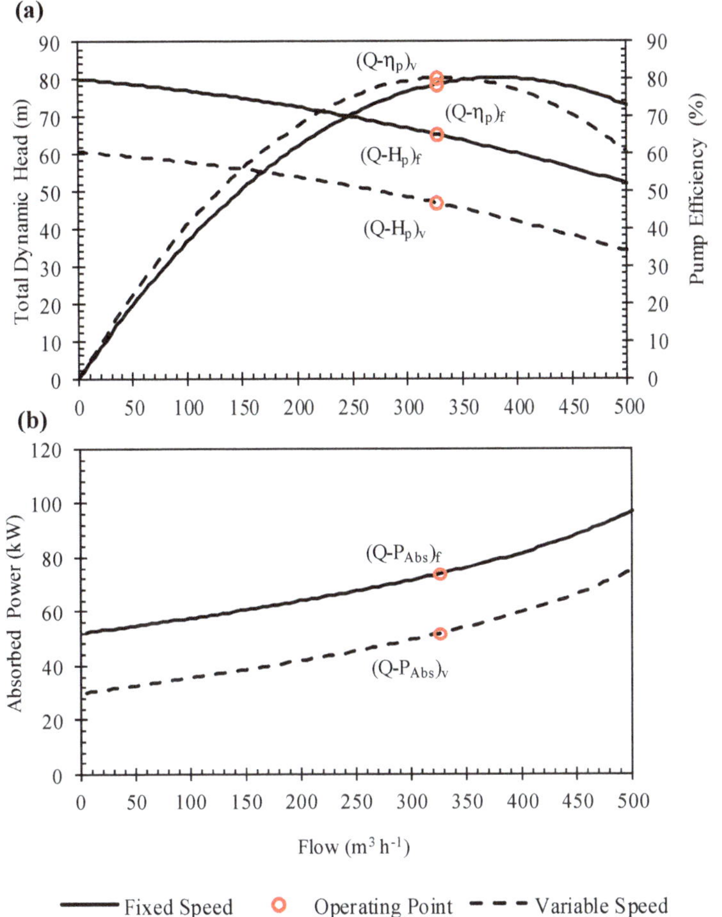

Figure 5. Pumping station characteristic curves for the 40° angular position lateral line with a fixed speed (f) and variable speed (v), (**a**) (Q-H) and (Q-η) curves, (**b**) (Q-P$_{Abs}$) curve.

In Figure 5, it can be seen that the displacement of curves (Q-H$_p$)$_v$ and (Q-P$_{Abs}$)$_v$ to (Q-H$_p$)$_f$ and (Q-P$_{Abs}$)$_f$ did not result in significant differences in pump efficiency (Q-η$_p$)$_{f,v}$. This result demonstrates that the use of the VSD in the irrigation system does not significantly interfere with the pump efficiency, whose high and low values were close to 80.26% (α = 0.85) and 78.54% (α = 1.00). This displacement also demonstrates the reduction of energy consumption (12.2%) through the influence of the VSD on the pumping station of the irrigation system under study. This same behavior of curve displacement was reported by Brar et al. [20] in a study on energy efficiency in center pivots. They determined that by reducing the speed of the pumping station's rotation through the VSD, it was possible to reduce its head pressure, thereby maintaining the efficiency of the pump. Also, as described by Córcoles et al. [27], the ratio (in kWh m^{-3}) can be higher at lower frequencies than the nominal value, thus presenting another source of energy saving.

3.3. Pressure Distribution Along the Lateral Line

After the hydraulic simulation of the center pivot, the pressure values at the top of the lateral line were obtained for each angular position at the nodes referring to the water outlets. Thus, the minimum pressure head value corresponds to the minor value of the pressure distribution. The pump speed computed at each angular position was the minimum pump speed that was able to keep pressure values at the downstream node of every PRV equal to the setting pressure (ie keeping every PRV at the Partially closed state). So, according to EPANET PRV rules described above, this condition was achieved when pressure values at every PRV upstream node were above the setting pressure. The minimum pressure head values and the location of the minimum pressure head along the lateral line are shown in Figure 6.

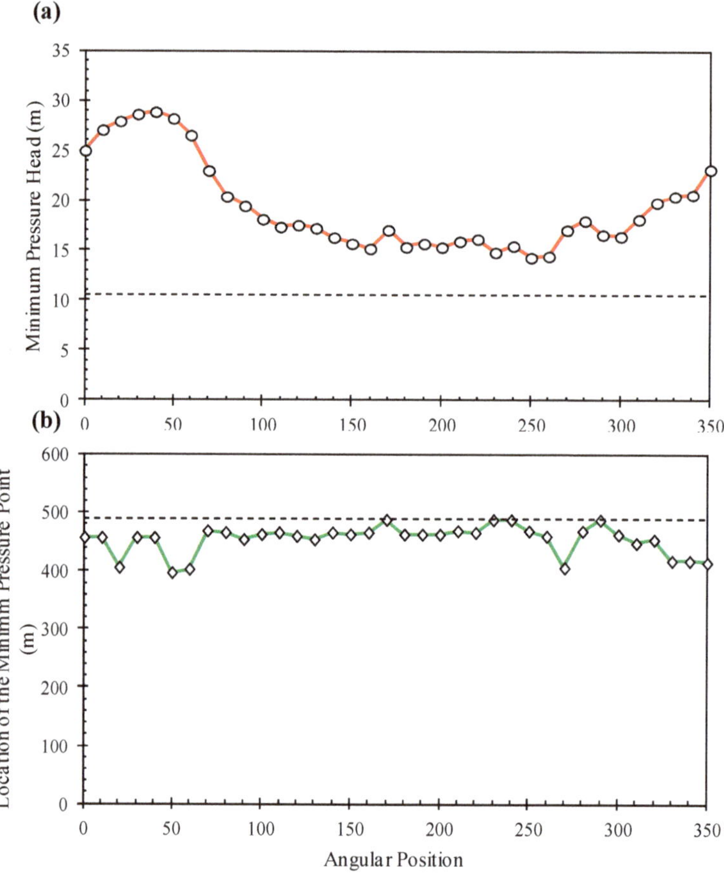

Figure 6. (**a**) Value and (**b**) location of the point of minimum pressure head in the different angular positions of the center pivot lateral line.

According to [13,20,21], the point of minimum pressure is located between the end of the lateral line and the center of the pivot according to the topographic changes of the irrigated area.

King et al. [13], studying the spatial distribution of pressure on the lateral portion of a center pivot (with a length of 392 m and a difference in the level of 18 m), reported that the point of minor pressure on the lateral line was not necessarily situated at the end. The location of that point was variable according to the different slopes of the irrigated area. This fact made it difficult to install pressure sensors to control the speed of the pumping station with a VSD.

Figure 6 also shows the location of the point of minimum pressure with respect to the end of the lateral line. In some angular positions, the minimum pressure value is located exactly at the end of the lateral line. This result shows the influence of the small topographic differences of the irrigated area on the location of the minimum pressure point and, consequently, on the potential reduction of energy consumption.

3.4. Energy Analysis

Pumping stations are designed for the lateral line's most critical angular position, i.e., the position where the lateral line has the largest positive level difference. Thus, the minimum pressure head is located at the lateral end. Pumping station values for the different lateral line angular positions and values for the fixed speed pump are shown in Table 1.

Table 1. Flow rate (Q), pumping pressure head (H_i), hydraulic power (P_H), speed of the pumping station (n), variable speed ratio (α), efficiencies (η), and specific energy consumption (CEE) in the different angular positions of the lateral line.

Angular Position	Q (m^3 h^{-1})	Hi (m)	P_H (kW)	n	α	η_p (%)	η_m (%)	η_v (%)	η_c (%)	η_t [1] (%)	CEE (kWh m^{-3})
Fixed	326.61	65.23	58.04	1750	1.00	78.54	94.13		98.60	72.90	0.244*
0°	326.61	50.72	45.12	1574	0.90	80.08	93.86	96.75	98.65	71.74	0.193
10°	326.61	48.64	43.27	1547	0.88	80.20	93.77	96.21	98.66	71.38	0.186
20°	326.61	47.77	42.50	1535	0.88	80.23	93.73	95.95	98.66	71.19	0.183
30°	326.61	47.07	41.88	1526	0.87	80.26	93.69	95.71	98.66	71.01	0.181
40°	326.61	46.84	41.67	1523	0.87	80.26	93.68	95.63	98.67	70.94	0.180
50°	326.61	47.49	42.25	1531	0.88	80.24	93.71	95.86	98.66	71.12	0.182
60°	326.61	49.16	43.74	1553	0.89	80.17	93.79	96.36	98.66	71.48	0.187
70°	326.61	52.69	46.88	1599	0.91	79.93	93.93	97.13	98.64	71.93	0.200
80°	326.61	55.29	49.20	1631	0.93	79.70	94.00	97.43	98.63	72.00	0.209
90°	326.61	56.24	50.04	1643	0.94	79.61	94.02	97.48	98.63	71.96	0.213
100°	326.61	57.57	51.22	1659	0.95	79.47	94.04	97.51	98.63	71.88	0.218
110°	326.61	58.33	51.90	1669	0.95	79.39	94.05	97.50	98.62	71.80	0.221
120°	326.61	58.15	51.73	1666	0.95	79.41	94.05	97.50	98.62	71.82	0.221
130°	326.61	58.45	52.01	1670	0.95	79.38	94.06	97.50	98.62	71.79	0.222
140°	326.61	59.38	52.83	1681	0.96	79.27	94.07	97.45	98.62	71.67	0.226
150°	326.61	60.00	53.38	1689	0.96	79.20	94.08	97.41	98.62	71.58	0.228
160°	326.61	60.49	53.82	1695	0.97	79.14	94.09	97.37	98.62	71.49	0.230
170°	326.61	58.63	52.16	1672	0.96	79.36	94.06	97.49	98.62	71.77	0.223
180°	326.61	60.33	53.67	1693	0.97	79.16	94.08	97.38	98.62	71.52	0.230
190°	326.61	59.99	53.37	1688	0.96	79.20	94.08	97.41	98.62	71.58	0.228
200°	326.61	60.35	53.69	1693	0.97	79.16	94.08	97.38	98.62	71.52	0.230
210°	326.61	59.75	53.16	1686	0.96	79.23	94.08	97.43	98.62	71.61	0.227
220°	326.61	59.52	52.96	1683	0.96	79.25	94.07	97.44	98.62	71.65	0.226
230°	326.61	60.86	54.15	1699	0.97	79.10	94.09	97.33	98.62	71.43	0.232
240°	326.61	60.21	53.57	1691	0.97	79.17	94.08	97.39	98.62	71.54	0.229
250°	326.61	61.37	54.60	1705	0.97	79.03	94.10	97.27	98.61	71.33	0.234
260°	326.61	61.24	54.48	1703	0.97	79.05	94.09	97.28	98.61	71.36	0.234
270°	326.61	58.61	52.15	1672	0.96	79.36	94.06	97.49	98.62	71.77	0.222
280°	326.61	57.68	51.32	1661	0.95	79.46	94.04	97.51	98.63	71.87	0.219
290°	326.61	59.07	52.56	1677	0.96	79.31	94.07	97.47	98.62	71.71	0.224
300°	326.61	59.24	52.70	1679	0.96	79.29	94.07	97.46	98.62	71.69	0.225
310°	326.61	57.53	51.19	1659	0.95	79.48	94.04	97.51	98.63	71.88	0.218
320°	326.61	55.85	49.69	1638	0.94	79.65	94.01	97.47	98.63	71.98	0.211
330°	326.61	55.22	49.13	1630	0.93	79.71	93.99	97.42	98.63	72.00	0.209
340°	326.61	55.05	48.98	1628	0.93	79.73	93.99	97.41	98.63	72.00	0.208
350°	326.61	52.50	46.71	1596	0.91	79.95	93.92	97.10	98.64	71.92	0.199

[1] Total efficiency * variable speed drives (VSDs) efficiency not considered.

In Table 1, it can be sheen that the i = 40° position (Figure 4) has the lowest pumping pressure head value (46.84 m) and, consequently, a lower specific energy consumption value (0.180 kWh m^{-3}). On the other hand, position I = 250° presents the highest values–61.37 m and 0.234 kWh m^{-3}, respectively. It should be noted that for these positions, the determination of the CEE with the variable speed of the pump is different from the CEE with a fixed speed, because, in this configuration, the value of VSD efficiency is not considered.

Specific energy consumption (kWh m^{-3}) values for the fixed speed (without VSD) and variable speed (with VSD) pumping stations were determined. The specific energy consumption using the VSD was computed as the average values considering all the angular positions of the lateral line. In general, due to the small variations in the topography of the irrigated area, the average value of the CEE with variable speed of the pumping station was close to the value of the CEE with a fixed speed (0.214 and 0.244 kWh m^{-3}, respectively), resulting in an energy reduction value of 12.2%. This percentage is lower than the values found by [18] (32% energy savings using variable speed well pumps in a center pivot system), [19] (27% to 35% of energy savings can be achieved using VSD in two Italian irrigation districts operating with three parallel horizontal axis pumps), and higher than those found by Brar et al. [20] (9.6% energy reduction is possible for 13.6 m difference in the irrigated area for a study containing 100 center pivots in Nebraska (USA), with each pivot containing a pumping station). However, these studies did not take into account the efficiency of VSD. The energy reduction value of this study is close to that found by King et al. [13] (7.5% to 15.8%), who considered the variable speed drive efficiency of a center pivot pumping unit. This result demonstrates that using VSD can reduce energy consumption in pumping units for water distribution.

In the present study, the center pivot system is equipped with a single pump. The use of VSDs in situations where multiple pumps supply several irrigation systems can result in greater energy savings, as can be seen in the studies conducted by [18,19].

The values of the absorbed power (the relation of the hydraulic power and total efficiency, in kW) at the pumping station at each of the angular positions of the center pivot's lateral line are shown in Figure 7.

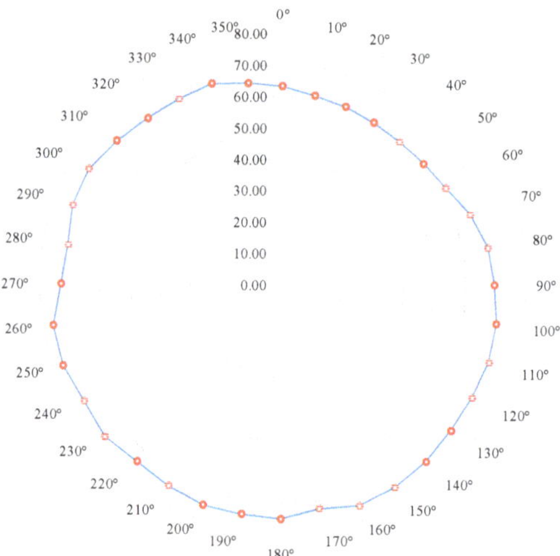

Figure 7. Absorbed power at the pumping station for the different angular positions of the lateral line.

In Figure 7, the lowest values of the absorbed power (51.92 kW) were reached in the angular positions where the lateral line assumes a downward slope (i = 40°) from the center to the end. In the same way, when the lateral line presents an ascending slope (i = 250°), it results in higher values of absorbed power (69.09 kW). This difference in values results in a reduction of 28.5% between these two angular positions. It should be noted that the total efficiency of these two positions is very close (70.9% (for i = 40°) and 71.3% (for i = 250°)). These results are directly related to the movement of the point of minimum pressure along the lateral line according to topographic variations of the irrigated area. The same type of behavior for the minimum pressure point was reported by Brar et al. [20] and

King et al. [13] in studies with energy conservation in the center pivot systems using variable speed drives at the pumping station.

3.5. Economic Analysis

To analyze the operation costs with the use of VSDs in this center pivot, the values of energy consumption and costs, considering the two configurations (fixed and variable speed) for an irrigation season with maize crop in the region of Albacete (Spain), are presented in Table 2.

Table 2. The consumption and cost of energy for the pumping stations with fixed and variable speeds.

Crop		Maize
GIWR (gross irrigation water requirement)	mm	627.00
Flow rate	$m^3\ h^{-1}$	326.61
Operation time	h	1440.00
EC_f	kWh	114739.03
EC_v	kWh	100632.47
Average cost	€ kW h^{-1}	0.20
C_f	€	22947.96
C_v	€	20126.49
Energy Savings	€ year^{-1}	2821.47

For an irrigated area with a maximum elevation difference of 15.3 m, and considering the use of a VSD with a fixed speed, the annual average energy savings is close to 14107.35 kWh (Table 2), representing an annual average energy cost savings of 2821.47€.

In a similar study, [13] presented an average annual saving of 18100.00 kWh, applying a requirement of 635 mm of gross water irrigation and an area with an elevation difference of 18 m. They concluded that the VSD installation would not be economically viable. This result can be explained due to the low performance of the utility's equipment and the low cost of energy (0.031 kWh^{-1}) in the region where the study was conducted, which resulted in a low average annual saving (564.30 €). The average annual energy savings (kWh) presented by [13] were similar to those determined in the present study. However, due to the unit cost of energy consumption, the financial savings were lower. This fact shows that the cost of energy, besides the topography of the irrigated area, must be taken into account for this type of analysis.

Working with several center pivots for the maize crop, Brar et al. [20] presented an average annual saving of 4155 kWh, applying a requirement of 284 mm of gross water irrigation to an average irrigated area of 49 ha with a height difference of 13.6 m. The average pumping cost was 0.098 kWh^{-1}, resulting in average annual savings of 407.40€. The difference of the values in this study can be explained by the smaller difference in elevation, their low energy cost, and their lower gross irrigation requirements.

4. Conclusions

This study presented a proposal to vary the pressure supplied to a center pivot irrigation system according to the angular position of the lateral line by adjusting the speed of the pumping station with a variable speed drive (VSD). A model was developed to simulate the possible reduction in energy consumption when using a VSD in a center pivot pumping station in Albacete (Spain). For topographic characterization, DEMs from satellite images (with 5×5 m spatial accuracy) were used, and it was concluded that the topographic precision for this type of study is essential to determine the required pressure values at each angular position of the center pivot's lateral line.

The present study showed a saving of 12.2% of energy when using speed control at the pumping station in comparison with the commonly used fixed speed. This reduction in energy allowed economic savings close to 2821.47 €, in an area irrigated by a center pivot with a difference in maximum elevation of 15.3 m and the major part of the lateral line on the ascending slope.

The VSPM tool, which added topographic, hydraulic, and energy characteristics to the irrigation system, facilitated the simulation of the center pivot irrigation system in the EPANET software, along with the analyses of the use of VSD in the speed control of the pumping station. This tool demonstrates the possible financial return when using a VSD in a center pivot pumping station that operates in areas of variable topography. This center pivot represents the greater part of the irrigated area in the ascending slope during the movement of the lateral line. Therefore, for an irrigated area where the lateral line remains in a descending slope during the greater part of its rotation, the energy savings would be greater. Thus, this tool seems to be useful for users to achieve better precision in energy and economic analyses of center pivot irrigation systems.

The use of VSD to control the speed of the pumping station of the irrigation system showed a greater reduction in energy consumption when the different angular positions of the lateral line along the irrigated area were in an ascending slope. In these positions, the point of minimum pressure tends to migrate from the extremity to the center of the pivot.

This study aimed to quantify the energy saving potential of VSDs, providing irrigation producers and professionals results that show the benefits of installing VSDs in the pumping stations of center pivot irrigation systems.

Author Contributions: Conceptualization, V.B.d.S.B., M.Á.M., and J.I.C.; methodology, V.B.d.S.B., A.C., M.Á.M., and J.I.C.; software, V.B.d.S.B., A.C., and M.Á.M.; validation, V.B.d.S.B., A.C., and M.Á.M.; formal analysis, V.B.d.S.B., J.I.C., A.C., and M.Á.M.; investigation, V.B.d.S.B., M.Á.M., J.I.C., and A.C.; writing—original draft preparation, V.B.d.S.B., J.I.C, M.Á.M., and A.C.; writing—review and editing, V.B.d.S.B., J.I.C., and M.Á.M.; visualization, V.B.d.S.B., J.I.C., M.Á.M., and A.C.; supervision, M.Á.M.

Funding: This research was funded by the Spanish Ministry of Education and Science (MEC).grant number AGL2017-82927-C3-2-R (Co-funded by FEDER). And, this research was financed in part by the Coordenação de Aperfeiçoamento de Pessoal de Nível Superior, Brasil (CAPES), process PDSE 88881.190146/2018-01.

Conflicts of Interest: The authors declare no conflict of interest.

Nomenclature

A_c	cross-sectional area of cable (mm^2)
C_c	electrical conductivity of copper (m Ω^{-1} mm^{-2})
CEE^f	specific energy consumption, considering the pumping station with fixed speed (kWh m^{-3})
CEE_i^v	specific energy consumption using the VSD, for each angular position i (kWh m^{-3})
$(CEE_i^v)_{av}$	average of the specific energy consumption in the different angular positions (kWh m^{-3})
C_{HW}	roughness coefficient of the pipe material of the Hazen-Williams equation
$\cos\phi$	power factor
D	pipe diameter of the lateral line
DEM	Digital Elevation Model
EC	Energy Consumption (kWh)
ER	Energy Reduction (%)
GIWR	gross irrigation water requirement (mm)
h_{arc}	maximum height of the lateral line arc (m)
h_E	height of the emitter relative to the ground (m)
hf	head loss (m)
H_i	pressure head that the pump must provide for each angular position i (m)
$H_{min(i)}$	minimum pressure along the lateral line for each angular position i (m)
H_p	pressure head at the fixed pumping speed (m)
H_{prv}	nominal pressure of the PRV (m)
H_{prv}^*	PRV pressure, including the minimum regulator requirement (m)
h_T	height of moving towers (m)
i	angular position of the lateral line ($0°, 10°, \ldots, 350°$)
j	number of moving towers
k_e	emitter discharge coefficient (m$^{2.5}$ s^{-1})
L	equivalent length of lateral line (m)
Lb	gross irrigation depth (mm day^{-1})
L_c	cable length (m)
L_{DP}	total length of the drop pipe (m)
L_E	spacing between emitters (m)
L_{FE}	distance between the last emitter and next tower of the span (m)
L_{IE}	spacing between the tower and the first emitter (m)
L_j	distance from the centre tower to the index tower j (m)
L_N	distance of the node referring to the water outlet in relation to the previous tower (m)
L_S	length of span (m)
N	water outlet in the span
N_O	number of water outlets in the span
P_H	hydraulic power (kW)
PNOA	Spanish National Program of aerial photogrammetry
$P_{Abs(i)}$	absorbed power (kW)
P_{Nom}	nominal power (kW)
PRV	Pressure Regulator Valve
Q	total flow rate of the irrigation system
q_x	flow of the outlet with order number x (m^3 h^{-1})
R_{inst}	radius of installation of the emitter, relative to the centre tower (m)
R_T	radius of rotation of the tower relative to the centre tower (m)
S_T	slope between towers n e n -1
Tg	rotation time (h)
T_o	operating time of irrigation system (h)
U	nominal voltage (V)
VSD	Variable Speed Drive
VSPM	Variable Speed Pivot Model
X_i^j, Y_i^j	geographical coordinates of the moving towers j (m)

$Z_{n=x}$ elevation of the node x (m)
Z_{Tn} tower elevation posterior to node n (m)
Z_{Tn-1} tower elevation previous to node n (m)
α ratio between the speed of the variable speed drive and the maximum speed as a fixed speed drive
β exponent of the pressure
γ water specific weight (N m^{-3})
Δh length of the drop pipe between the lateral line and the tower (m)
η_c cable efficiency
η_m motor efficiency
η_p pump efficiency
η_t total efficiency of the pumping station
η_v VSD efficiency

References

1. Alexandratos, N.; Bruinsma, J. *World Agriculture Towards 2030/2050: The 2012 Revision*; ESA Work. Pap. No. 12–03; FAO: Rome, Italy, 2012; p. 153.
2. Pereira, L.S. Water, Agriculture and Food: Challenges and Issues. *Water Resour. Manag.* **2017**, *31*, 2985–2999. [CrossRef]
3. World Business Council for Sustainable Development. *Water, Food and Energy Nexus Challenges*; World Business Council for Sustainable Development: Geneve, Switzlerand, 2014.
4. Tarjuelo, J.M.; Rodriguez-diaz, J.A.; Abadía, R.; Camacho, E.; Rocamora, C.; Moreno, M.A. Efficient water and energy use in irrigation modernization: Lessons from Spanish case studies. *Agric. Water Manag.* **2015**, *162*, 67–77. [CrossRef]
5. AQUASTAT Database. *FAO's Global Water Information System: Area Equipped for Irrigation*; FAO: Rome, Italy, 2014.
6. Frizzone, J.A.; Rezende, R.; Camargo, A.P.; Colombo, A. *Irrigação por Aspersão: Sistema Pivô Central*, 1st ed.; Editora UEM: Maringá, PR, Brazil, 2018.
7. Keller, J.; Bliesner, R.D. *Sprinkle and Trickle Irrigation*; Van Nostrand Reinholh: New York, NY, USA, 1990.
8. Folegatti, M.V.; Pessoa, P.C.S.; Paz, V.P.S. Avaliação do desempenho de um Pivô Central de Grande Porte e Baixa Pressão. *Sci. Agric.* **1998**, *55*, 119–127. [CrossRef]
9. Gilley, J.R.; Watts, D.G. Possible Energy Savings in Irrigation. *J. Irrig. Drain. Div.* **1977**, *103*, 445–457.
10. Moreno, M.A.; Planells, P.; Córcoles, J.I.; Tarjuelo, J.M.; Carrión, P.A. Development of a new methodology to obtain the characteristic pump curves that minimize the total cost at pumping stations. *Biosyst. Eng.* **2008**, *102*, 95–105. [CrossRef]
11. Moreno, M.A.; Medina, D.; Ortega, J.F.; Tarjuelo, J.M. Optimal design of center pivot systems with water supplied from wells. *Agric. Water Manag.* **2012**, *107*, 112–121. [CrossRef]
12. Barbosa, B.D.S.; Colombo, A.; de Souza, J.G.N.; da Baptista, V.B.; de Araújo, A.C.S. Energy Efficiency of a Center Pivot Irrigation System. *Eng. Agrícola* **2018**, *38*, 284–292. [CrossRef]
13. King, B.A.; Wall, R.W. Distributed Instrumentation for Optimum Control of Variable Speed Eletric Pumping Plants with Center Pivots. *Appl. Eng. Agric.* **2000**, *16*, 45–50. [CrossRef]
14. Kranz, W.L.; Irmak, S.; Martin, D.L.; Yonts, C.D. Flow Control Devices for Center Pivot Irrigation Systems. *Univ. Neb. Linc. Ext. Inst. Agric. Nat. Resour.* **2007**, *888*, 1–3.
15. Alandi, P.P.; Pérez, P.C.; Álvarez, J.F.O.; Moreno, M.Á.; Tarjuelo, J.M. Pumping Selection and Regulation for Water-Distribution Networks. *J. Irrig. Drain. Eng.* **2005**, *131*, 273–281. [CrossRef]
16. Khadra, R.; Moreno, M.A.; Awada, H.; Lamaddalena, N. Energy and Hydraulic Performance-Based Management of Large-Scale Pressurized Irrigation Systems. *Water Resour. Manag.* **2016**, *30*, 3493–3506. [CrossRef]
17. Fernández García, I.; Moreno, M.A.; Rodríguez Díaz, J.A. Optimum pumping station management for irrigation networks sectoring: Case of Bembezar MI (Spain). *Agric. Water Manag.* **2014**, *144*, 150–158. [CrossRef]
18. Hanson, B.R.; Weigand, C.; Orloff, S. Variable-frequency drives for electric irrigation pumping plants save energy. *Calif. Agric.* **1996**, *50*, 36–39. [CrossRef]

19. Lamaddalena, N.; Khila, S. Energy saving with variable speed pumps in on-demand irrigation systems. *Irrig. Sci.* **2012**, *30*, 157–166. [CrossRef]
20. Brar, D.; Kranz, W.L.; Lo, T.; Irmak, S.; Martin, D.L. Energy Conservation Using Variable-Frequency Drives for Center-Pivot Irrigation: Standard Systems. *Trans. ASABE* **2017**, *60*, 95–106. [CrossRef]
21. Scaloppi, E.J.; Allen, R.G. Hydraulics of Center Pivot Laterals. *J. Irrig. Drain. Eng.* **1993**, *119*, 554–567. [CrossRef]
22. Moreno, M.A.; Córcoles, J.I.; Tarjuelo, J.M.; Ortega, J.F. Energy efficiency of pressurised irrigation networks managed on-demand and under a rotation schedule. *Biosyst. Eng.* **2010**, *107*, 349–363. [CrossRef]
23. Rossman, L.A. *EPANET 2: User Manual*; National Risk Management Research Laboratory Office of Research and Development, U.S. Environmental Protection Agency: Cincinnati, OH, USA, 2000.
24. Valiantzas, J.D.; Dercas, N. Hydraulic Analysis of Multidiameter Center-Pivot Sprinkler Laterals. *J. Irrig. Drain. Eng.* **2005**, *131*, 137–146. [CrossRef]
25. Bernier, M.A.; Bourret, B. Pumping energy and variable frequency drives. *ASHRAE J.* **1999**, *41*, 37–40.
26. Alazba, A.A.; Asce, M.; Mattar, M.A.; Elnesr, M.N.; Amin, M.T. Field Assessment of Friction Head Loss and Friction Correction Factor Equations. *J. Irrig. Drain. Eng.* **2012**, *138*, 166–176. [CrossRef]
27. Córcoles, J.; Gonzalez Perea, R.; Izquiel, A.; Moreno, M. Decision Support System Tool to Reduce the Energy Consumption of Water Abstraction from Aquifers for Irrigation. *Water* **2019**, *11*, 323. [CrossRef]

Article

Evaluation of the Dual Crop Coefficient Approach in Estimating Evapotranspiration of Drip-Irrigated Summer Maize in Xinjiang, China

Fengxiu Li and Yingjie Ma *

College of Water Conservancy and Civil Engineering, Xinjiang Agricultural University, Urumqi 830052, China; fengxiu_l1@163.com
* Correspondence: xj-myj@163.com; Tel.: +86-135-799-98634

Received: 22 April 2019; Accepted: 14 May 2019; Published: 21 May 2019

Abstract: A dual crop coefficient approach was validated experimentally to estimate evapotranspiration of drip-irrigated summer maize with partial mulch and no mulch in an arid region in Aksu, Xinjiang, China, during 2016–2017. In this study, five treatments were established based on fixed or variable irrigation cycles. Summer maize transpiration and evapotranspiration were estimated by the dual crop coefficient approach. Evapotranspiration was validated, and a positive regression with those values was obtained using the water balance method, with a root mean square error (RMSE) of 10 mm. The estimated transpiration also had a positive regression with measurements obtained by the stable carbon isotope technique, with a RMSE of 20 mm. By analyzing the RMSE, regression coefficients, and concordance index, we suggest that the dual crop coefficient approach is an effective method to estimate and partition evapotranspiration. Across the entire growing season for partially mulched summer maize, the estimated crop transpiration accounted for 78.7% and 76% of the total evapotranspiration in 2016 and 2017, respectively. For non-mulched summer maize, the estimated crop transpiration accounted for 64.9% of the total evapotranspiration over the entire growing season, which implied that the soil evaporation was about 12% higher than that of the partially mulched treatments. Water consumption with partial mulching was reduced by about 10%, compared with non-mulching, which indicated that mulching improved the use of water during irrigation.

Keywords: summer maize; drip irrigation; evapotranspiration; crop transpiration; the stable carbon isotope technique

1. Introduction

Evapotranspiration (ET) includes soil evaporation (E) and crop transpiration (T). As an important term in both water and land surface energy balance equations [1], ET plays an important role in energy and water balance. The Food and Agricultural Organization (FAO) use the FAO-56 Penman-Monteith reference evapotranspiration (ET_0) and crop coefficient method to estimate cropland ET [2]. The crop coefficient method is a semi-empirical model recommended by the FAO. The crop coefficient (K_c) is multiplied by ET_0 to obtain ET. The crop coefficient approach consists of single and dual coefficient approaches. The dual crop coefficient approach can partition ET into E and T. It can also be used to estimate the effect of rainfall, irrigation, and use of mulch on soil water.

The dual crop coefficient approach has been used widely in many regions [3]. For example, Bodner et al. [4] found that the integration of stress compensation into the FAO crop coefficient approach provided reliable estimates of water losses under dry conditions. López-Urrea et al. [5] calculated the dual crop coefficient of irrigated sorghum biomass and found that the method aided in monitoring and estimating the spatially distributed water requirements of the sorghum biomass at field and regional scales. By investigating the crop water demands for winter wheat and summer maize, Liu et al. [6] found a good agreement between predictions from the dual crop coefficient approach and measurements from a lysimeter in the North China Plain. Zhao et al. [7] calculated the single crop coefficient and the dual crop coefficient of winter wheat and summer maize, respectively, and found the dual crop coefficient approach to be more precise than the single crop coefficient approach, particularly when the crops incompletely covered the ground. Shrestha et al. [8] investigated the basal crop coefficient (K_{cb}) values of mulched erect and vine crops in a sub-tropical region using the dual crop coefficient approach and they found that some K_{cb} values of watermelon and pepper recommended by FAO-56 were not applicable. Ding et al. [9] developed a modified dual crop coefficient approach to estimate and partition ET of mulched maize in the Shiyang River Basin of the Gansu Province and found that the modified dual crop coefficient approach had high precision for maize ET under mulching. The results suggested that the modified dual crop coefficient approach predicted E and T accurately. Tomomichi et al. [10] investigated the seasonal variation in the basal crop coefficient K_{cb} and the soil evaporation coefficient K_e of sorghum and quantified the relationship between crop coefficient and LAI (Leaf Area Index). Majnooni-Heris et al. [11] determined the crop coefficient and evapotranspiration ratio of canola using the dual crop coefficient approach. Current studies that estimate and partition crop ET using the dual crop coefficient approach are mostly carried out in wet and semi-humid regions, and these are validated by water balances derived using a large lysimeter and a stem flow meter under completely covered or bare ground conditions. However, in arid areas, the application of the dual crop coefficient approach under the partially mulched drip irrigation to estimate and partition crop ET are still limited, compared with the non-mulched condition.

The stable carbon isotope technique is a new approach in the study of plant physiology and ecology, and its reliability and stability have been shown in previous studies [12–14]. Anyia et al. [15] evaluated the application of carbon isotope discrimination as a selection criterion for improving water use efficiency (WUE) and productivity of barley (*Hordeum vulgare* L.) under field and drought stress conditions in a greenhouse. Chen et al. [16] validated the stability of a leaf carbon isotope discrimination as a measure of WUE across years and locations in Alberta, Canada, based on selected barley genotypes. Chen et al. [17] investigated WUE and water consumption in different growth stages of walnut-woad/semen cassia intercrop systems using the stable carbon isotope technique and a sap flow meter, and they found that the intercropping systems consumed less water than the mono-cropping systems. He et al. [18] measured the sap flow in walnut trees and the stable carbon isotope composition of different components of a walnut-wheat intercropping system and mono-cropped wheat, and they calculated WUE and water consumption. Total water consumption of mono-cropped wheat was higher than that of intercropped wheat. The stable carbon isotope technique is reliable and stable, but the employment of this technique to validate the dual crop coefficient approach is still limited.

Experiments were conducted in the Aksu, Xinjiang province during 2016–2017 to evaluate the FAO-56 dual crop coefficient approach for estimating evapotranspiration of summer maize with partial mulch and with no mulch. This study aimed to (1) establish a suitable dual crop coefficient model with the condition of partial mulch or non-mulch of summer maize in arid region, combining with the ecological environment of the experimental area and measured data to parameterize the dual crop coefficient model. (2) Using the water balance method to evaluate the estimated ET of the model. At the same time, the crop water consumption of summer maize was calculated by the method of a stable carbon isotope technique, and it was compared with T, predicted by the dual crop coefficient method. (3) The values of ET, E, and T, with the condition of partial mulch or non-mulch of summer maize, were simulated by the model in an arid region. The changes of ET, E, and T were analyzed in this paper. (4) Provide a scientific basis for improved water management on farmland in the region.

2. Materials and Methods

2.1. Study Site

This study was conducted in 2016 and from June–October 2017, at the Xinjiang Agricultural University's Linguo Experimental Station, located in Hongqipo, Wensu County, Aksu Prefecture, China (41°16′ N, 80°20′ E). The site was warm and arid and had mean annual sunshine of 2800–3000 h and 200–220 frost-free days each year. The mean annual precipitation was 80.4 mm with large evaporation. During the growing season for summer maize, rainfall in 2016 was 75.9 mm and rainfall in 2017 was 64.8 mm. The groundwater depth was >10 m. The values of field capacity and permanent wilting point were 28% (volumetric water content) and 6% (volumetric water content), respectively. Soils were mostly sandy or silt loam (Table 1).

Table 1. Description of soils at the Xinjiang Agricultural University's Linguo Experimental Station in Hongqipo, Wensu County, Aksu Prefecture, China.

Depth cm	Bulk Density g·cm^{-3}	Particle Size Distribution %				Soil Texture
		0–0.002 mm	0.002–0.05 mm	0.05–2 mm	>2 mm	
0–20	1.38	7.0	56.5	36.5	0	Silt loam
20–40	1.42	7.2	67.9	24.9	0	Silt loam
40–60	1.40	2.9	15.8	81.3	0	Sandy loam
60–80	1.38	0.1	1.7	98.2	0	Fine sand
80–100	1.35	0.2	8.0	91.8	0	Fine sand

2.2. Experimental Design

The seeds of New Maize No. 9 were sown in 3 m × 2.2 m × 2 m plots. Each plot was spot-seeded manually with a plant spacing of 0.3 m, 0.4 m, 0.3 m, and 0.6 m in sequence, and 0.25 m between rows (Figure 1). The drip-irrigated treatment included a two-pipe and four-row system. The drip tube with 0.1 m emitter intervals was placed between two rows of maize, with a dripper discharge rate of 0.8 L h^{-1}. Each plot consisted of eight rows of maize (Figure 1). Water meters were installed in each plot to monitor the amount of irrigation water (measurement precision was 0.001 m^3).

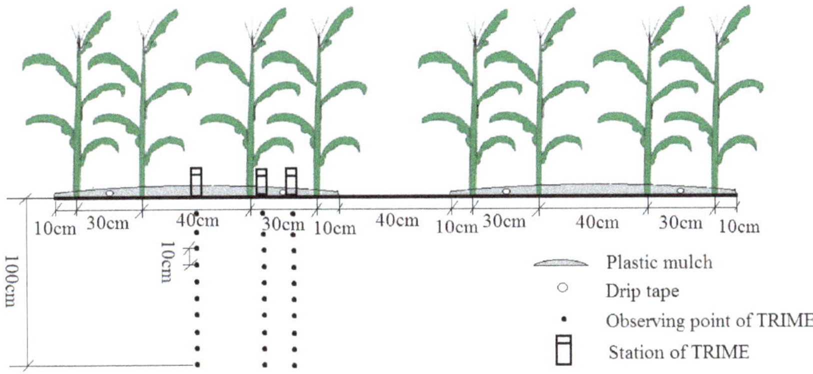

(a) With partial mulch

Figure 1. *Cont.*

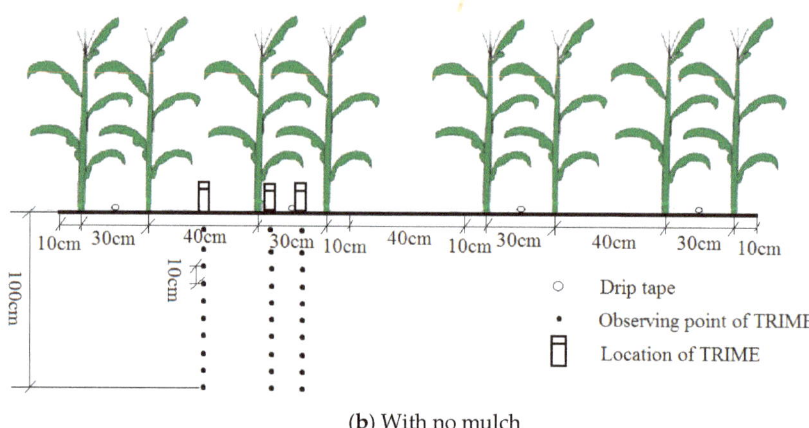

(b) With no mulch

Figure 1. Plot design for estimating evapotranspiration of drip-irrigated summer maize in Xinjiang, China. (**a**) With partial mulch; (**b**) With no mulch. (TRIME was the TRIME TDR system to determine the soil volumetric water content).

The experiments with partial mulch were conducted in 2016. In 2017, plots with partial mulch and without mulch were tested simultaneously. Five treatments were set up that included fixed irrigation cycles (W1, W2, and W3) and variable irrigation cycles (W4 and W5). Each treatment included three replicates. The fertilization and field management methods of each treatment were the same. To ensure seedling emergence, all the plots were irrigated once with approximately 125 mm water before sowing. The seeds were sown on 20 June (Table 2). The WUE values in Table 2 were calculated based on yields and water consumption.

Table 2. Description of experiments in 2016 and 2017 on mulched and non-mulched plots in Xinjiang, China.

Year	Treatment	Irrigation Cycle	Irrigation Amount/mm	WUE/kg·hm^{-2}·mm^{-1}
2016	W1	8 days (Y)	316 (Y)	41 (Y)
	W2	8 days (Y)	256 (Y)	41 (Y)
	W3	8 days (Y)	196 (Y)	44 (Y)
	W4	10 days (Y)	256 (Y)	47 (Y)
	W5	6 days (Y)	256 (Y)	47 (Y)
2017	W1	8 days (Y/N)	181 (Y/N)	38 (Y) 25 (N)
	W2	8 days (Y/N)	151 (Y/N)	35 (Y) 29 (N)
	W3	8 days (Y/N)	121 (Y/N)	35 (Y) 28 (N)
	W4	10 days (Y/N)	151 (Y/N)	39 (Y) 25 (N)
	W5	6 days (Y/N)	151 (Y/N)	42 (Y) 32 (N)

(Y) With partial mulch; (N) without mulch.

For the 2016 experiments with partial mulch, drip irrigation was started on 20 July and ended on 14 September. Maize was harvested on 14 October. The W1, W2, and W3 irrigation cycles were 8 d, with irrigation rates of 45, 37.5, and 30 mm, respectively, and total irrigation amounts of 316, 256, and 196 mm, respectively. The W4 and W5 irrigation cycles were 10 d and 6 d, respectively. The irrigation rates were 49.5 and 30 mm, respectively, and the total irrigation amount was 256 mm for both W4 and W5.

For the 2017 experiments with both partial mulch and no mulch, drip irrigation was started on 23 July and ended on 1 September, due to a pump failure. The maize was harvested on 3 October. The W1, W2, and W3 irrigation cycles were 8 d. The corresponding amount of water for irrigation was determined as 120% ET, 100% ET, and 80% ET for the W1, W2, and W3 irrigation cycles, which equated to 181, 151, and 121 mm, respectively. The ET in each irrigation cycle was calculated according to ET = K_c × ET_0, and the crop coefficient (K_c) of each growth period, used values from Liang [19]. The irrigation cycles of W4 and W5 were 10 d and 6 d, respectively, and the irrigated water was determined as 100% ET, and the total irrigation amount was 151 mm. There were one or two fewer irrigation events for each treatment in 2017 compared with 2016, which resulted in a reduced amount of irrigation water in 2017 compared with 2016.

2.3. Measurements and Methods

2.3.1. Soil Water Content

The soil volumetric water content was determined by the TRIME TDR system [7,20]. Three TRIME tubes were placed in the middle of the two drip irrigation belts, next to the maize and between the rows in each plot. The data were recorded before each irrigation. The measured depth was 100 cm. The soil water content was collected at every 10 cm of depth for a total of 10 layers (Figure 2). Two lines in Figure 2 are 100% field capacity and 60% field capacity, respectively. In 2016 and 2017, the volumetric moisture content of soil declined with the whole growth period under partial mulched or non-mulched summer maize. The values of volumetric moisture content are almost between 60–100% field capacity. The volumetric moisture content of some treatments was lower than 60% field capacity at the beginning of September, in 2017. It was caused by drip irrigation was ended on 1 September.

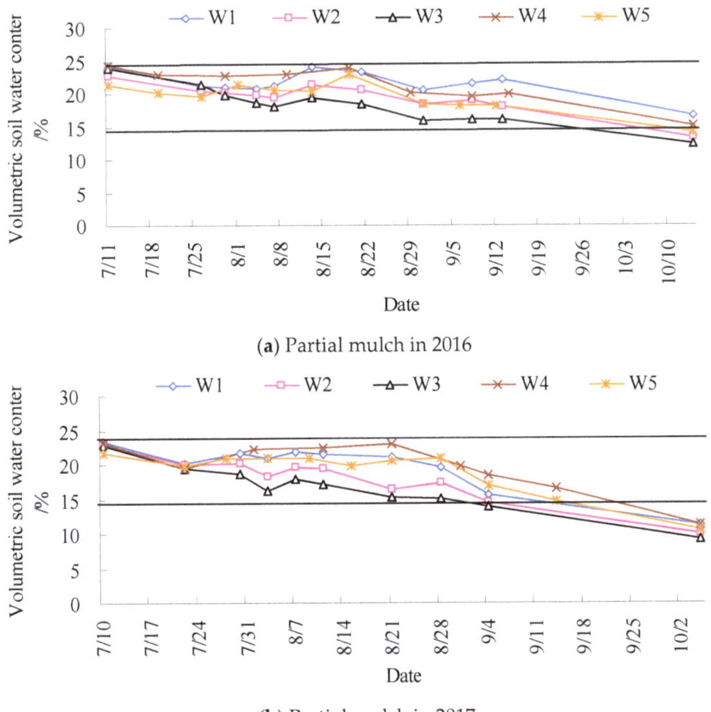

(**a**) Partial mulch in 2016

(**b**) Partial mulch in 2017

Figure 2. *Cont.*

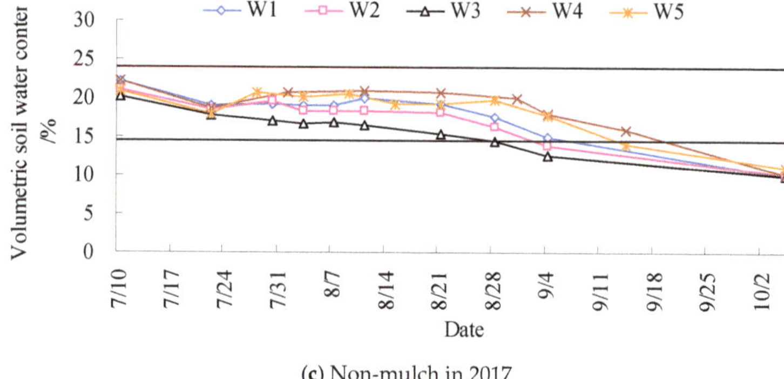

(**c**) Non-mulch in 2017

Figure 2. Volumetric soil water content for W1–W5 treatments throughout the season for drip-irrigated summer maize in Xinjiang, China. (**a**) Partial mulch in 2016; (**b**) Partial mulch in 2017; (**c**) Non-mulch in 2017.

2.3.2. Calculation of Water Consumption

In this study, the water balance equation was used to calculate the evapotranspiration of partial mulched/non-mulched crops. The calculation of ET followed the equation below:

$$ET = Pr + I + SWS - R - D + K \tag{1}$$

where ET (mm) is crop water consumption (evaporation transpiration), Pr (mm) is rainfall during the growth period, I (mm) is irrigation amount, SWS (mm) is the difference in soil water storage during sowing and harvesting, R (mm) is surface runoff, D (mm) is deep drainage, and K (mm) is groundwater recharge. Because the groundwater depth was at 10 m, R, D, and K were negligible and set to zero. Because the mulch intercepted effective rainfall, a factor of 0.25 was added to the effective rainfall for non-mulch treatments to calculate water consumption for the partially mulched summer maize, based on the percentage (25%) of the plots that were not mulched.

2.3.3. Meteorological Data

All meteorological data was obtained from a weather station (Watchdog) 300 m away from the experimental plots. Data included solar radiation, temperature, relative humidity, wind speed, atmospheric pressure, and rainfall. The data were recorded every 1 h.

2.3.4. Determination Water Use Based on the Stable Carbon Isotope Technique

The stable carbon isotope technique was used to calculate water use of summer maize. After the measurement of physiological index had been completed, the leaves in different treatments were sampled during the 6-leaf stage, 12-leaf stage, silking stage, filling stage, and mature stage, and healthy leaves (no pest or disease) were sampled at a similar height. In the laboratory, the leaf samples were dried in an oven at 70 °C for 48 h and sieved (80 mesh). The samples were sealed, stored, and sent to a laboratory (Department of Renewable Resources, Xinjiang Agricultural University, China) for determination of $\delta^{13}C$ values. The values of $\delta^{13}C$ of leaves were measured combining the Flash EA 2000 elemental analyzer (Thermo Electron, Waltham, MA, USA) with the Delta V Advantage isotope mass spectrometer (Thermo Finnigan, Bremen, German). The measurement error did not exceed 0.2‰. Observations were conducted to determine the actual water use of summer maize based on methods used as Farquhar and He et al. [12–16,21–24]. The calculation of WUE followed the equation below:

$$WUE = \frac{(1 - \varphi)Ca(b - \delta a + \delta p)}{(b - a)1.6VPD} \tag{2}$$

where WUE (mmol C/mol H_2O) is the water use efficiency, a and b (empirical coefficient [18]) are isotopic shunt coefficients of CO_2 diffusion and carboxylation, respectively (a = 4.4‰, b = 30‰ [14]), Ca is the atmospheric CO_2 concentration, δa and δp [12] are values of $\delta^{13}C$ of air and plant material, and the carbon isotope composition of air was taken as −9‰ (Brugnoli and Farqhuar 2000). The the diffusion ratio of water vapor and CO_2 in air was 1.6, φ is the ratio of carbon consumed by nocturnal respiration of leaves and respiration of other organs during the entire growth period of a plant (φ = 0.3) [25], and VPD is the difference in vapor pressure between the inside and outside of the leaf blade. The VPD can be calculated according to the average daily meteorological data (air temperature, air humidity, etc.) on the day of sampling [26,27]

$$VPD = E - e$$

$$E = 0.611 \times 10^{17.502T/(240.97+T)}$$

$$RH = \frac{e}{E} \times 100\%$$

$$VPD = 0.611 \times 10^{17.502T/(240.97+T)} \times (1 - RH) \ (kPa) \tag{6}$$

where T is blade temperature, RH is the relative humidity of the atmosphere, 0.0611 is the saturated vapor pressure of the horizontal plane when T = 0 °C, e is the actual vapor pressure, and E is the saturated vapor pressure at the same temperature.

$$WUE = \frac{DW \times CC}{WU} \tag{7}$$

At the same time, WUE is the ratio between the total amount of carbon assimilated by plants and water use (WU, kg·m^{-2}) over a period of time, DW (g) is dry weight biomass of each organ, and CC (mg·g^{-1}) is carbon content.

2.3.5. Measurement of Physiological Indices

(1) Plant height: During the summer maize 6-leaf stage, 12-leaf stage, the silking stage, the filling stage, and the mature stage, three plants were selected at each stage for each treatment, and the plant height (cm) was measured with a ruler (1 mm).

(2) Dry mass: During the summer maize 6-leaf stage, 12-leaf stage, the silking stage, the filling stage, and the mature stage, three plants were selected at each stage for each treatment. The aboveground parts, including the leaves, stems, and ears, were dried at 105 °C initially, then dried at 80 °C until they reached a consistent weight.

2.4. Dual Crop Coefficient Approach

We calculated the evapotranspiration of summer maize in the experimental plots using the dual crop coefficient approach, according to the FAO-56 Equation as follows:

$$ET_c = (K_s K_{cb} + K_e) \ ET_0 \tag{8}$$

where ET_c (mm) is the actual crop evapotranspiration, ET_0 (mm) is the reference crop evapotranspiration that was calculated from the Penman-Monteith equation based on meteorological data [2], K_{cb} (-) is basal crop coefficient, and K_s (-) is the water stress coefficient. Both K_{cb} and K_s were calculated based on the K_e [2]. K_e (-) is the soil evaporation coefficient that was used to describe the soil evaporation component of crop evapotranspiration. Miao and Wen et al. [28,29] calculated K_e [2], which is generally calculated as

$$K_e = K_r \left(K_{c(\max)} - K_{cb} \right) \leq f_{ew} K_{c(\max)} \tag{9}$$

where $K_{c(\max)}$ (-) is the maximum value of K_c after irrigation or rainfall, K_r (-) is the soil attenuation coefficient, and f_{ew} (-) is the percentage of exposed and wet parts of the soil surface. Detailed descriptions of the calculation of the parameters $K_{c(\max)}$ and K_r is available in FAO-56 [2].

This experiment included two treatments: Partial mulch and no mulch. The K_e of the no mulch treatments was calculated by Equation (9). The calculation of the soil evaporation coefficient for the partial mulch treatments consisted of two parts: Membrane hole evaporation and bare soil evaporation [29]. Based on the ratio of the area between covered and the bare soils in the partial mulch treatments, K_e was calculated as:

$$K_e = \frac{3}{4}K_{e_1} + \frac{1}{4}K_{e_2} \tag{10}$$

where K_{e_1} (-) and K_{e_2} are the membrane pore evaporation coefficient and the bare soil evaporation coefficient, respectively, which were calculated from Equation (9).

The f_{ew} [2] under the drip irrigation condition was calculated as follows:

$$f_{ew} = \min\left(1 - f_c, \left(1 - \frac{2}{3}f_c\right)f_w\right) \tag{11}$$

where f_c (-) is the proportion of vegetation cover to the surface soil area and f_w is the moisture ratio at the soil surface.

In the calculation of K_{e_1}, the f_w [29] in Equation (11) was calculated as:

$$f_w = \alpha N A_h / A_{total} \tag{12}$$

where α (-) is the membrane effective area factor, N (-) is the number of membrane holes, A_h (m^2) is the area of a single membrane hole, and A_{total} (m^2) is the total area of the membrane.

In the calculation of K_{e_2}, f_w was set to 1.

2.5. Statistics

The regression coefficient (b) [30], coefficient of agreement (d) [30], and root mean square error (RMSE) [30] were used to evaluate the model's applicability. They were calculated as follows:

(1) The regression coefficient through the origin (b)

$$b = \frac{\sum\limits_{i=1}^{n} O_i \times P_i}{\sum\limits_{i=1}^{n} O_i^2} \tag{13}$$

(2) The coefficient of agreement (d)

$$d = 1 - \frac{\sum\limits_{i=1}^{n} (O_i - P_i)^2}{\sum\limits_{i=1}^{n} \left(\left|P_i - \overline{O}\right| + \left|O_i - \overline{O}\right|\right)^2} \tag{14}$$

(3) The root mean square error (RMSE)

$$RMSE = \sqrt{\frac{1}{m}\sum\limits_{i=1}^{m} (O_i - P_i)^2} \tag{15}$$

where O_i and P_i are the measured and estimated value of i, respectively, and $\overline{O}$ is the average of O_i (I = 1, 2, ... , n).

The closer the values of b and d were to 1.0, the higher was the agreement between the estimates and the measurements. A lower RMSE value indicated a better fit.

3. Results

3.1. Parameterization of the Dual Crop Coefficient Approach

Using the parameters in Table 3, the dual crop coefficient approach was used to estimate the evapotranspiration of the partial mulched summer maize in 2016 and the non-mulched summer maize in 2017. Under mulching, the comparisons between evapotranspiration were estimated by the model and the practical crop evapotranspiration was calculated by the water balance method for the five treatments (Figure 3a). Under the non-mulched condition, the comparisons between the estimated evapotranspiration of W1, W3, and W5 by the model, and the practical crop evapotranspiration calculated by the water balance method, are shown in Figure 3b. The comparisons between measured and estimated ET are shown in Table 4.

Table 3. Parameters for the dual crop coefficient approach that was used to estimate evapotranspiration of summer maize on plots in Xinjiang, China.

Variables	Parameters	Range	Value	Sources
	$Kcb_{(ini)}$	0.15	0.15/0.15	
Basal crop coefficient (with/without mulch)	$Kcb_{(mid)}$	1.15	1.20/1.12	calibration
	$Kcb_{(end)}$	0.50	0.87/0.50	
Depth of the surface soil layer	Ze	0–0.15	0.10	FAO-56 [2]
Readily evaporable water	REW	8–11	9	calibration
Total evaporable water	TEW	18–25	20	calibration
Effective area coefficient of membrane hole	α	2–8	6	Wen et al. [29]

$Kcb_{(ini)}$, $Kcb_{(mid)}$, and $Kcb_{(end)}$ are basal crop coefficients in initial, middle, and late stages.

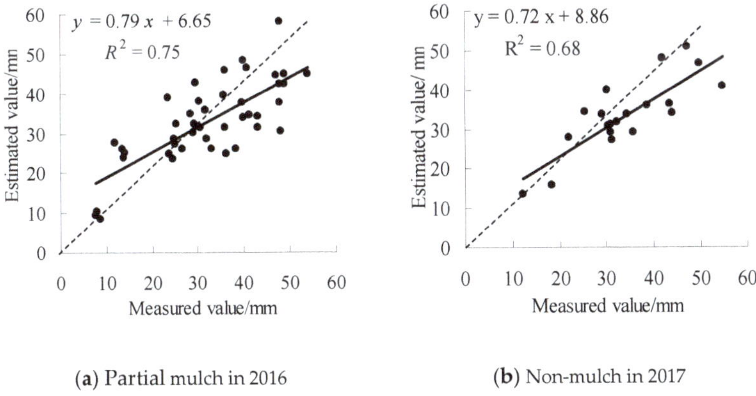

(**a**) Partial mulch in 2016   (**b**) Non-mulch in 2017

Figure 3. Regression between observed and estimated evapotranspiration in 2016 and 2017 on partial mulched and non-mulched plots in Xinjiang, China. (**a**) Partial mulch in 2016; (**b**) non-mulch in 2017.

The measured and estimated ET values under the partial mulching treatment in 2016 and the non-mulching treatment in 2017 were close to the 1:1 line, with determination coefficients (R^2) of 0.75 and 0.68, respectively (Figure 3a,b). The RMSE of the estimated ET values of mulched summer maize for the five treatments in 2016 ranged from 6.95 to 10.66 mm, and the regression coefficient (b) varied from 0.91 to 1.06 (Table 4). The concordance index (d) varied from 0.98 to 0.99. In 2017, the RMSE of estimated ET values of non-mulched summer maize for the three treatments (W1, W3, and W5) were similar to those of the mulched treatments and ranged from 7.71 to 10.43 mm. The regression coefficient (b) varied from 0.99 to 1.18, and (d) varied from 0.90 to 0.96, which indicated that there was a good fit between measurements and estimates.

Table 4. Statistics of observed and estimated evapotranspiration in 2016 and 2017 on partial mulched and non-mulched plots in Xinjiang, China.

Year	Treatment	RMSE/mm	b	d
2016 With Partial Mulch	W1	8.12	0.93	0.99
	W2	7.52	0.91	0.99
	W3	7.59	0.92	0.99
	W4	6.95	0.99	0.99
	W5	10.66	1.06	0.98
2017 Without Mulch	W1	7.71	0.99	0.96
	W3	10.94	1.18	0.90
	W5	10.43	0.99	0.92

3.2. Model Evaluation

3.2.1. Model Evaluation Based on the Water Balance Method

The calibrated model parameters were put into the dual crop coefficient model to calculate crop ET for five treatments for different irrigation periods in 2017. We compared the model outputs with the measured ET. Figure 4a showed the relationship between the measured and estimated ET values of the five treatments in the mulched summer maize in 2017. Figure 4b demonstrated the relationship between the measured and estimated ET values of the two treatments (W2 and W4) in the non-mulched summer maize in 2017. The statistics are shown in Table 5.

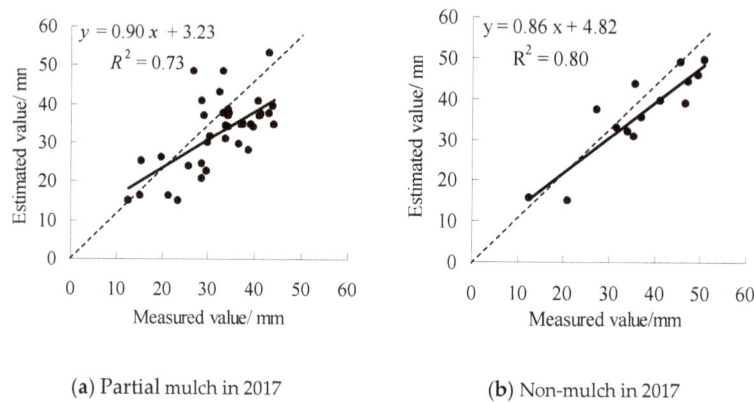

(**a**) Partial mulch in 2017   (**b**) Non-mulch in 2017

Figure 4. Regression between observed and estimated evapotranspiration in 2017 on partial mulched and non-mulched plots in Xinjiang, China. (**a**) Partial mulch in 2017; (**b**) Non-mulch in 2017.

Compared with the measurements from the water balance method, the model accurately estimated ET of summer maize during the growing period in the experimental region. The measured and estimated ET values of the mulched summer maize in the growing season were close to the 1:1 line, with an R^2 of 0.73 (Figure 4a). The RMSE of the measured and estimated ET values in each treatment during the growing season ranged from 4.59 to 12.56 mm, with (b) of 1.03–1.12, and (d) of 0.86–1.00. The measured and estimated ET values of the non-mulched summer maize in the growing season were also close to the 1:1 line, with an R^2 of 0.80 (Figure 4b). The RMSE of the measured and estimated ET values of W2 and W4 treatments were 7.84 mm and 6.88 mm, with (b) of 1.04 and 0.98, and (d) of 0.96 and 0.95, respectively. Our results suggested that there were good agreements between predicted and measured ET for summer maize during the growing season after parameterizing the model correctly.

Table 5. Statistics of observed and estimated evapotranspiration in 2017 on partial mulched and non-mulched plots in Xinjiang, China.

Year	Treatment	RMSE/mm	b	d
2017 With Partial Mulch	W1	4.59	1.03	0.99
	W2	11.37	1.06	0.89
	W3	12.56	1.06	0.86
	W4	10.15	1.12	0.99
	W5	11.75	1.03	1.00
2017 Without Mulch	W2	7.84	1.04	0.96
	W4	6.88	0.98	0.95

3.2.2. Model evaluation based on stable carbon isotope techniques

Figure 5a showed the partial mulched regression between the measured and simulated crop transpiration (T) of summer maize with five treatments in 2017 using stable carbon isotope method. Since only non-mulched treatments, i.e., W1, W2 and W5, were used the stable carbon isotope technique to calculate crop transpiration, Figure 5b showed the regression between the measured and estimated crop T of the three treatments in summer maize during each growth periods. The statistics are shown in Table 6.

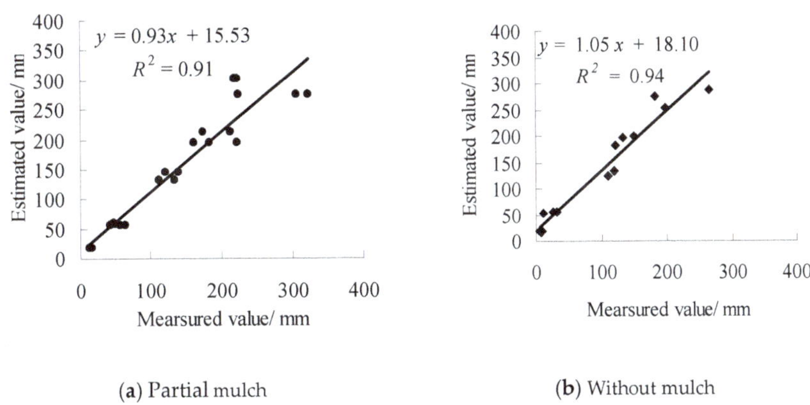

(**a**) Partial mulch   (**b**) Without mulch

Figure 5. Regressions between observed and estimated transpiration in 2017 on mulched and non-mulched plots in Xinjiang, China. (**a**) Partial mulch; (**b**) without mulch.

Table 6. Statistics of observed and estimated transpiration for mulched and non-mulched plots in 2017 in Xinjiang, China.

Treatment	Partial Mulch	RMSE/mm	b	d
W1	Yes	17.83	1.14	0.98
	No	26.48	1.29	0.94
W2	Yes	22.62	1.23	0.97
	No	23.07	1.22	0.95
W3	Yes	14.21	1.14	0.98
W4	Yes	18.02	0.91	0.98
W5	Yes	25.44	0.82	0.97
	No	13.05	1.05	0.99

Compared with the measurements from stable carbon isotopes, the model accurately estimated summer maize crop transpiration (T) during the growing season. There was good agreement between the predicted and measured crop T during the growing season (Figure 5, Table 6). For the partial mulched and non-mulched treatments, the determination coefficients (R^2) were 0.91 and 0.94, respectively. The RMSE was in the range of 14.21–25.44 mm and 13.05–26.48 mm, with a coefficient (b) of 0.82–1.23 and 1.05–1.29, and (d) of 0.97–0.98 and 0.94–0.99 for partial mulched and non-mulched plots, respectively. The evaluations of the model suggested that after the model was calibrated, it accurately estimated summer maize change in T during the growing season. This further shows the effectiveness of the stable carbon isotope technique in quantifying the T of summer maize in arid regions.

3.3. Evapotranspiration Dynamics of Summer Maize

The measurements of five treatments under mulching conditions in 2016 suggested that the dual crop coefficient approach underestimated ET in the rapid growth and late growth stages of summer maize, but overestimated ET in the late growth stage (Table 7). The reason may be that the pump failure on 1 September 2017 led to early water emergence of summer maize, which resulted in a slight deviation between the measured ET and the simulated ET value. The highest water consumption of summer maize was in the middle growth stage, which accounted for about 40% of the total ET over the entire growing season. The second highest water consumption was during the rapid growth stage, which accounted for about 35% of the total ET over the entire growing season. Under the non-mulched condition, water consumption of summer maize during the rapid growth phase was larger than during other periods, which accounted for about 45% of the total ET, followed by the middle growth phase, where water consumption accounted for about 35% of the total ET. Compared with the two methods of partial mulch or no mulch, the proportion of ET differed in the different time periods. The reason is that summer maize under the mulching condition entered the middle growth stage earlier, which shortened the rapid growth stage and extended the crops middle growth stage. This conclusion is similar to that of Wen et al. [29].

Table 7. Values of observed and estimated evapotranspiration at different growth stages of summer maize on partially mulched and non-mulched plots in Xinjiang, China.

Year	Growth Stage	W1 Observations (O)/mm	W1 Estimations (E)/mm	W2 O/mm	W2 E/mm	W3 O/mm	W3 E/mm	W4 O/mm	W4 E/mm	W5 O/mm	W5 E/mm
2016 With PartialMulch	Initial	60.0	69.3	58.0	63.7	63.0	58.9	56.0	62.2	55.0	58.9
	Rapid	165.5	143.0	149.2	130.0	150.8	118.0	135.6	124.9	132.3	119.4
	Middle	183.4	219.2	176.4	199.2	152.0	179.2	167.8	188.9	129.6	179.2
	Late	58.4	31.2	50.5	28.9	39.9	25.5	40.8	26.9	67.6	25.5
	Whole	467.3	462.7	434.1	421.8	405.6	381.7	400.2	403.0	384.5	383.1
2017 With PartialMulch	Initial	69.5	70.0	62.6	67.1	55.7	63.0	56.0	68.5	65.6	63.0
	Rapid	129.5	132.5	146.0	131.2	146.1	120.5	132.5	123.7	116.6	120.5
	Middle	176.2	174.9	152.4	174.8	119.7	159.1	146.5	159.5	140.1	159.5
	Late	14.3	13.3	8.3	13.2	8.9	12.0	11.1	12.1	6.8	12.1
	Whole	389.6	390.7	369.2	386.3	330.4	354.7	346.1	363.7	329.0	355.0
2017 Without Mulch	Initial	81.5	68.4	81.2	67.6	63.0	66.2	71.0	53.6	76.2	53.6
	Rapid	191.9	192.9	169.3	186.6	154.1	175.9	164.8	169.1	143.9	173.2
	Middle	137.6	141.6	122.7	134.6	105.4	121.4	115.3	138.8	140.0	149.3
	Late	26.1	14.1	18.7	13.3	14.0	12.6	24.8	12.6	16.9	20.2
	Whole	437.1	417.0	392.0	402.1	336.5	376.0	375.9	374.1	377.0	396.3

3.4. Partitioning of Evapotranspiration of Summer Maize

The temporal patterns of ET were similar under different treatments with the condition of partial mulched. For example, the estimated ET for W1 had similar patterns under partial mulching during the two growing seasons (Figure 6a,b). In the initial growth stage, ET values were small. As the crops grew, ET gradually increased in the middle growth stage and decreased gradually in the late stage.

During the two growing seasons, the pattern of estimated T was consistent with ET. It was small at the initial stage, increased during the development stage, reached a maximum during the middle stage, and then decreased during the late stage. However, the E dynamics were the opposite of T; it was large at the initial stage, and decreased gradually as the crop grew during the middle and late stages. The dynamics of E changed little in the late growth stage compared with the middle growth stage.

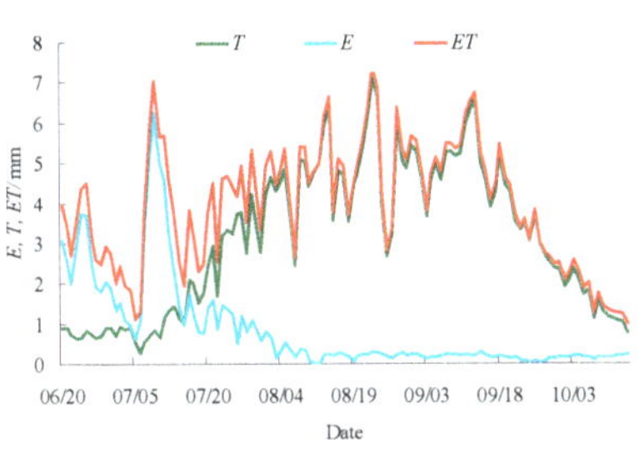

(**a**) 2016—with partial mulch

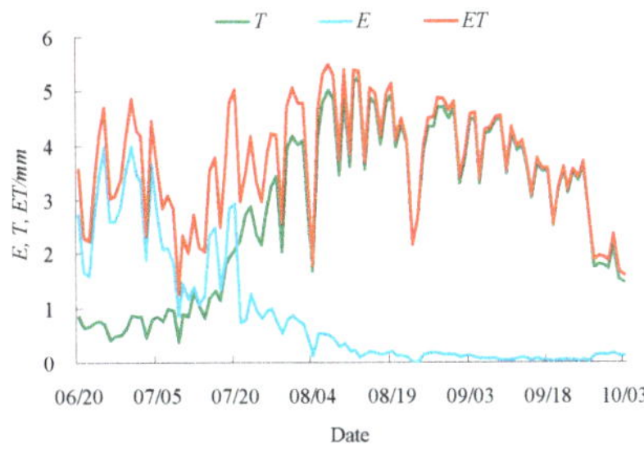

(**b**) 2017—with partial mulch

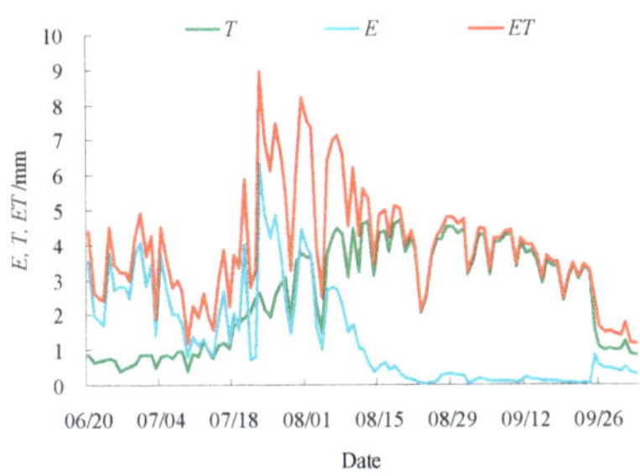

(**c**) 2017—with no mulch

Figure 6. Seasonal variations in evapotranspiration (ET), evaporation (E), and transpiration (T) of summer maize for treatment W1 during the growing season on partially mulched and non-mulched plots in Xinjiang, China. (**a**) 2016—with partial mulch; (**b**) 2017—with partial mulch; (**c**) 2017—with no mulch.

ET with no mulching was high during the rapid growth stage because evaporation was high in this period (Figure 6c). T was low at the initial stage. During the rapid growth stage, T gradually increased and exhibited a relatively stable pattern during the middle stage. However, the E dynamics were the opposite of T. The E value in the initial and rapid growth stages were high, and it gradually became lower after the middle stage. The magnitude of the change was relatively stable.

The dual crop coefficient approach was used to estimate the evaporation (E) and leaf transpiration (T) under partially mulched and non-mulched conditions, and it was used to estimate the ratio of leaf transpiration to crop evapotranspiration (T/ET) and the ratio of evaporation to evapotranspiration (E/ET) during a crop's growing season (Table 8). In 2016 and 2017, under the partially mulched condition with summer maize, the estimated T/ET for 2016 and 2017 were 78.7% and 76.0%, respectively, and the E/ET were 21.3% and 24.0%, respectively. In 2017, under the non-mulched condition, the estimated T/ET was 64.89% over the entire growing season, which was about 12% lower than that of the partial mulched treatments. The E/ET was 35.11%, which was about 12% higher than that of the partial mulched treatments. This conclusion is different from the results of Ding et al. [10]. The reason may be that Ding et al. simulated the situation of complete film mulching, but in the current experiment, we used partial film mulching, which resulted in a higher evaporation.

Table 8. Variations in *T, E, T/ET*, and *E/ET* of summer maize on partial mulched and non-mulched plots during different growth periods in Xinjiang, China.

Stage	2016 With Partial Mulch				2017 With Partial Mulch				2017 With no Mulch			
	T/mm	E/mm	T/ET/%	E/ET/%	T/mm	E/mm	T/ET/%	E/ET/%	T/mm	E/mm	T/ET/%	E/ET/%
Initial	15.3	54.0	22.1	77.9	15.0	55.0	21.4	78.6	15.0	53.4	21.9	78.1
Rapid	109.4	33.6	76.5	23.5	99.2	33.4	74.8	25.2	109.6	83.3	56.8	43.2
Middle	211.0	8.2	96.3	3.7	170.6	4.3	97.6	2.4	136.3	5.3	96.2	3.8
Late	28.3	2.9	90.7	9.3	12.2	1.0	92.3	7.7	9.7	4.4	68.7	31.3
Whole	364.0	98.7	78.7	21.3	297.0	93.7	76.0	24.0	270.6	146.4	64.9	35.1

4. Conclusions

(1) The dual crop coefficient approach accurately estimated the ET during different summer maize growth periods. The RMSE was around 10 mm under the partially mulched conditions in 2016 and 2017. The averaged regression coefficient (b) was about 1. The consistency index (d) for 2016 and 2017 was in the range of 0.97–1 and 0.86–1, respectively. In 2017, the RMSE was about 6 mm under non-mulched conditions. The regression coefficient (b) was about 1. The coefficient of agreement (d) was in the range of 0.68–0.97, which was consistent with the observed values.

(2) The dual crop coefficient approach accurately partitioned summer maize evapotranspiration. The estimated transpiration over the entire growing season of summer maize under partially mulched conditions in 2016 and 2017 accounted for 78.7% and 76.0% of ET, respectively, and the evaporation accounted for 21.3% and 24.0% of ET, respectively. In 2017, the estimated transpiration over the entire maize growing season under non-mulched conditions accounted for 64.89% of ET, which was about 12% lower than that of the partially mulched treatments. The evaporation of non-mulched treatments accounted for 35.1% of ET, which was about 12% higher than that of the partially mulched treatments.

(3) In 2017, the average water consumption of summer maize under partially mulched treatments was about 350 mm, but the water consumption under the non-mulched conditions was about 380 mm. The water consumption over the entire growing season for the partially mulched maize was about 30 mm, which was about 10% lower than that of the non-mulched conditions, which suggested that partially mulched treatments can be used to improve water use efficiency.

Author Contributions: Conceptualization, Y.M. and F.L.; Methodology, Y.M.; Software, F.L.; Validation, Y.M. and F.L.; Formal Analysis, F.L.; Investigation, F.L.; Data Curation, F.L.; Writing-Original Draft Preparation, F.L.; Writing-Review & Editing, Y.M. and F.L.; Funding Acquisition, Y.M.

Funding: This research was funded by University Innovation Team Project of Xinjiang, China, grant number XJEDU2017T004 and Water Conservancy Science and Technology Project of Xinjiang, China, grant number 2019G01.

Acknowledgments: Fengxiu, Li acknowledges the support from the University Innovation Team Project of Xinjiang, China (Grant No. XJEDU2017T004) and the Water Conservancy Science and Technology Project of Xinjiang, China (Grant No. 2019G01). We are also grateful for the comments and suggestions of the editors and reviewers.

References

1. Xu, C.Y.; Singh, V.P. Evaluation of three complementary relationship evapotranspiration models by water balance approach to estimate actual regional evapotranspiration in different climatic regions. *J. Hydrol.* **2005**, *308*, 105–121. [CrossRef]

2. ASCE-EWRI. *The ASCE Standardized Reference Evapotranspiration Equation*; Report 0-7844-0805-X, ASCE Task Committee on Standardization of Reference Evapotranspiration; American Society of Civil Engineers: Reston, VA, USA, 2005.

3. Zhang, B.Z.; Liu, Y.; Xu, D.; Zhao, N.; Lei, B.; Rosa, R.D.; Paredes, P.; Paço, T.A.; Pereira, L.S. The dual crop coefficient approach to estimate and partitioning evapotranspiration of the winter wheat-summer maize crop sequence in North China Plain. *Irrig. Sci.* **2013**, *31*, 1303–1316. [CrossRef]

4. Bodner, G.; Loiskandl, W.; Kaul, H.P. Cover crop evapotranspiration under semi-arid conditions using FAO dual crop coefficient method with water stress compensation. *Agric. Water Manag.* **2007**, *93*, 85–98. [CrossRef]

5. López-Urrea, R.; Martínez-Molina, L.; de la Cruz, F.; Montoro, A.; González-Piqueras, J.; Odi-Lara, M.; Sánchez, J.M. Evapotranspiration and crop coefficients of irrigated biomass sorghum for energy production. *Irrig. Sci.* **2016**, *34*, 287–296. [CrossRef]

6. Liu, Y.; Luo, Y. A consolidated evaluation of the FAO-56 dual crop coefficient approach using the lysimeter data in the North China Plain. *Agric. Water Manag.* **2010**, *97*, 31–40. [CrossRef]

7. Zhao, N.; Liu, Y.; Cai, J.; Paredes, P.; Rosa, R.D.; Pereira, L.S. Dual crop coefficient modelling applied to the winter wheat–summer maize crop sequence in North China Plain: Basal crop coefficients and soil evaporation component. *Agric. Water Manag.* **2013**, *117*, 93–105. [CrossRef]

8. Shrestha, N.K.; Shukla, S. Basal crop coefficients for vine and erect crops with plastic mulch in a sub-tropical region. *Agric. Water Manag.* **2014**, *143*, 29–37. [CrossRef]

9. Ding, R.; Kang, S.; Zhang, Y.; Hao, X.; Tong, L.; Du, T. Partitioning evapotranspiration into soil evaporation and transpiration using a modified dual crop coefficient model in irrigated maize field with ground-mulching. *Agric. Water Manag.* **2013**, *127*, 85–96. [CrossRef]

10. Tomomichi, K.; Makio, K. Determination of a crop coefficient for evapotranspiration in a sparse sorghum field. *Irrig. Drain.* **2006**, *55*, 165–175.

11. Niaghi, A.R.; Vand, R.H.; Asadi, E.; Majnooni-Heris, A. Evaluation of single and dual crop coefficient methods for estimation of wheat and maize evapotranspiration. *Adv. Environ. Biol.* **2015**, *9*, 963–971.

12. Farquhar, G.D.; Ehleringer, J.R.; Hubick, K.T. Carbon isotope discrimination and photosynthesis. *Annu. Rev. Plant Physiol. Plant Mol. Biol.* **1989**, *40*, 503–537. [CrossRef]

13. Yan, C.R.; Han, X.G.; Chen, L.Z. Water use efficiency of six woody species in relation to micro-environmental factors of different habitats. *Acta Ecol. Sin.* **2001**, *21*, 1952–1956.

14. Farquhar, G.D.; O'leary, M.H.; Berry, J.A. On the relationship between carbon isotope discrimination and the intercellular carbon dioxide concentration in leaves. *Aust. J. Plant Physiol.* **1982**, *9*, 121–137. [CrossRef]

15. Anyia, A.O.; Slaski, J.J.; Nyachiro, J.M.; Archambault, D.J.; Juskiw, P. Relationship of carbon isotope discrimination to water use efficiency and productivity of barley under field and greenhouse conditions. *J. Agron. Crop Sci.* **2007**, *193*, 313–323. [CrossRef]

16. Chen, J.; Chang, S.X.; Anyia, A.O. The physiology and stability of leaf carbon isotope discrimination as a measure of water-use efficiency in barley on the Canadian prairies. *J. Agron. Crop Sci.* **2011**, *197*, 1–11. [CrossRef]

17. Chen, P.; Meng, P.; Zhang, J.; He, C.; Jia, C.; Li, J. Water use of intercropping system of tree and two herbal medicine in the low hilly area of north China. *J. Northeast Univ.* **2014**, *42*, 52–56, 78. (In Chinese)

18. He, C.X.; Meng, P.; Zhang, J.S.; Gao, J.; Sun, S.J. Water use of walnut-wheat intercropping system based on stable carbon isotope technique in the low hilly area of North China. *Acta Ecol. Sin.* **2012**, *32*, 2047–2055. (In Chinese)

19. Liang, W.Q. *Study on Measuring Crop Evapotranspiration and Crop Coefficients Changes for Winter Wheat and Maize*; North West A&F University: Xianyang, China, 2012. (In Chinese)

20. Kang, S.; Cai, H.; Zhang, J. Estimation of maize evapotranspiration under water deficits in a Semiarid Region. *Agric. Water Manag.* **2000**, *43*, 1–14.

21. Smith, D.M.; Allen, S.J. Measurement of sap flow in plant stems. *J. Exp. Bot.* **1996**, *47*, 1833–1844. [CrossRef]

22. Köstner, B.; Granier, A.; Cermák, J. Sap flow measurements in forest stands: Methods and uncertainties. *Annu. Sci.* **1998**, *55*, 13–27. [CrossRef]

23. Wullschleger, S.D.; Meinzer, F.C.; Vertessy, R.A. A review of whole-plant water use studies in trees. *Tree Physiol.* **1998**, *18*, 499–512. [CrossRef]

24. Farquhar, G.D.; Richards, R.A. Isotopic composition of plant carbon correlates with water-use efficiency of wheat genotypes. *Aust. J. Plant Physiol.* **1984**, *11*, 539–552. [CrossRef]

25. Evans, J.R. Nitrogen and photosynthesis in the flag leaf of wheat (*Triticum aestivum* L.). *Plant Physiol.* **1983**, *72*, 297–302. [CrossRef]

26. Abbate, P.E.; Dardanelli, J.L.; Cantarero, M.G.; Maturano, M.; Melchiori, R.J.M.; Suero, E.E. Climatic and water availability effects on water-use efficiency in wheat. *Crop Sci.* **2004**, *44*, 474–484. [CrossRef]

27. Hu, J.; Moore, D.J.; Riveros-Iregui, D.A.; Burns, S.P.; Monson, R.K. Modeling whole tree carbon assimilation rate using observed transpiration rates and needle sugar carbon isotope ratios. *New Phytol.* **2010**, *185*, 1000–1015. [CrossRef]

28. Xu, D.; Mermoud, A. Modeling water use, transpiration and soil evaporation of spring wheat–maize and spring wheat–sunflower relay intercropping using the dual crop coefficient approach. *Agric. Water Manag.* **2016**, *165*, 211–229.

29. Wen, Y.Q.; Yang, J.; Shang, S.H. Analysis on evapotranspiration and water balance of cropland with plastic mulch in arid region using dual crop coefficient approach. *Trans. Chin. Soc. Agric. Eng.* **2017**, *33*, 138–147. (In Chinese)

30. Yu, C.; Chao, L.B.G.; Gao, R.; Zhu, Z. Validity examination of simulated results of crop water requirements. *Trans. Csae* **2009**, *25*, 13–21. (In Chinese)

Article

Pump-as-Turbine Selection Methodology for Energy Recovery in Irrigation Networks: Minimising the Payback Period

Miguel Crespo Chacón [1], Juan Antonio Rodríguez Díaz [2], Jorge García Morillo [2] and Aonghus McNabola [1,*]

1 Department of Civil, Structural and Environmental Engineering, Trinity College Dublin, Dublin D02 PN40, Ireland; crespocm@tcd.ie
2 Department of Agronomy, University of Córdoba, International Campus of Excellence ceiA3, 14071 Córdoba, Spain; jarodriguez@uco.es (J.A.R.D.); jgmorillo@uco.es (J.G.M.)
* Correspondence: amcnabol@tcd.ie; Tel.: +353-1896-3837

Received: 9 December 2018; Accepted: 11 January 2019; Published: 16 January 2019

Abstract: In pressurized irrigation networks, energy reaches around 40% of the total water costs. Pump-as-Turbines (PATs) are a cost-effective technology for energy recovery, although they can present low efficiencies when operating outside of the best efficiency point (BEP). Flow fluctations are very important in on-demand irrigation networks. This makes flow prediction and the selection of the optimal PAT more complex. In this research, an advanced statistical methodology was developed, which predicts the monthly flow fluctuations and the duration of each flow value. This was used to estimate the monthly time for which a PAT would work under BEP conditions and the time for which it would work with lower efficiencies. In addition, the optimal PAT power for each Excess Pressure Point (EPP) studied was determined following the strategy of minimising the PAT investment payback period (PP). The methodology was tested in Sector VII of the right bank of the Bembézar River (BMD), in Southern Spain. Five potential sites for PAT installation were found. The results showed a potential energy recovery of 93.9 MWh and an annual energy index per irrigated surface area of 0.10 MWh year^{-1} ha^{-1}. Renewable energy will become increasingly important in the agriculture sector, to reduce both water costs and the contribution to climate change. PATs represent an attractive technology that can help achieve such goals.

Keywords: hydropower; irrigation networks; combinatorial analysis; statistical analysis; pump-as-turbine; payback period

1. Introduction

The energy consumption embodied in pressurized water networks represents around 2–3% of global energy consumption [1]. Furthermore, it has been estimated that the energy consumption associated with the water sector will increase by 23% in 2020 and by 63% in 2050, above the 2000 levels in the USA [2]. The energy dependency of the water sector is reflected in the cost percentage represented by production and supply, which have risen to 80% of the operating budget [3]. In pressurized irrigation networks, energy reaches around 40% of the total water costs, and therefore water and energy efficiency cannot be considered independently in agriculture [4]. Several measures have been studied to reduce energy consumption in irrigation networks, such as critical hydrant detection [4], irrigation sectoring [5], the optimization of energy consumption in pumped systems [6–9], or solar energy production [10].

To counteract this rising energy dependency, several authors have also highlighted the potential for energy recovery from urban water supply networks using micro-hydropower (MHP) turbines at points of excess pressure. Water networks are commonly sub-optimal in terms of their use of energy and water resources, due to changes in elevation, demand, water pressure and leakage rates across many kilometres of pipelines. Recent research has studied the application of MHP turbines in water supply and wastewater infrastructure to reduce pressure to desired levels and recover energy in the form of electricity [11–14]. In pressurized irrigation networks, the irrigation devices (drippers, sprinklers) continue to evolve towards greater efficiency in the consumption of water and energy. This results in a lower working pressure requirement in some areas of an irrigation network, triggering the potential for available energy recovery. In addition, due to changes in elevation and demand across a typical irrigation network, areas of excess pressure are unavoidable, unless a network is situated in an area with uniform gradient and demand distribution.

The use of pump-as-turbines (PATs) for energy recovery have been shown to be cost-effective—potentially just 10% of the cost of conventional MHP turbines [15–19] at sites with small power output capacity [20,21]. Nonetheless, PATs have the disadvantage of relatively low efficiency, which can reduce further with large flow fluctuations. It has been shown that the efficiency of the PAT can reduce to approximately 70% of the maximum efficiency when the flow is 20% below the Best Efficiency Point (BEP) flow rate [21]. Different investigations have also analysed the use of PATs in irrigation networks for energy recovery. Nonetheless, the flow fluctuations in the irrigation sector are more pronounced since the demand will depend on the irrigation requirements of the crops cultivated and the yearly agronomic parameters, such as rainfall and potential evapotranspiration. The water demand is also concentrated in just a few months of the year in certain cases. Some previous research has analysed the potential energy recovery available, considering average flows and pressure in irrigation networks, along with the most probable exceedance flow and average head [22]. The maximum potential found was 270.5 MWh in a total surface of 4000 ha. However, flow fluctuations were not taken into account to quantify the energy recovered. Previous investigators have also used the farmers' habits, historical records and the characteristics of the network to estimate the flow over time in any line of the network and considering fixed turbine parameters [23]. A total energy of 188.23 MWh was estimated to be recovered in 290.2 ha, where flow fluctuations were considered but PAT performance was assumed to be constant [23]. Additional research in this field was focused on maximising energy recovery through different objective functions such as: selection of best energy converter, operation in best efficiency conditions, varying the rotational speed, and providing the required flow in each situation. The maximum energy recovered was estimated as 58.18 MWh year^{-1}, increasing the energy recovery by 141% and 184% when comparing to constant rotational speed PATs [24]. Finally, the use of PATs for energy recovery in irrigation networks was also studied in an area of 68 ha in Portugal. The potential estimated to be recovered was 2.12 MWh [25].

The fluctuations in irrigation water demand are particularly important in on-demand irrigation networks. This kind of infrastructure allows greater flexibility to the farmers since water is available at any time every day and year, and the flow circulating at any point of the network depends on the number of downstream hydrants that are open [26]. Therefore, depending on the combination of open and closed hydrants, the flow and head at an issue point vary greatly. When analysing MHP installations, these variations will directly affect the energy recovery.

To characterise a network and the different monthly flow values, statistical methods are commonly used based on the probability of each hydrant being open or closed. Several methods have been used to calculate the monthly open hydrant probability. Rodriguez-Diaz et al. [26] stated that the gamma function adjusts better to this demand than other distributions, but local farmers' practices and the desired constraints of the network have to be taken into account. However, Clément's methodology [27] requires fewer initial data and several previous investigators concluded that the methodology provides good approximate values that can be used to design on-demand irrigation networks [26,28].

In this paper, an advanced statistical methodology is developed to determine the power available for energy recovery through radial PATs in on-demand irrigation networks. The methodology is applied to the common scenario where no flow data are recorded in the irrigation network, and seeks to minimise the PAT investment payback period. The methodology is developed and applied in a real case study in Southern Spain. This methodology uses statistical methods to estimate the variability of flows and heads during the irrigation season. It also provides a useful tool to select the PAT with the lowest payback period for pre-selected locations.

2. Materials and Methods

2.1. Methodology

The proposed methodology is based on the characterisation of the monthly behaviour of the network through a statistical experiment known as a Bernoulli Experiment. The experimental results define the value domain of the flow, considered a random variable Q, and their occurrence probabilities each month. The objective was to determine the PAT power for each selected Excess Pressure Point (EPP), while minimising the PAT installations payback period (PP). Experimental curves approximating the head recovered and the relative PAT efficiency, both depending on the flow rate together with the flow–head (Q–H) curve of the system, were used to estimate the power ranges and energy recovered. The methodology was defined as a general strategy for reducing the investment risks for PAT installation in irrigation networks. The methodology diagram can be seen in Figure 1 and is divided into five main steps, explained below:

2.1.1. Location of Excess Pressure Points and Calculation of Downstream Open/Closed Hydrants Combinations

The first stage in Figure 1 was to simulate the network's hydraulic performance and find the excess pressure points along it, considering a pre-set hypothesis, such as design hypothesis or 100% of hydrants open. Within this first step, the next boundary condition was applied: BEP head (H_{BEP}) is equal to the head available for each EPP in the first simulation under the hypothesis used, ensuring no lack of pressure in any scenario. H_{BEP} had the same value for every scenario analysed within each EPP. Novara et al. [19] presented a Q-H space to locate the BEP conditions for a large set of PATs, showing several points where the head could reach up to 3 m for certain flows. Considering this space, the minimum head (excess pressure) for a point to be evaluated as an EPP was fixed at 3 m above the service pressure.

The flow fluctuations depended on the crop irrigated by each hydrant and their water requirements along the irrigation season. These fluctuations defined the values of the domain of the random variable Q, and was analysed through a Bernoulli Experiment. Hence, in each EPP, the range of possible values for Q was determined depending on the amount of possible combinations of downstream open/closed hydrants. Supposing a number of hydrants n, the number of possible combinations C, was calculated as defined by Equations (1) and (2), for a random combination of open hydrants a, with $0 \leq a \leq n$. Finally, the Q-H curve for each branch to be analysed, was obtained from the hydraulic model. In the scheme shown in Figure 2, a random combination a for an EPP is represented.

$$C = \binom{n}{0} + \binom{n}{1} + \binom{n}{2} \ldots + \binom{n}{n} \tag{1}$$

$$C_a^n = \binom{n}{a} = \frac{n!}{a!(n-a)!} \tag{2}$$

2.1.2. Open Hydrant Probability Calculation

This step aimed to calculate the monthly probability of each hydrant to be open. To obtain these probabilities, the formula proposed by Clément [27] was used. The distribution of crops irrigated by each hydrant and their monthly irrigation requirements were needed. For this, the method proposed by Allen et al. [29] was used through the CROPWAT software. Hence, the monthly water requirements matrix, IN_{ij} (l ha^{-1} month^{-1}), was obtained, with i referring to the hydrant and j to the month.

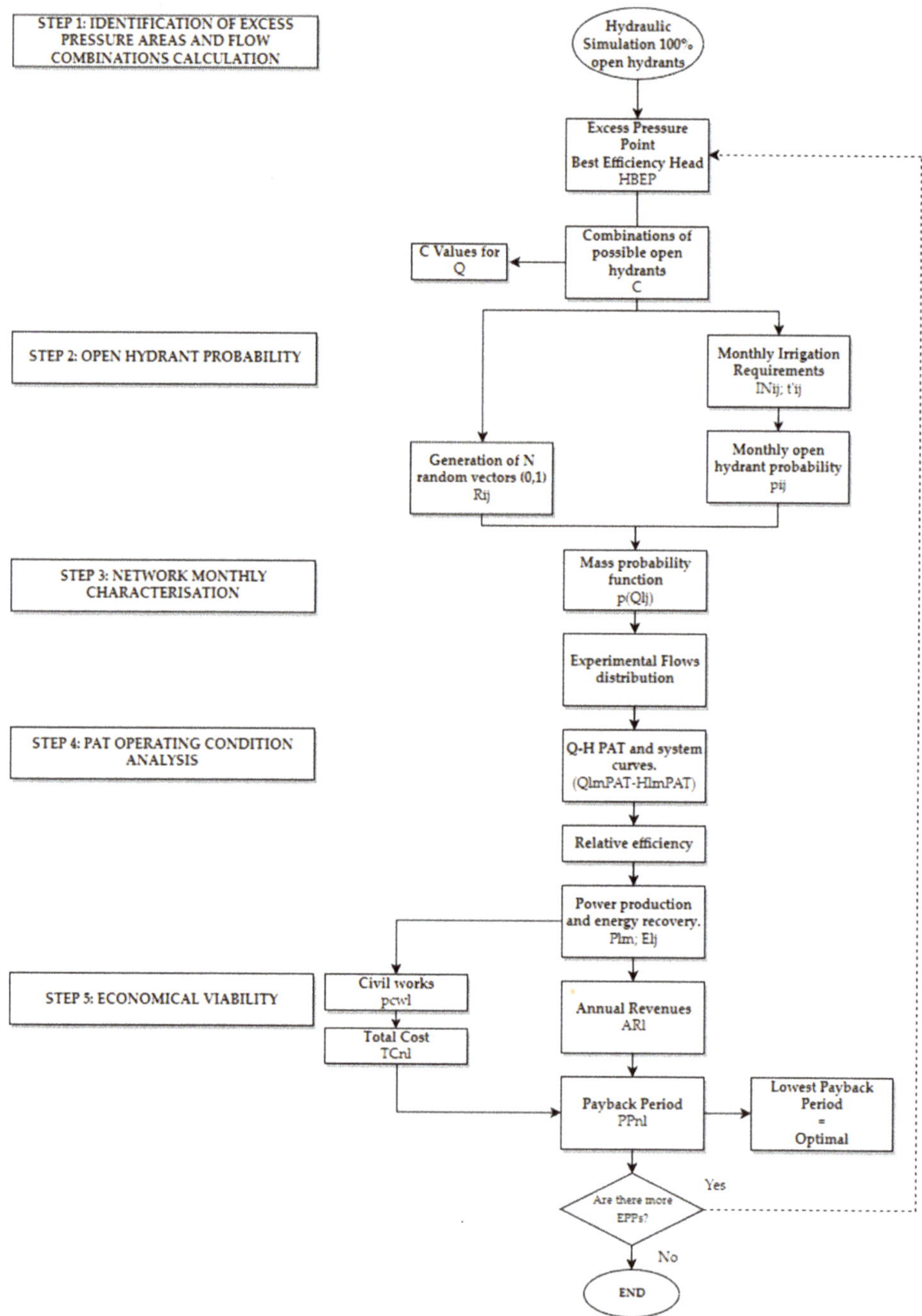

Figure 1. Flowchart of the methodology.

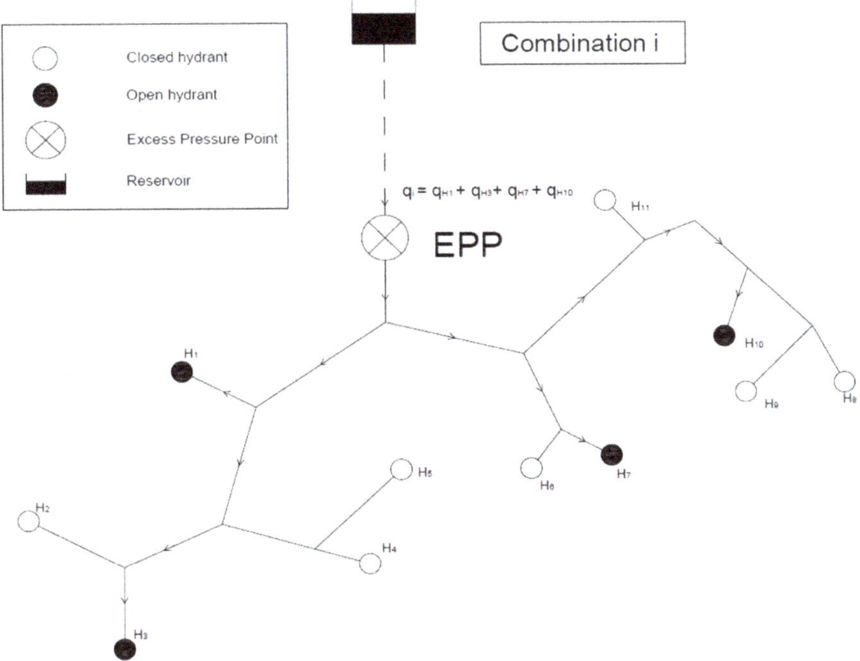

Figure 2. Random combination of downstream open hydrant for a general EPP.

Clément defined that one hydrant has two possible working states: open, with a probability of p, and closed, with a probability of *1-p* [27,30]. Thus, the monthly probability of an open hydrant (p_{ij}), defined in Equation (3), was estimated as the relationship between monthly irrigation hours required by the crops, associated to each hydrant, t'_{ij} (hours month^{-1}), and the monthly water availability, T'_{ij} (hours month^{-1}) for each hydrant i in each month j. These were calculated following Equations (4) and (5), respectively. Finally, *hours$_i$* refers to the daily water availability (hours) per hydrant and *days$_j$* (days month^{-1}) to the number of days in the month j. q_{max} is the design flow allowed per unit of irrigated area.

$$p_{ij} = \frac{t'_{ij}}{T'_{ij}} \tag{3}$$

$$t'_{ij} = \frac{1}{3600} \frac{IN_{ij}}{q_{max}} \tag{4}$$

$$T'_{ij} = hours_i\, days_j \tag{5}$$

2.1.3. Monthly Characterisation of the Network: Mass Probability Function, $p_X(x)$ Calculation

The Bernoulli Experiment involved repeated independent trials of an experiment, called Bernoulli Trials (BTs), with two possible outcomes, arbitrarily called success (S) and failure (F) [31]. Knowing that the trials are independent and assigning the value 1 to S and 0 to F, the combinations of open and closed hydrants downstream of the EPPs were obtained, depending on the results of the trials. Therefore, every BT had two possible outcomes, $X = 1$ is understood as success, and the issue hydrant is open. On the other hand, if the result was $X = 0$, then the result is failure and the issue hydrant is closed. Depending on the number of possible downstream open hydrant combinations (C), a number of BTs, N, is defined, since the greater the number of hydrants the greater the number of combinations, and so the greater the domain of Q. N will be at least double the number of combinations, in order to obtain every possible combination. Thus, every BT consisted of the generation of N random vectors, R_i, with values between [0, 1], and its comparison with monthly probability of each hydrant to be open. The results obtained in each BT followed Equations (6) and (7):

$$If\ R_i >\ p_{ij} \rightarrow X = 0 \tag{6}$$

$$If\ R_i \leq\ p_{ij} \rightarrow X = 1. \tag{7}$$

The aim of the BTs was to generate matrices with dimensions [N × j], which contained all the possible monthly values of the domain of the random variable Q, depending on the different combinations of open and closed hydrants. With these matrices, the behaviour of the network downstream of the EPPs could be characterized on a monthly basis.

The Bernoulli Experiment was run integrating the EPANET engine into Python (v. 2.7.15) through its Dynamic Link Library (DLL). Bernoulli distributions were obtained after each trial. These distributions are directly related with binomial distribution. Binomial distribution is defined by the number of independent trials carried out, N, and the probability of success, p. When the number of trials is 1, then the binomial distribution is called a Bernoulli distribution. Therefore, the results obtained for every EPP composed the domain of Q. Analysing these results, the monthly flow values and their occurrence probability could be calculated. Hence, for each EPP, 12 (monthly) binomial distributions were obtained.

The probability mass function of a discrete random variable X conveys the same information as a table of probabilities of simple events for the possible values of X [31]. Thus, after calculating every possible flow value Q_l, the next step was to calculate how often these values occur monthly. The mass probability function for the whole domain was obtained dividing the times, n_{lj}, that each value Q_l was repeated in each month j, by the number of total BTs (N). The occurrence probability of each flow value was obtained following Equation (8):

$$p\left(Q_{lj}\right) = \frac{n_{lj}}{N}. \tag{8}$$

A comparison between the monthly experimental volume and the monthly theoretical volume required was made in each EPP. Its aim was to check how good the experiment fitted with the theoretical values. To calculate the monthly experimental volumes per unit of irrigated surface, the monthly flow distributions and their probabilities were required. Applying Equation (9), the monthly experimental volumes were calculated as:

$$EV_j = 3600\ T'_{ij}\frac{q_{max}}{q_M} \sum_{l=1}^{C} Q_l\ p\left(Q_{lj}\right), \tag{9}$$

where EV_j is the monthly experimental volume; q_{max} is the design flow allowed per unit of irrigated area; q_M is the maximum flow circulating through the EPP.

2.1.4. PAT Operating Conditions Analysis

Every possible flow circulating through the EPP pipe was considered as a possible BEP flow of a theoretical PAT. Hence, l different scenarios were defined, each of them corresponding to one PAT, whose BEP was (Q_{IBEP}, H_{BEP}). To regulate the PAT inlet conditions and keep the network service conditions downstream, the bypass scheme proposed by Lydon et al. [32], has been considered, in which there are two control valves, one before the PAT and the other in parallel. The regulation used was hydraulic regulation (HR), since previous investigations concluded that HR is generally more efficient than electric regulation, showing larger efficiencies when the working conditions vary from the design values. In addition, they were also shown to be less expensive [13]. This can be observed in Figure 3.

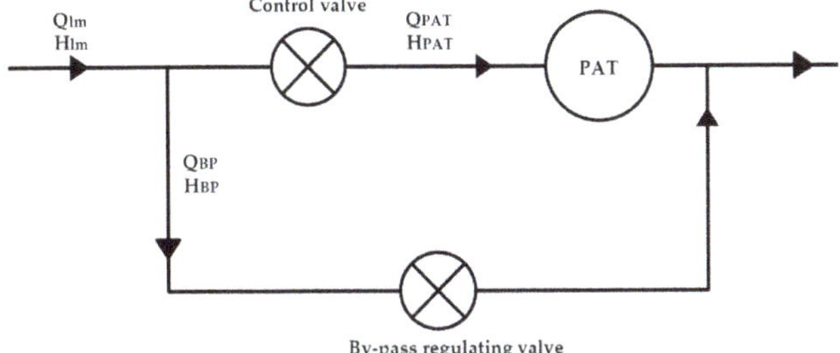

Figure 3. Typical PAT installation scheme.

The methodology followed to estimate the flows running through the turbine, simulated the interaction between the Q–H system and PAT curves. The two operating rules fixed were: (i) the flow demanded downstream of the EPP would fully circulate through the turbine if its value is lower than or equal to the maximum flow to be turbined Q_{lMAX} in each scenario l (This value was calculated obtaining the intersection between both, PAT and system Q-H curves); (ii) if the flow demanded downstream is greater than the maximum fixed for each scenario Q_{lMAX}, this flow would be diverted to the bypass. To obtain the amount of flow diverted in each scenario l for each flow demanded downstream m, Q_{lmBP}, the interaction between both system and PAT curves is required again.

Operating Rules:

(i) $\quad if\ Q_{lm} \leq Q_{lMAX} \left\{ \begin{array}{l} Q_{PAT} = Q_{lm} \\ Q_{lmBP} = 0 \end{array} \right\}$

(ii) $\quad if\ Q_{lm} > Q_{lMAX} \left\{ \begin{array}{l} Q_{PAT} = Q_{lmPAT} \\ Q_{lmBP} = Q_{lm} - Q_{lmPAT} \end{array} \right\}$

The methodology assumes that the selection of a pump to operate as a turbine with the specified BEP can be carried out independently, using the approaches described in Lydon et al. [32]. The method then adopts the approach proposed by Barbarelli et al. [32] to estimate the PAT characteristic curves (head & flow). Barbarelli, proposed an alternative curve to the curve suggested by Derakshan and Nourbakhsh [33], to obtain the relative head for any given machine, based on 12 pumps tested as turbines. This curve followed Equation (10). Thus, all the relative heads were obtained for every flow demanded downstream Q_{lm} in the scenario l. The value of the head recovered by the PAT H_{lm} was calculated multiplying the relative heads by the BEP head, H_{BEP}. With these heads quantified for every PAT associated to every scenario l, all the Q-H curves for the specific system were obtained. These equations had the form of Equation (11), where the coefficients changed for each hypothetical PAT tested:

$$\frac{H_{lm}}{H_{BEP}} = 0.922 \left(\frac{Q_{lm}}{Q_{lBEP}} \right)^2 - 0.406 \left(\frac{Q_{lm}}{Q_{lBEP}} \right) + 0.483 \tag{10}$$

$$H_{lmPAT} = a Q_{lmPAT}^2 + b Q_{lmPAT} + c, \tag{11}$$

where H_{lmPAT} is the head recovered for a certain flow Q_{lmPAT} running through the PAT. In Figure 4, this interaction and intersection between a potential PAT and the system curve for a random site is displayed. To calculate the amount of flow turbined and the amount diverted through the bypass, every possible flow value greater than Q_{lMAX}, was introduced in the system curve, obtaining the head available ($H_{lm-System}$) in the system for such a flow (Q_{lm}). Using this head ($H_{lm-System}$) in the PAT curve as the head recovered (H_{lmPAT}), the flow circulating through the PAT was fixed. Applying this sequence to every possible flow greater than Q_{lMAX}, all the pairs (Q_{lmPAT}, H_{lmPAT}), for which the device could work, were calculated. Consequently, in scenario l, the portion of flow diverted through the bypass for values greater than Q_{lMAX} was the difference between the flow demanded downstream and the flow turbined ($Q_{lm} - Q_{lmPAT}$).

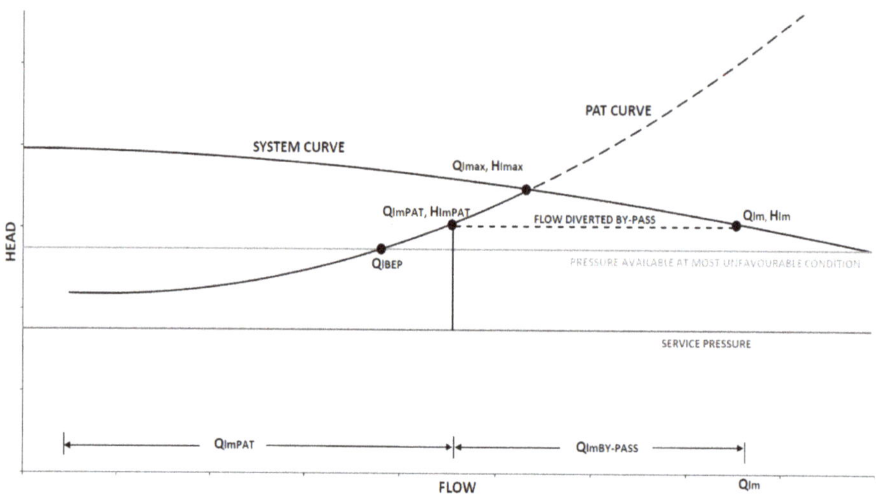

Figure 4. Representation of a potential PAT flow-head curve for a hypothetical site, and working pairs (Q_{lmPAT}, H_{lmPAT}) for a random flow Q_{lm} greater than the maximum Q_{lMAX} in the Q-H space.

Each of these pairs (Q_{lmPAT}, H_{lmPAT}) had an associated relative efficiency, under which the PAT operates. Novara et al. [34] proposed a model, through the extrapolation 116 measured PAT characteristic curves, estimating the behaviour of their relative efficiency depending on the flow rate. Thus, the mechanical relative efficiency was obtained for each flow, Q_{lmPAT}, in scenario l, following Equation (12). As a conservative estimate the maximum efficiency was fixed at 55% (0.65 PATs + generator efficiency and 0.85 to take into account the hydraulic regulation losses) [12]. For very low flow rates, this relative efficiency has negative values, for which the device should be off or the flow would be diverted. The power production for each scenario l, whose BEP is (Q_{lBEP}, H_{lBEP}), for each pair (Q_{lm}, H_{lm}), was obtained as per Equation (13). For very low flow rates, this relative efficiency has negative values, for which the device should be off and no flow would be turbined.

$$\eta_{lm} = 0.5197 \left(\frac{Q_{lmPAT}}{Q_{lBEP}} \right)^3 - 2.3328 \left(\frac{Q_{lmPAT}}{Q_{lBEP}} \right)^2 + 3.0931 \left(\frac{Q_{lmPAT}}{Q_{lBEP}} \right) - 0.2757 \tag{12}$$

$$P_{lm} = 0.55\, Q_{lmPAT}\, H_{lmPAT}\, \gamma\, \eta_{lm} \tag{13}$$

where Q_{lBEP} is the value of the BEP flow for each scenario l; γ is the specific weight of the water; η_{lm} is the relative efficiency value for each pair for scenario l. Lastly, to estimate the monthly energy recovered, the powers produced by each PAT for each pair (Q_{lmPAT}, H_{lmPAT}) in scenario l were used, together with the monthly mass probability function and the monthly available time. The monthly available time matrix was reduced to a single vector, since it was an on-demand irrigation network, where every hydrant had 24 h availability every day of the year. Its calculation followed Equation (14):

$$E_{lj} = P_{lm}\, p\left(q_{lj} \right)\, T'_j. \tag{14}$$

2.1.5. Economic Viability

The last stage of the methodology was to assess the economic viability of each scenario studied. Payback Period (PP) was used here to determine the period needed to recover the investment made, neglecting the time value of money.

To quantify the total installation cost, three different main components have been considered: electromechanical (PAT + generator), civil works and additional works. Regarding the electromechanical part, different investigations have given different approaches. Ramos et al. [15] estimated the cost of a PAT to vary between 200–400 €/kW for nominal power lower than 40 kW. Carravetta et al. [13] proposed the sum of nominal power of the turbine, 230 €/kW, and the maximum PAT power accounting for the cost of the generator, 115 €/kW. De Marchis et al. [35] proposed a cost per power unit of 2000 €/kW for PAT plus generator. In this research, a cost model, which estimates the unitary price for PAT and generator, has been used. This model estimates different kinds of radial PATs, including generator with one, two or three pairs of magnetic poles (pp) [19]. The cost per kW of the centrifugal PATs coupled with induction generators with the number of pp mentioned is related to the parameter $Q_{lBEP}\sqrt{H_{BEP}}$. Thus, the electromechanical cost has been calculated for every possible BEP flow value and the BEP head was fixed, using Equations (15)–(17):

$$C_{PP1l} = 11,589.32\ Q_{lBEP}\sqrt{H_{BEP}} + 1380.79 \tag{15}$$

$$C_{PP2l} = 12,864.77\ Q_{lBEP}\sqrt{H_{BEP}} + 949.43 \tag{16}$$

$$C_{PP3l} = 15,484.97\ Q_{lBEP}\sqrt{H_{BEP}} + 1172.72. \tag{17}$$

In addition to the PAT costs, other works have to be added, such as civil work and the cost of the bypass. Regarding the civil works, a new approach has been developed within this research, explained in Appendix A. The percentage of the civil works costs depending on the power installed was calculated using Equation (18), proposed in this research. For additional works, such as electric connection or maintenance, 20% of the total costs has been considered. Following Equation (19), the total costs for the installations were obtained as:

$$p_{cwl} = 1 \cdot 10^{-7} P_{lBEP}^4 - 2 \cdot 10^{-5}\ P_{lBEP}^3 + 0.0011\ P_{lBEP}^2 - 0.0349\ P_{lBEP} + 0.6714 \tag{18}$$

$$TC_{nl} = \frac{CC_{PPnl}}{(1 - p_{cwl})\ 0.8}, \tag{19}$$

where p_{cwl} is the percentage of the civil works over the total installation cost in scenario l, and C_{PPnl} is the total installation cost for each flow value and number of polar pairs, n, of the electromechanical devices. To complete the economic analysis, the calculation of the annual revenues (AR) and the PP was carried out. For the first term, the total energy produced in each scenario has been multiplied by the income rate, in case of selling to the grid, or the energy tariff in case of auto consumption. This rate will depend on the country where the installation is made. Thus, applying Equations (20) and (21) the AR and PP for each scenario was calculated:

$$AR_l = \sum_{j=1}^{12} E_{lj} r_j \tag{20}$$

$$PP_{nl} = \frac{C_{PPnl}}{AR_l}, \tag{21}$$

where E_{lj} is the monthly energy recovered in each scenario l; the vector r_j represents the money received or saved per kilowatt every month. Finally, analysing all the PPs associated to scenario l and every potential PAT, n, the selected scenario would be the one whose PP is the lowest, considering the respective BEP power to be installed. It has to be highlighted, that for MHP technology, the PP has to be lower than 10 years [14] to be considered economically viable in the water sector. Thus, of all the points studied, those with $PP \geq 10$ years were discarded.

2.2. Study Area

Sector VII of the right bank of the Bembezar River (BMD) is a pressurized water distribution network located in Seville (Spain). The network is composed of pipes with diameters between 150 and 800 mm. It contains 162 hydrants that irrigate a total surface of 920 hectares. The main crops cultivated in the district are: citrus (56%), maize (32%), cotton (9%) and sunflower (3%) [36]. The hydrants are distributed in levels that vary between 47 m and 97 m.

A pumping station is located at 86 m.O.D (meters above the ordenance Datum). and is composed of two kinds of pumps. The first type has a power of 90 kW, and there are three of these units. The second type has a power of 270 kW, and there are two of these units. The network was designed to supply 1.2 l/s/ha on demand, so water is continuously available to farmers (24 h per day). The network was designed for 100% of open hydrants simultaneity. The methodology has been applied for the 2017 irrigation season, for which the agronomic parameters (rainfall and evapotranspiration) have been considered. The values of these parameters for the 2017 irrigation season were 440 mm and 1210 mm, respectively.

3. Results

3.1. Location of Excess Pressure Areas and Calculation of Downstream Open/Closed Hydrant Combinations

In this first stage, hydraulic simulation following the design hypothesis, 100% of the hydrants of the network set as open, was conducted, using EPANET [37]. As a result, five points have been identified as potential EPPs, with an available excess pressure of 19.1 m, 13.9 m, 19.8 m, 18 m and 14.3 m, respectively. In this first assessment, the BEP head for the turbine was fixed, since the first simulation has been carried out under the most unfavourable conditions. In this way, it was ensured that the service pressure reaching the hydrants located downstream was always greater than or equal to the minimal pressure required, 35 m in this case. The location of these points can be seen in Figure 5, noting that each EPP was located on a separate branch of the network.

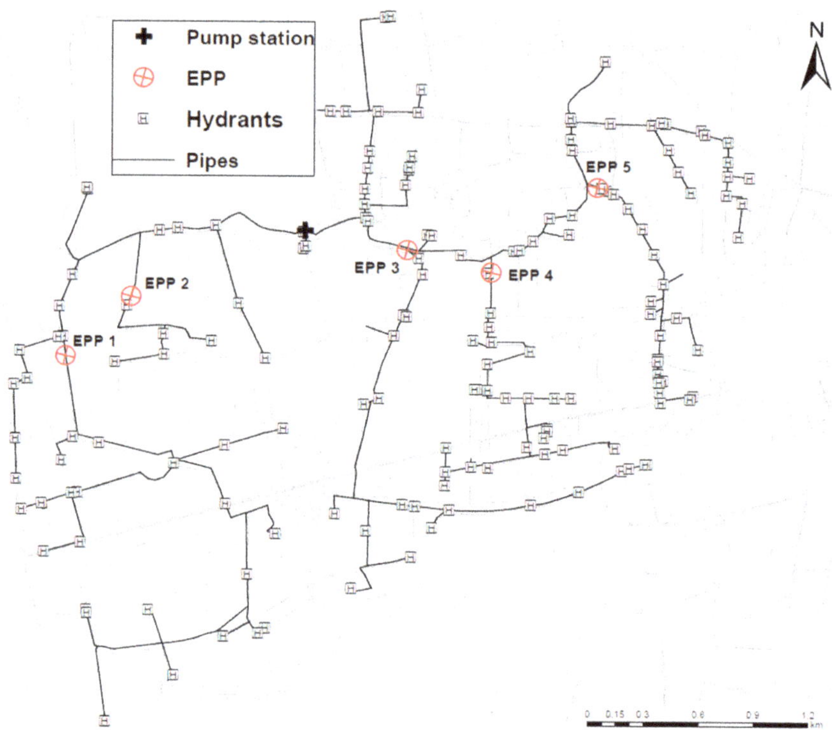

Figure 5. EPPs in the sector VII of the right bank of the Bembezar River irrigation district.

The number of hydrants located downstream of each EPP was then counted, to obtain the number of possible open hydrant combinations, following Equations (1) and (2).

3.2. Open Hydrant Probability Calculation

As there is no record available of the actual open hydrant time, it has to be estimated by means of the formula proposed by Clément (1966) [27]. Regarding the crop distribution, the crops irrigated by each hydrant were also not available, and just the percentage of total land for each crop was known. Therefore, this general distribution has been applied to each hydrant. These two first steps of the second stage could be replaced by actual information in case the irrigation network studied had these data available. Thus, firstly, the monthly crop water requirements were calculated. The required irrigation time per hydrant and month was then calculated using Equation (4). Specifically, this network is an on-demand irrigation network, so the water availability is 24 h every day. Then, the monthly probability of open hydrant was calculated for each hydrant using Equation (3). Since the crop distribution per hydrant was not available, the general percentage of crops mentioned in the description of the case study has been applied to each hydrant, assuming all of them had the same open hydrant probability, as shown in Table 1. The characteristics of each EPP before running the experiment and the input information are shown in Table 2.

Table 1. Monthly open hydrant probability by crops depending on the surface occupied, and total monthly open hydrant finally applied to every hydrant during the irrigation season.

Crop	Surface Percentage	Monthly Open Hydrant Probability (%)							
		March	April	May	June	July	August	September	October
Citrus	56	0.3	4.1	14.7	25.5	28.1	24.4	13.0	1.0
Maize	32	0.0	0.0	7.6	23.8	26.7	14.6	0.0	0.0
Cotton	9	0.0	0.0	1.8	6.0	7.1	4.2	0.0	0.0
Sunflower	3	0.0	0.0	1.1	2.5	2.4	0.3	0.0	0.0
Total (%)	**100**	**0.3**	**4.1**	**25.2**	**57.8**	**64.3**	**43.5**	**13.0**	**1.0**

Table 2. Summary of the EPPs found, downstream hydrants, number of possible flow values, flow range and monthly and yearly number of Bernoulli Trials run conducted.

EPP	Downstream Hydrants	Flow Values	Q Range (l/s)	Bernoulli Trials	Total Simulations
1	23	8,388,608	0–297	17,000,000	204,000,000
2	5	32	0–82	15,000	180,000
3	21	2,097,152	0–179	5,000,000	60,000,000
4	26	67,108,864	0–101	140,000,000	1,680,000,000
5	21	2,097,152	0–75	5,000,000	60,000,000

3.3. Monthly Characterisation of the Network: Mass Probability Function, $p_X(x)$ Calculation

Once the monthly open hydrant matrices were defined for every hydrant of the network, several BTs were run, in order to characterise the behaviour of the network across the year. Thus, analysing the results obtained for each EPP it can be seen that the flow values varied from 0–$297\,\mathrm{l\,s^{-1}}$, 0–$82\,\mathrm{l\,s^{-1}}$, 0–$179\,\mathrm{l\,s^{-1}}$, 0–$101\,\mathrm{l\,s^{-1}}$ and 0–$75\,\mathrm{l\,s^{-1}}$, respectively. From these results, a distribution of the flows along the irrigation season was obtained, and the monthly behaviour of the network could be characterised by analysing the 12 monthly binomial distributions. The mass probability functions were calculated using Equation (8).

In Figure 6 the mass probability functions corresponding to the months of irrigation season for EPP3 can be seen. The mass probability function illustrates the monthly occurrence probability of every flow of the domain of Q. Thus, in Figure 7 it can be seen how the occurrence probabilities change along the months represented. Higher probabilities can be seen for lower flows in months where the irrigation requirements are lower, and higher probabilities for greater flows in months with more irrigation requirements. The >60 million experimental flows obtained for the irrigation season are displayed for EPP3 in Figure 8.

The monthly experimental volumes were calculated using Equation (9). The variations between the theoretical and experimental values for EPP3 can be seen in Figure 9. The annual variations found between the theoretical and experimental volumes in the five EPPs were −0.0873%, 0.3867%, 0.0816%, −0.08024% and 1.2287%, respectively.

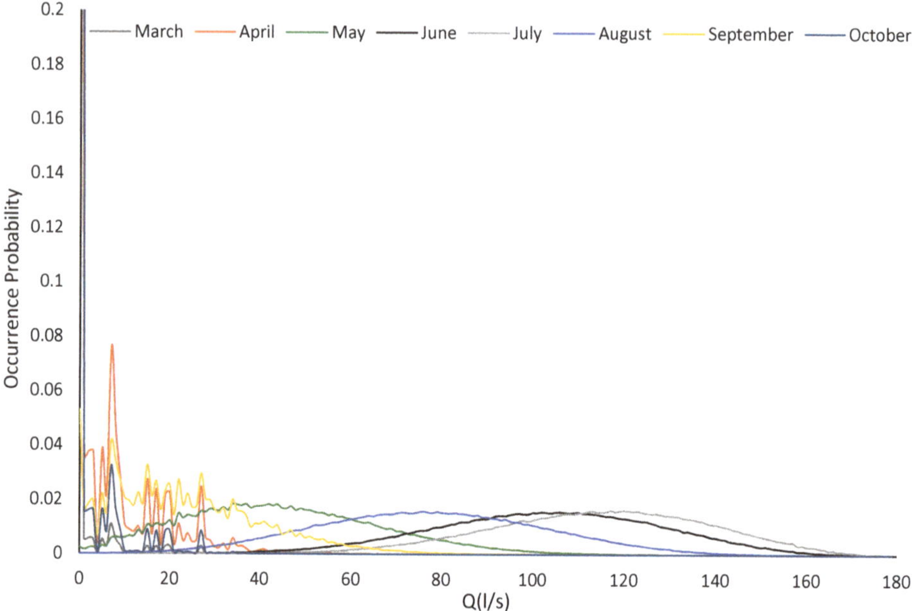

Figure 6. Mass function, $p(q)$, for the possible flow values in March, April, May, June, July, September and October for EPP 3.

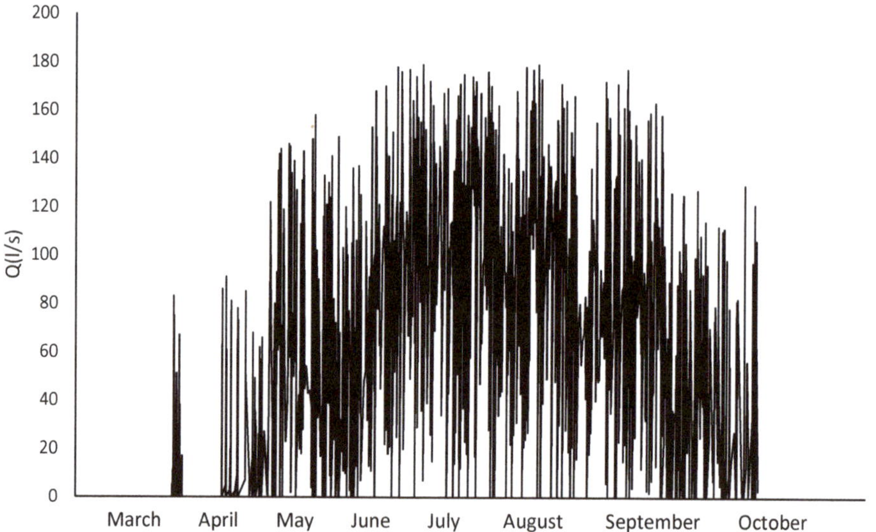

Figure 7. Experimental flow distribution during the irrigation season for EPP 3.

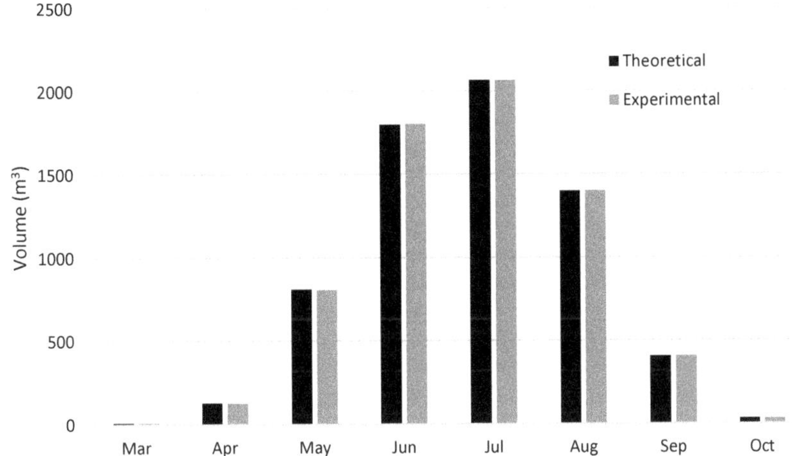

Figure 8. Theoretical irrigation volume requirements and experimental irrigation volume requirements for EPP 3.

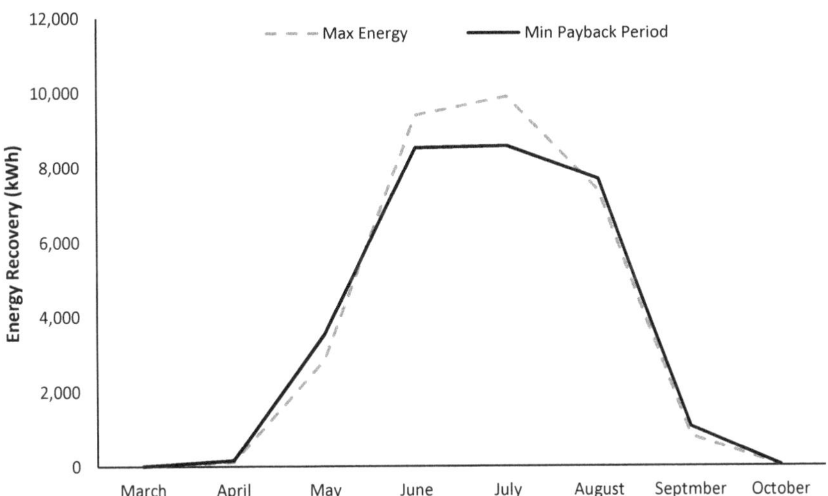

Figure 9. Comparison between the highest energy-producing scenario and the lowest PP scenario in EPP3.

3.4. PAT Operating Conditions Analysis

Every experimental Q-H and PAT curve had to be defined. Every Q-H system curve was obtained from the hydraulic model. For the different EPPs, an average of 16 million experimental curves were tested per EPP using Equation (10). Calculating the intersection between every PAT curve and the system curve, the maximum flow allowed to run through each device was obtained in each scenario l. Once these maximum flows were defined, the space of (Q_{lmPAT}, H_{lmPAT}) for each PAT associated to each scenario l were also defined for each possible value of Q. The relative efficiencies that every pair of (Q_{lmPAT}, H_{lmPAT}) would produce in each PAT were calculated using Equation (12), depending on the BEP flow of each device. With all the flows and heads for which the PATs would operate under, and the relative efficiencies associated to these values, the power produced in each circumstance was estimated.

3.5. Economic Viability

The cost associated to each scenario and every PAT evaluated was calculated using Equations (15)–(17). To estimate the civil works associated to each scenario, Equation (18) was applied. Depending on the power, the percentage represented by the civil works varies from low values, close to 10% of the total installation cost for greater powers, up to high values close to 70% for lower powers.

For the five EPPs analysed in the network studied, the energy that would be recovered varies within the range 0.9–43.3, 2.4–6.9, 1.1–30.4, 0.5–11.2 and 0.4–5.6 MWh, respectively. To calculate the annual revenues, several authors have used different values in their research. Perez-Sanchez et al. [23] fixed a price of 0.0842 kWh^{-1}, whilst Garcia Morillo et al. [22] applied the monthly average of the Spanish tariff based on six periods. In this case, the application LUMIOS (REE, 2014) [38], developed by the Spanish Electrical Grid, which provides the monthly average tariff for a selected period, has been used to calculate the monthly tariff for 2017.

The values, in kWh^{-1}, for the months in which energy is produced, were: April (0.111242), May (0.112542), June (0.113439), July (0.113044), August (0.113056) and September (0.113611). These tariffs were considered since the energy recovered has been assumed to be for self-consumption instead of selling it to the grid, as in many cases, there are no grid connection points close to the installation and it would be considered as saved energy. Thus, this connection could make the installation much more expensive, and was not considered as a viable solution for energy production in the irrigation sector.

Using Equations (20) and (21) to calculate the annual revenues and the payback period, respectively, for each scenario, and following the boundary conditions imposed for the payback period, the optimal PAT for each EPP was obtained, or the EPP was rejected. Thus, for the five EPPs, the summarised results can be seen in Table 3. Two of them were considered as viable individually for a PAT installation for energy recovery, two of them were rejected because of their PP exceeded 10 years, and one could be considered as potentially viable for being just in the border of the 10 years for returning the investment. These three EPPs would recover a sum of 81.4 MWh. Nevertheless, considering the five EPPs as a single investment, the PP would be 6.4 years, increasing the energy recovered to 93.9 MWh for the whole set.

Table 3. Summary of the results obtained for each EPP and for the set, showing the optimal scenario, BEP flow, BEP power of the optimal scenario, number of polar pairs of the electromechanical device, total installation costs, energy recovered in the optimal scenario and its PP.

EPP	Optimal Scenario	BEP Flow (l/s)	BEP Power (kW)	Polar Pairs	Cost (€)	Energy (MWh)	PP (Years)
1	2,743,236	88	9.1	1	16,438	40.8	3.5
2	13	39	2.9	2	12,339	6.9	15.8
3	631,784	54	5.8	2	14,207	29.5	4.2
4	30,122,847	46	4.5	2	13,352	11.1	10.6
5	1,051,433	36	2.8	2	12,278	5.6	19.4
Total	-	-	25.1	-	68,614	93.9	6.4

The civil works, which were calculated following Equation (20), for the optimal solution of each EPP represented 43.5%, 58%, 50.3%, 53.5% and 58.2% of the total cost, respectively.

4. Discussion

Flow fluctuations are very significant in irrigation networks, since the irrigation requirements vary along the irrigation season, depending on the crops. Furthermore, the farmers' irrigation habits are not standardized in on-demand irrigation networks. Due to the difficulty of accessing data in this sector, one method to obtain the performance of the network is a statistical analysis based on the crop water requirements. Applying Clément's methodology and Bernoulli Experiments to an on-demand irrigation network, an estimation of the data along the network can be obtained. Characterising the network through their application makes the estimation of the flow fluctuations possible, approximating the monthly probability that each flow has of occurring and the probability to be exceeded. This analysis estimated the different values that could run through a specific pipe during the whole irrigation season. All of these flows have been evaluated as BEP flows simulating as many theoretical PATs as the number of flow values there were for each EPP. However, only one machine can be selected for installation, and this methodology allows us to select a PAT whose BEP gives the best return on investment from all possible flow/head combinations across the irrigation season.

A limitation of this methodology is that a general PAT performance curve has been considered for all of the possible PATs that could be installed, underestimating in some cases and overestimating in others, the energy that could be recovered using each specific PAT studied. This general performance curve has been developed from the characteristic curves of 116 different PATs, extrapolating them and obtaining a general curve [34]. Therefore, this methodology, applied to the general PAT performance curve, helped us to study all the possible scenarios for an energy recovery installation, pre-selecting the possible power outputs and choosing the best one for the payback period. However, a deeper investigation would be necessary in each EPP site once it is established by this methodology that the economic viability is predicted to be favourable.

Another limitation of the methodology is the fact that while many theoretical PAT BEPs were analysed among the possible combinations of flow across the irrigation season, a finite number of pumps exist in the market. The PAT curves were obtained using Equation (10), which is based on experimental data from 12 pumps tested as turbines [32]. In reverse these pumps function as PATs and are considerably cheaper than traditional turbines due to mass production. Therefore not every theoretical PAT is in existence in the marketplace and to retain cost-competitiveness, in practice we would need to select the closest available machine to the selected theoretical one for a specific EPP. The current methodology may under- or over-estimate economic viability at specific EPPs as a result of this limitation.

Regarding the domain of the random variable Q, formed by the number of possible combinations of downstream open and closed hydrants, it will be greater as the number of downstream hydrants increases. This means that the possible flow values will increase as the number of downstream hydrants does, having a larger probability of flow fluctuations as the quantity of hydrants increases. In the present case study, four of the EPPs had more than 20 hydrants downstream, where, >2 million possible flows could occur.

The consideration of relative efficiencies in this paper are very important. For the flow fluctuations, as their occurrence probabilities change significantly along the irrigation season, the energy that would be produced using other methodologies that just account for the energy recovered under BEP for average flows and heads, would not give realistic results. Thus, the variation of the PATs efficiency depending on the flow rate variation, allows a more realistic power output capacity to be installed in a specific EPP. Different variables could be considered when the viability of a PAT installation is being examined, such as the energy maximization. Nonetheless, if the variable to be maximised was the energy, then, the optimal scenarios would divert to higher BEP powers, where the PP would rise to levels that could make the investment unviable. In spite of this, maximising the energy has been used in other research, and this methodology selects the best PAT for which the investment would be recovered the soonest. This would be more inviting for farmers to install.

A comparison between the energy recovered in the scenario with the lowest PP and the scenario producing the greatest energy is displayed in Figure 9 (EPP3). The energy would increase up to 30.2 MWh. This would amount to 2.3% more energy recovery. However, the PP would increase by 4.4%.

The civil works accounted for here differed from the civil works used by other authors. Some authors stated that 65% of the total installation corresponded to the cost of the PAT and the generator, and the other 35% was accounted for other works [39]. In other cases, the civil works were considered to be around 25% of the total installation costs [21]. Both considerations linked the civil work costs to the power to be installed and many previous papers were also not considering costs in the irrigation setting. Nonetheless, for a random installation, the percentage of costs represented by civil works will change with the power. Thus, different power values of PAT for the same point will not change the civil works to be conducted, but the percentage of these will change, being lower as the power increases. Therefore, for this research, an estimation of the general civil works to be carried out in these kind of installations in irrigation networks has been calculated, which contains general works and the main elements to be carried out. The result is a curve relating the power of the installation with the percentage represented by the civil works within the total costs of the installation. This gives a more realistic weight to the civil works than the previously used.

The energy prices to be considered depend on the location of the irrigation network and energy use. In this case, the energy recovered has been considered to be auto-consumed by the farmers or the irrigation district itself. Hence, the energy recovered could be considered as a savings on the energy consumption.

The case studies pump station accounts for a power consumption of 1080 kW. The power estimated to be potentially viable, was 25.1 kW. This amount represents 2.3% of the total power of the pump station. However, the power production of 25.1 kW represents the average power output across the year, where peak production of up to 45.8 kW would be reached in some stages of the irrigation season. The five PATs would be able to recover 93.9 kWh in an irrigation season. If the nominal power of the whole set is compared with the unitary pumps' power, the PATs' power amounts to 28.0% of the total power for the first type and 9.3% for the second type. Depending on the stage of the irrigation season, the number of pumps working changes. Therefore, this could translate to an important energy savings in those stages where a lower number of pumps work. The index of the annual energy recovered per irrigated surface area was 0.10 MWh year^{-1} ha^{-1}. This shows that the potential available in this specific network is not large. Previous investigations showed values of 0.65 and 0.08 MWh year^{-1} ha^{-1} [22,23]. However, these values cannot be compared, since each index will partially depend on the topography in which the network is built. In addition, previous estimates did not considered both flow variations and turbine efficiency variations.

Finally, the application of MHP for energy recovery together with other potential energy saving measures proposed in other investigations would have a positive effect, reducing the energy dependency of the activity. For instance, the optimisation of pump stations would not remove the excess pressure in every area of a network. The excess pressure due to change in elevation, among others, would still exist, and therefore the application of MHP would be a viable solution for both reducing the excess pressure and energy dependency in the network.

5. Conclusions

The use of renewable energy sources in the agricultural sector will increase in the next few years. The percentage of crop water costs related to the energy comprises an important percentage of the total costs paid by farmers. In addition, the environmental pressure to reduce the greenhouse gases emissions will be a critical driver in this issue. PAT installations in these infrastructures have been shown to be viable solution to improve the sustainability and economic viability of this sector, due to their low cost in comparison with other technologies such as traditional turbines in the case of hydropower.

Reducing operating costs by this amount will result in lower food prices for consumers and potential for greater crop yields (avoiding deficit irrigation). As a result, the incorporation of MHP energy recovery in irrigation networks has an important role to play in the water-energy-food nexus, lowering GHG emissions, lowering food prices, reducing energy consumption and increasing crop yields.

This research develops a new methodology to optimise the PAT power to install at pre-selected sites in irrigation networks, where no data are recorded, minimising the payback period of the investment and combining combinatorial and statistical analysis. Three constraint conditions were fixed to achieve this goal. There can be no lack of pressure in the network after the installation with these constraints applied. The installation PP had to be lower than 10 years. Moreover, the scenario with the lowest PP was selected, whose power is the basis of the PAT selected.

The energy recovery for the set including the five EPPs, summed to 93.9 MWh. The energy savings estimated in this paper could comprise important economical savings for farmers. Future works will study the validation of this methodology with actual measured data, and its use in irrigation networks where there is no access to actual data, to assess the potential in this sector and the percentage represented by energy saved over the total energy consumed.

Author Contributions: All authors have participated in every step of this research. J.A.R.D. and J.G.M. provided information about the irrigation networks and supervised the probabilistic approach and hydraulic analysis. M.C.C. proposed the methodology to minimize the payback period and reduce the risk investment as well as the methods and materials used. A.M. supervised the research and document preparation.

Funding: This research was partially funded by the European Regional Development Fund (ERDF) through the Interreg Atlantic Area Programme 2014–2020, as part of the REDAWN project (Reducing the Energy Dependency in the Atlantic Area from Water Networks), and the Spanish Ministry of Economy and Competitiveness through the AGL2017-82927-C3-1-R Project.

Conflicts of Interest: The authors declare no conflict of interest. The funders had no role in the design of the study; in the collection, analyses, or interpretation of data; in the writing of the manuscript; or in the decision to publish the results.

Abbreviations

a	Random scenario
AR_l	Annual revenues generated for the installation designed for the flow l
BT	Bernoulli Trial
C	Combinations of downstream open hydrants
C_{PPnl}	Total cost for a PAT with n magnetic polar pairs generator installation for the flow l
$days_j$	Monthly days
EV_j	Monthly experimental volumes
E_{lj}	Monthly energy recovered in the scenario l
H_{BEP}	Best efficiency point head
H_{lmPAT}	Head recovered for flow m in scenario l
IN_{ij}	Monthly water requirements per hydrant
N	Number of simulations
n	Number of downstream hydrants
n_{lj}	Monthly number of repeating times of the flow value l
p_{ij}	Monthly open hydrant probability
P_{lm}	Power for the scenario produced for the inlet flow m in the scenario l
$p_X(x)$	Mass probability function
$p\left(q_{lj}\right)$	Monthly mass probability function of the flow value l
p_{cw}	Percentage of civil works in the total installation cost
PP_{nl}	Payback period for an PAT installation design for the flow l with a generator with n polar pairs
q_{max}	Design flow of the network
q_M	Maximum flow running in the pipe EPP studied
q_l	Value of the flow in scenario l
Q	Random variable Flow
Q_{lBEP}	BEP flow value in scenario l
Q_{lmPAT}	Flow running through the PAT when m flow is demanded in the scenario l
R_j	Monthly random vector [0-1]
r_j	Monthly energy tariff
t'_{ij}	Monthly hydrant irrigation time required
T'_{ij}	Monthly hydrant water availability
η_{max}	Maximum PAT performance
η_{lm}	Relative performance of the flow value m in the PAT designed for flow value l
γ	Water specific weight

Sub-indexes

i	Hydrant sub-index
j	Month sub-index
l	Flow values for main scenarios sub-index
m	Flow values for relative scenarios sub-index
n	Generator magnetic polar pair sub-index

Appendix A

This appendix contains a description of the civil works considered in this research. Previous investigations stated that the civil work costs could be taken into account as a fixed percentage of the total installation, independent of the PAT costs and power. This consideration does not match reality since the civil works to be made will depend on the power to be installed, and the cost. Therefore, for general cases, the percentage represented by the civil works would depend on the power of the PAT, or rather, the PAT cost. The method proposed in this paper is based on the estimation of the parties that would be involved in a general PAT installation in irrigation networks. Concrete foundation for the PAT, earthworks, materials and construction of the bypass, backfilling and protection house have been considered as general parties for a general PAT installation in these infrastructures. In some cases, the parties involved would be more expensive. Nonetheless, this approach provides a better estimation of the civil works, since they will be almost the same for any specific point, independently of the power to be installed. Thus, it can be said that the lower is the power the higher percentage will be represented by the civil works in the total costs. A brief bill of quantities (BOQ), whose unitary prices have been fixed from the Spanish Price Generator for Construction Database [40], is given in Table A1. From this BOQ, which shows the percentage represented by the civil works over the total installation costs, Figure A1 has been developed.

Table A1. Summary of the parties accounted in the civil work costs.

	Civil Works				
CW.1	Manual trench excavation ($20 \times 2 \times 1.5$ m)	m^3	76	49.45	3758.20
CW.2	Bypass: Supply + fixing 300 mm ductile iron pipes	lm	18	96.35	1734.30
CW.3	Reinforced concrete slab 10 cm	m^2	8	€16.2	€129.8
CW.4	Protection House: Concrete blocks ($40 \times 20 \times 10$ cm) supply and fixing ($4 \times 2 \times 2.5$ m)	m^2	30	€41.78	€1253.40
CW.5	Manual backfilling: Same material excavation	m^3	76	€3.54	€269.04
Total					**€7144.78**

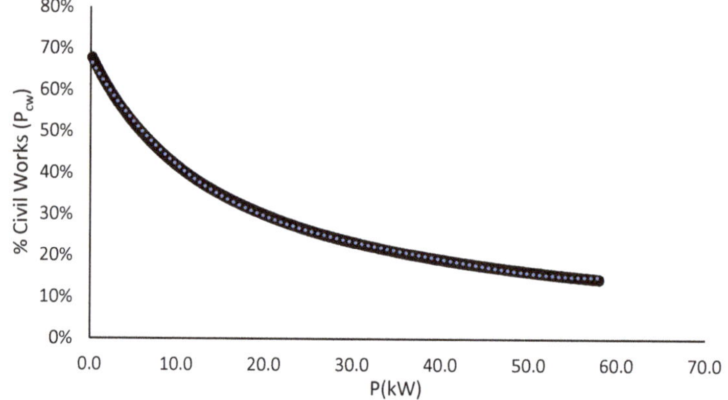

Figure A1. Cost percentage represented by the civil works depending on the PAT power.

References

1. Nogueira, M.; Perrella, J. Energy and hydraulic efficiency in conventional water supply systems. *Renew. Sustain. Energy Rev.* **2014**, *30*, 701–714. [CrossRef]
2. Copeland, C.; Carter, N.T. *Energy-Water Nexus: The Water Sector's Energy Use*; Congressional Research Service: Washington, DC, USA, 2017.
3. Lofman, D.; Petersen, M.; Bower, A. Water, Energy and Environment Nexus: The California Experience. *Int. J. Water Resour. Dev.* **2002**, *18*, 73–85. [CrossRef]
4. Rodríguez-Díaz, J.A.; Pérez-Urrestarazu, L.; Camacho-Poyato, E.; Montesinos, P. The paradox of irrigation scheme modernization: More efficient water use linked to higher energy demand. *Span. J. Agric. Res.* **2011**, *9*, 1000–1008. [CrossRef]
5. Navarro Navajas, J.M.; Montesinos, P.; Camacho-Poyato, E.; Rodríguez-Díaz, J.A. Impacts of irrigation network sectoring as an energy saving measure on olive grove production. *J. Environ. Manag.* **2012**, *111*, 1–9. [CrossRef] [PubMed]
6. Moreno, M.; Corcoles, J.; Tarjuelo, J.; Ortega, J. Energy efficiency of pressurised irrigation networks managed on-demand and under a rotation schedule. *Biosyst. Eng.* **2010**, *107*, 349–363. [CrossRef]
7. Fernández García, I.; Rodríguez Díaz, J.A.; Camacho Poyato, E.; Montesinos, P. Optimal operation of pressurized irrigation networks with several supply sources. *Water Resour. Manag.* **2013**, *27*, 2855–2869. [CrossRef]
8. Pardo, M.A.; Manzano, J.; Cabrera, E.; García-Serra, J. Energy audit of irrigation networks. *Biosyst. Eng.* **2013**, *115*, 89–101. [CrossRef]
9. Jimenez-Bello, M.A.; Royuela, A.; Manzano, J.; Prats, A.G.; Martínez-Alzamora, F. Methodology to improve water and energy use by proper irrigation scheduling in pressurised networks. *Agric. Water Manag.* **2015**, *149*, 91–101. [CrossRef]
10. Mérida García, A.; Fernández García, I.; Camacho Poyato, E.; Montesinos Barrios, P.; Rodríguez Díaz, J.A. Coupling irrigation scheduling with solar energy production in a smart irrigation management system. *J. Clean. Prod.* **2018**, *175*, 670–682. [CrossRef]
11. Power, C.; Coughlan, P.; McNabola, A. The applicability of hydropower technology in wastewater treatment plants. In Proceedings of the International Water Association World Congress on Water, Climate and Energy, Dublin, Ireland, 13–18 May 2012.
12. Carravetta, A.; del Giudice, G.; Fecarotta, O.; Ramos, H. Energy production in water distribution networks: A PAT design strategy. *Water Res. Manag.* **2012**, *26*, 3947–3959. [CrossRef]
13. Carravetta, A.; del Giudice, G.; Fecarotta, O.; Ramos, H.M. PAT design strategy for energy recovery in water distribution networks by electrical regulation. *Energies* **2013**, *1*, 411–424. [CrossRef]
14. Corcoran, L.; Coughlan, P.; McNabola, A. Energy recovery potential using micro hydropower in water supply networks in the UK & Ireland. *Water Sci. Technol.* **2013**, *13*, 552–560. [CrossRef]
15. Ramos, H.M.; Borga, A.; Simão, M. New design solutions for low-power energy production in water pipe systems. *Water Sci. Eng.* **2009**, *4*, 69–84.
16. Carravetta, A.; Fecarotta, O.; Del Giudice, G.; Ramos, H. Energy recovery in water systems by PATs: A comparisons among the different installation schemes. *Procedia Eng.* **2014**, *70*, 275–284. [CrossRef]
17. Power, C.; McNabola, A.; Coughlan, P. Development of an evaluation method for hydropower energy recovery in wastewater treatment plants: Case studies in Ireland and the UK. *Sustain. Energy Technol. Assess.* **2014**, *7*, 166–177. [CrossRef]
18. Power, C.; McNabola, A.; Coughlan, P. Micro-hydropower energy recovery at waste-water treatment plants: Turbine selection and optimization. *J. Energy Eng.* **2016**, *1*, 143. [CrossRef]
19. Novara, D.; Carravetta, A.; McNabola, A.; Ramos, H.M. A cost model for Pumps As Turbines in run-of-river and in-pipe energy recovery micro-hydropower applications. *J. Water Res. Plan. Manag.* **2019**. under review.
20. Fecarotta, O.; Aricò, C.; Carravetta, A.; Martino, R.; Ramos, H.M. Hydropower potential in water distribution networks: Pressure control by PATs. *Water Resour. Manag.* **2015**, *29*, 699–714. [CrossRef]
21. Lydon, T.; Coughlan, P.; McNabola, A. Pump-as-turbine: Characterization as an energy recovery device for the water distribution network. *J. Hydraul. Eng.* **2017**, *143*. [CrossRef]

22. García Morillo, J.; McNabola, A.; Camacho, E.; Montesinos, P.; Rodríguez Díaz, J.A. Hydropower energy recovery in pressurized irrigation networks: A case of study of an Irrigation District in the South of Spain. *Agric. Water Manag.* **2018**, *204*, 17–27. [CrossRef]

23. Pérez-Sánchez, M.; Sánchez-Romero, F.J.; Ramos, H.M.; López-Jiménez, P.A. Modelling Irrigation Networks for the Quantification of Potential Energy Recovering: A Case Study. *Water* **2016**, *8*, 234. [CrossRef]

24. Perez-Sanchez, M.; Sanchez-Romero, F.J.; Ramos, H.M.; Lopez-Jimenez, P.A. Optimization Strategy for Improving the Energy Efficiency of Irrigation Systems by Micro Hydropower: Practical Application. *Water* **2017**, *9*, 799. [CrossRef]

25. Tarragó, E.F.; Ramos, H. Micro-Hydro Solutions in Alqueva Multipurpose Project (AMP) towards Water-Energy-Environmental Efficiency Improvements. Bachelor's Thesis, Universidade de Lisboa, Lisboa, Portugal, 2015.

26. Rodríguez-Díaz, J.A.; Camacho-Poyato, E.; López-Luque, R. Model to Forecast Maximum Flows in On-Demand Irrigation Distribution Networks. *J. Irrig. Drain. Eng.* **2007**, *133*, 222–231. [CrossRef]

27. Clément, R. Calcul des débits das les réseaux d'irigation fonctionant a la demande. *La Houille Blanche* **1966**, *20*, 553–576. [CrossRef]

28. Monserrat, J.; Poch, R.; Colomer, M.A.; Mora, F. Analysis of Clément's first formula for irrigation distribution networks. *J. Irrig. Drain. Eng.* **2004**, *130*, 99–105. [CrossRef]

29. Allen, R.G.; Pereira, L.S.; Raes, D.; Smith, M. *Crop Evapotranspiration: Guidelines for Computing Crop Water Requirements*; Irrigation and Drainage Paper No. 56; FAO: Rome, Italy, 1998.

30. Lamaddalena, N.; Sagardoy, J.A. *Performance Analysis of On-demand Pressurized Irrigation Systems*; Irrigation and Drainage Paper 59; FAO: Rome, Italy, 2000.

31. Olkin, I.; Gleser, L.J.; Derman, C. *Probability Models and Applications*, 1st ed.; Macmillan: New York, NY, USA, 1980.

32. Barbarelli, S.; Amelio, M.; Florio, G. Experimental activity at test rig validating correlations to select pumps running as turbines in microhydro plants. *Energy Convers. Manag.* **2017**, *149*, 781–797. [CrossRef]

33. Derakhshan, S.; Nourbakhsh, A. Experimental study of characteristic curves of centrifugal pumps working as turbines in different specific speeds. *Exp. Therm. Fluid Sci.* **2008**, *290*, 800–807. [CrossRef]

34. Novara, D.; McNabola, A. A model for the extrapolation of the characteristic curves of Pumps as Turbines from a datum Best Efficiency Point. *Energy Convers. Manag.* **2018**, *174*, 1–7. [CrossRef]

35. De Marchis, M.; Fontanazza, C.M.; Freni, G.; Messineo, A.; Milici, B.; Napoli, E.; Notaro, V.; Puleo, V.; Scopa, A. Energy recovery in water distribution networks. Implementation of pumps as turbine in a dynamic numerical model. *Proc. Eng.* **2014**, *70*, 439–448. [CrossRef]

36. Fernández García, I.; Rodríguez Díaz, J.A.; Camacho Poyato, E.; Montesinos, P.; Berbel, J. Effects of modernization and medium-term perspectives on water and energy use in irrigation districts. *Agric. Syst.* **2014**, *131*, 56–63. [CrossRef]

37. Rossman, L.A. *EPANET. Users Manual*; US Environmental Protection Agency (EPA): Cincinnati, OH, USA, 2000.

38. LUMIOS. Available online: https://www.esios.ree.es/es/lumios?rate=rate1&start_date=03-09-2018T14:02&end_date=04-09-2018T14:02 (accessed on 7 May 2018).

39. Lydon, T.; Coughlan, P.; McNabola, A. Pressure management and energy recovery in water distribution networks: Development of design and selection methodologies using three pump-as-turbine case studies. *Renew. Energy* **2017**, *114*, 1038–1050. [CrossRef]

40. CYPE Ingenieros, S.A. Generador de Precios de la Construcción. Spain. Available online: http://www.generadordeprecios.info/ (accessed on 15 April 2018).

Article

Modeling Sugar Beet Responses to Irrigation with AquaCrop for Optimizing Water Allocation

Margarita Garcia-Vila [1,*], Rodrigo Morillo-Velarde [2] and Elias Fereres [1,3]

[1]   Agronomy Department, University of Cordoba, 14007 Córdoba, Spain; ag1fecae@uco.es
[2]   Research Association for Sugar Beet Crop Improvement, 47012 Valladolid, Spain; rmorillovelarde@icloud.com
[3]   Institute for Sustainable Agriculture, CSIC, 14004 Córdoba, Spain
*   Correspondence: g82gavim@uco.es; Tel.: +34-957-499-231

Received: 31 July 2019; Accepted: 10 September 2019; Published: 14 September 2019

Abstract: Process-based crop models such as AquaCrop are useful for a variety of applications but must be accurately calibrated and validated. Sugar beet is an important crop that is grown in regions under water scarcity. The discrepancies and uncertainty in past published calibrations, together with important modifications in the program, deemed it necessary to conduct a study aimed at the calibration of AquaCrop (version 6.1) using the results of a single deficit irrigation experiment. The model was validated with additional data from eight farms differing in location, years, varieties, sowing dates, and irrigation. The overall performance of AquaCrop for simulating canopy cover, biomass, and final yield was accurate (RMSE = 11.39%, 2.10 t ha^{-1}, and 0.85 t ha^{-1}, respectively). Once the model was properly calibrated and validated, a scenario analysis was carried out to assess the crop response in terms of yield and water productivity to different irrigation water allocations in the two main production areas of sugar beet in Spain (spring and autumn sowing). The results highlighted the potential of the model by showing the important impact of irrigation water allocation and sowing time on sugar beet production and its irrigation water productivity.

Keywords: modelling; AquaCrop; calibration; sugar beet; irrigation water allocation; water productivity

---

1. Introduction

Sugar beet (*Beta vulgaris*), together with sugarcane (*Saccharum officinarum*), are the main sources of commercial sucrose worldwide. Even though the sugar beet cropped area has been decreasing over recent decades, total production remains stable due to increasing yields [1]. The highest average fresh root yield has been recorded in Spain (90 t ha^{-1}), despite it not being ranked among the world's 10 largest producing countries, which have yields ranging from 39 t ha^{-1} to 88 t ha^{-1} [1]. This is the result of a significant effort made over recent decades to increase crop productivity through the adoption of optimum farming practices, together with plant breeding efforts, to introduce new, adapted varieties, which was led by the Spanish Research Association for Sugar Beet Crop Improvement (AIMCRA, in its Spanish acronym). Although only about one-fourth of the world's area devoted to sugar beet is irrigated, the fraction of irrigated area varies greatly from region-to-region, between 20% and 100% of the sugar beet area [2]. Because of the different environments where both crops are grown, sugar beet consumes from 500 to 800 mm of water [2], while the annual evapotranspiration of sugarcane ranges between 800 and 2000 mm [3]. This lower water consumption of sugar beet makes it suitable for water-limited environments (e.g., the Mediterranean region of limited rainfall concentrated in the winter and a rainless summer), where governments are fostering sugar beet production as an alternative to sugarcane. Thus, one would expect an increase of the sugar beet area in those countries in the coming years.

Sugar beet is sensitive to water deficits, particularly in the early growing stages [2,4], and there is a positive linear relationship between water use and root production [2,5]. Water stress is the major cause of sugar beet yield loss in many regions, even in areas where the crop is not normally irrigated or is supplementary irrigated [6]. Furthermore, yield losses due to water stress are expected to increase in the future, which becomes a serious new problem in many areas, such as North-East France and Belgium [7]. Under this scenario, optimum water usage, irrespective of its source (irrigation or rainfall), becomes a key target for sugar beet producers.

The response of crop to water availability is complex, but their characterization through field experiments have been conducted for many years [8]. However, the empiricism, time consumption, and the amount of resources needed limit their implementation for designing optimum irrigation management. Given these constraints, crop simulation models are a promising alternative used to simulate the crop response to different water supply scenarios. There are several examples of crop models used to simulate the sugar beet production under different environments and management practices. Simple models such as the one presented by Reference [9] or Reference [5] (PLANTGRO), both based on the approach in Reference [10], have been proposed. Under the production function approach, a crop response factor relates the normalized yield to its potential value against the normalized evapotranspiration or transpiration to their potential values. These models have limited capacity to predict yield response to water since they do not include the water stress effects on the different crop eco-physiological processes. To overcome these limitations, more complex process-based models have also been adapted to simulate sugar beet production, such as Broom's Barn crop growth model [6,11], CSM-CERES-Beet model, or DSSAT [12], STICS [13,14], Greenlab [14,15], LNAS [14], and PILOTE [14,16]. In addition, the AquaCrop model [17], in an attempt to develop a simple, versatile, and robust water-driven model, has been calibrated and validated for sugar beet [18–22]. This model can simulate the yield response to water more accurately with a relatively small number of parameters than other models, which makes it more attractive for simulations under water-limited conditions or for irrigation scheduling. For these reasons, AquaCrop has been implemented to assess the sugar beet production under different scenarios [23,24]. Nevertheless, as for any model application, AquaCrop must be accurately calibrated and validated. In the case of sugar beet, earlier calibration and validation efforts in AquaCrop have yielded uncertain results [18–22]. One reason may be that important modifications in the quantification of soil water stress have been introduced in the new model versions (v6.0 and v6.1), which makes it necessary to carry out a new calibration and validation process.

Once the model is properly parameterized, it can be used for multiple purposes, both at the plot scale (e.g., irrigation scheduling, as in Reference [25]) or farm scale (e.g., optimization of the irrigation and cropping patterns, as in Reference [26]) and at basin (e.g., integrated assessment modelling, as in Reference [27]) or regional level (e.g., crops responses to climate change [23]). Ultimately, AquaCrop can be a useful tool for supporting decision-making at different levels, especially in terms of irrigation water management for a crop such as sugar beet, which is facing water scarcity challenges in most of the production areas. In this regard, the application of the model for better irrigation water allocation, an issue faced by water authorities, would be relevant in a water scarce environment. A proper irrigation water allocation must enhance the irrigation water productivity, taking into account the availability of a scarce resource, as water, and ensuring the farmers' profitability.

Thus, the objectives of this work were first to calibrate and validate the AquaCrop model (version 6.1) for sugar beet grown under different levels of irrigation. Additionally, a scenario analysis was carried out to assess the crop response in terms of yield and water productivity to different irrigation water allocations in the two main production areas of sugar beet in Spain (spring and autumn sowing).

2. Materials and Methods

2.1. Calibration and Validation of AquaCrop for Sugar Beet

2.1.1. Datasets

A field experiment was performed in 2012 at the experimental station of the Research Association for Sugar Beet Crop Improvement (AIMCRA, in its Spanish acronym), Valladolid, Central Spain (41°39′ N, 4°41′ W, 690 m a.s.l.) to calibrate the model. The climate in the area is typically Mediterranean with an annual average precipitation of around 400 mm, and about 1200 mm average annual reference evapotranspiration (ETo). The soil of the experimental area is a Gleyic Cambisol of 1.2 m depth with uniform clay loam texture. Sugar beet was planted on 21 February, 2012, at a final density of 14.5 plants m^{-2} (row spaced 0.5 m apart), and the selected cultivar was *Amalia KWS*.

The experimental design was a randomized complete block with 144 m^2 plots with four irrigation treatments replicated four times. The four treatments were: control (100% ETc), two sustained deficit irrigation treatments (70% and 55% of the control), and a rainfed treatment. In order to enable proper crop germination and establishment, all the treatments were fully irrigated to cover the crop needs until June 1st. The irrigation system used was a sprinkler that adapted the nozzle orifice size to the irrigation treatment (3.6–4.4 mm). In order to avoid nutritional stresses, 220 Kg N ha^{-1}, 100 Kg P_2O_5 ha^{-1}, 100 kg K_2O ha^{-1}, and 50 Kg MgO ha^{-1} were applied in all treatments. Pests and diseases were carefully controlled, and no weeds were allowed to develop in the field.

Green canopy cover was monitored weekly at 5 points per replication by taking zenithal photographs above the canopy with a digital camera. Images were analyzed using the Green Crop Tracker v.1.0 software, developed by Agriculture and AgriFood Canada [28]. Biomass (leaves and roots) sampling (one row of 1.4 m long per plot) was performed monthly in every treatment, and oven dried (at 70 °C) to assess the time course of biomass. Final yield was determined by harvesting 7 m^2 per replicate plot in every treatment and drying the roots at 70 °C.

To validate the model, eight commercial farms from part of the AIMCRA trials were selected. AIMCRA technicians collect detailed data on the soil, crop, and irrigation management, as well as the final fresh yield. The conditions in the selected farms cover the variability found in the main sugar beet production areas in Spain, i.e., weather and soil variability (different years and locations), different varieties, sowing dates, plant density, and irrigation. A summary of the information on the selected farms is provided in Table 1. Another important aspect taken into account during the farms selection was that yield had not been limited by nutritional stresses pests and/or diseases. Dry yield was estimated considering a dry matter content of 20%, the average value obtained in the calibration experiment.

Table 1. Experimental data sets and farmers' fields measurements used for the calibration and validation of AquaCrop for sugar beet. n is the number of irrigation treatments.

Data Sets	Location	Year	n	Cultivar	Sowing Date	Plant Density (Plants m^{-2})	Soil Texture
Calibration							
AIMCRA	Valladolid (Valladolid)	2012	4	Amalia KWS	21 February	14.5	Clay loam
Validation							
FARM1	Villabañez (Valladolid)	2012	1	Amalia KWS	9 March	10.5	Silty clay
FARM2	Pozáldez (Valladolid)	2012	1	Ludwina KWS	1 March	10.6	Clay
FARM3	Herrera de Pisuerga (Palencia)	2012	1	Ludwina KWS	16 March	10.4	Loam clay
FARM4	Donhierro (Segovia)	2012	1	Geraldina	28 February	10.8	Loam
FARM5	Villamarciel (Valladolid)	2013	1	Amalia KWS	25 April	9.6	Loamy sand
FARM6	Lebrija (Sevilla)	2014	1	Brahms	28 October	11.1	Clay
FARM7	Lebrija (Sevilla)	2014	1	Sanlucar	16 October	10.5	Clay
FARM8	Lebrija (Sevilla)	2014	1	Portal	24 October	10.9	Clay

2.1.2. Calibration and Validation Procedures

AquaCrop v6.1 has been used in this study and its algorithms, calculation procedures, and parameters are described in detail in the model reference manual [29], while a brief description of the concepts and basic development can be found in Reference [17] and Reference [30]. The calibration process was performed by adjusting the conservative parameters [31] comparing the simulated with the observed values of green canopy cover (CC), dry biomass (B), and dry yield at harvest (Y), in that order, as recommended by Reference [31]. This approach is in line with the calculation scheme of AquaCrop, which estimates crop yield in four steps operating on a daily basis: (1) calculation of CC, (2) calculation of transpiration (T) proportional to CC, (3) conversion of T into B through a normalized water productivity (WP *) factor, and (4) calculation of Y as the product of B and the harvest index (HI). The crop water stress parameters were adjusted after satisfactory results for the control treatment were achieved. The global sensitivity analysis performed by Reference [32] was taken into account during the calibration process. The validation procedure was carried out by comparing the simulated with the observed values only of Y, since there was no data on CC or B.

The input climate data required by AquaCrop (minimum and maximum air temperature, rainfall, and ETo calculated by the FAO Penman-Monteith equation) were obtained from a weather station located at the experimental station, in the case of the calibration, and from weather stations nearby for the validation. Input data on plant density, phenology, and development were obtained from the experimental results. The soil hydraulic properties, i.e., soil water content at permanent wilting point, field capacity, and saturation, and hydraulic conductivity at saturation, were estimated from the soil texture using pedo-transfer functions [33]. The initial soil water content was considered at field capacity due to the pre-sowing irrigation carried out in all cases.

AquaCrop performance was evaluated by linear regression between the observed and simulated values of CC, B, and Y, and the slope, intercept, and coefficient of determination (r^2) were determined. The goodness of fit was also assessed by the following two statistics: root mean square error (RMSE, Equation (1)) and the index of agreement (*d*) of Reference [34] (Equation (2)).

$$RMSE = \sqrt{1/n \sum_{i=1}^{n} (O_i - S_i)^2} \tag{1}$$

$$d = 1 - \frac{\sum_{i=1}^{n} (O_i - S_i)^2}{\sum_{i=1}^{n} \left(|S_i - \bar{O}| + |O_i - \bar{O}| \right)^2} \tag{2}$$

where O_i and S_i are the observed and simulated values, respectively, and *n* is the number of observations. The model fit improves as RMSE, an indicator of the absolute model uncertainty, approaches zero while *d*, which is a stable and bounded index, approaches unity.

2.2. Simulating the Crop Response to Different Irrigation Water Allocations

Once AquaCrop was calibrated and validated for sugar beet, the model was used to optimize the irrigation schedules and to simulate yield and irrigation water productivity (WP) under different irrigation water allocation (IWA) scenarios in the two main production areas of Spain: Central-North area, where the sowing is carried out in spring, and the Southern area of autumn sowing. The two locations selected for the simulations were Valladolid and Sevilla, which are both locations representative of the production areas. The calibrated crop parameter values were used for the simulations, and AquaCrop was run in the growing degree day mode, which uses a base temperature and an upper or optimum temperature (the temperature above which crop development no longer increases with an increase in air temperature [29]).

The sowing dates used in the simulations (2 April and 6 November for North and South area, respectively) were the average sowing dates obtained from the AIMCRA database. The input soil parameters were those of the AIMCRA experiment for the North area, while a clay soil was used to represent the South area. The initial soil water content was considered at field capacity, which was consistent with the rainfall patterns and the pre-sowing irrigation practices in the study areas. The simulated irrigation method was sprinkler with an application efficiency of 80%, which is the average value obtained by AIMCRA in previous irrigation performance assessments.

To account for climate variability in the crop response to irrigation, a synthetic 30 years series of climate data for each study area was generated using the stochastic weather generator CLIMAGEN [35], based on SIMMETEO [36]. The climate data series were generated on the basis of the statistical characteristics of the observed weather at Valladolid and Sevilla locations. Using these climate data series, AquaCrop was run to simulate and analyze various IWA scenarios in both production areas, focusing on the potential yield and related irrigation WP (defined as the ratio of the dry root yield to the gross applied irrigation water). The current IWA for sugarbeet in each area (reference scenario), of 600 mm and 400 mm for the North and South area, respectively, was analyzed. Two alternative scenarios were also assessed: an increase of 100 mm in the irrigation water supply (i.e., 700 mm and 500 mm for the North and South area, respectively), and a 25% reduction in IWA under water scarcity (i.e., 450 mm and 300 mm for the North and South area, respectively). Under these three scenarios, AquaCrop was used to generate an optimum irrigation schedule to achieve maximum yield (potential yield). The irrigation strategy followed avoided water stress in the early growing stages (most sensitive period), while it allowed moderate water deficits toward the end of the season [2]. The simulated yield and irrigation WP gaps between the reference scenario and the two alternative IWA scenarios were analyzed.

3. Results

3.1. Calibration and Validation of AquaCrop for Sugar Beet

The calibration of the crop parameters involved in the simulation of the time course of CC was carried out first. Figure 1 presents a comparison between observed and simulated CC for the four irrigation treatments of the AIMCRA experiment. A very good matching of simulated CC against measured CC was found for all treatments, with the exception of the rainfed treatment (Figure 1d). It appears that, despite the improvements introduced in AquaCrop v6.0 and v6.1 to account for the effects of a light rain on the level of soil water stress of deep rooted crops [29], the model was not able to stop the early senescence despite a rainfall event (11 mm) which occurred on day 156 after sowing and slowed down the canopy senescence in the field (Figure 1d). The rainfall event was insufficient to reduce the simulated water depletion in the top soil up to the threshold for triggering early senescence. Nevertheless, a proper fit of water stress response parameters for early senescence can be observed by analyzing the simulated senescence process in the 70% Control and 55% Control treatments. The calibrated values of the crop parameters for sugar beet can be found in Table 2. The overall performance of AquaCrop for simulating CC is shown in Table 3, with an RMSE of 11.39% and *d* value of 0.999.

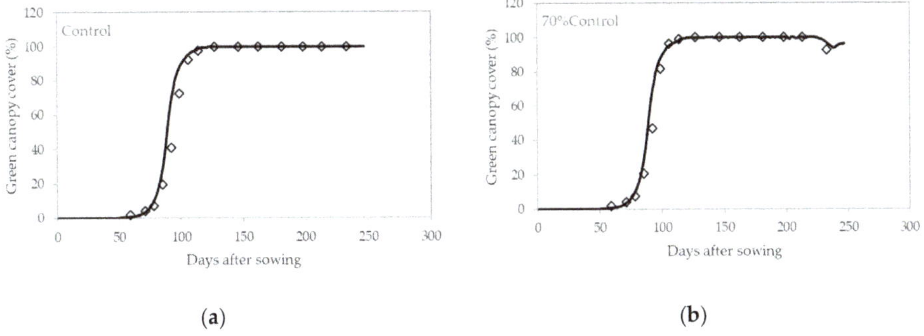

(a)   (b)

Figure 1. *Cont.*

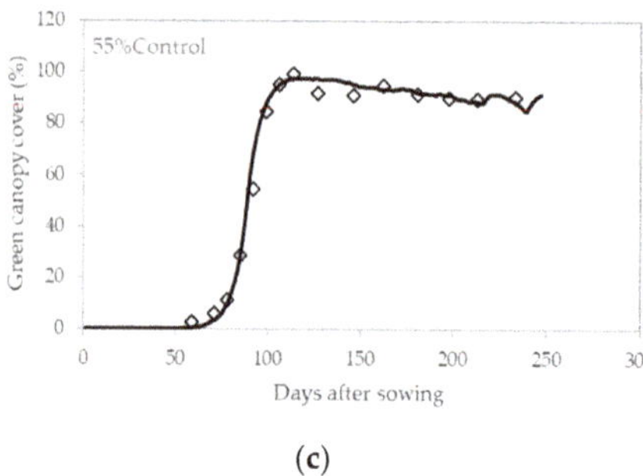

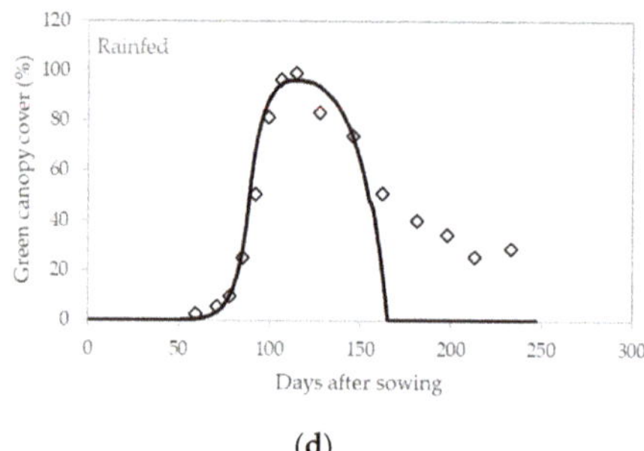

Figure 1. Observed and simulated seasonal course of green canopy cover for the irrigation treatments of the AIMCRA experiment used for model calibration: (**a**) control, (**b**) 70% control, (**c**) 55% control, and (**d**) rainfed.

Table 2. Calibrated values for selected AquaCrop crop parameters for sugar beet. The AquaCrop default values for sugar beet and the values calibrated in selected previous studies are shown. Information about the calibration process (model version, location, observed yield range, and evaluated variables) is presented. CC, B, and Y are green canopy cover, biomass, and yield, respectively.

	Calibration	Default	Stricevic et al., (2011) [18]	Alishiri et al., (2014) [19]	Van Straaten (2017) [20]	Malik et al., (2017) [21]
AquaCrop version	6.1	4.0	4.0	4.0	5.0	-
Location	Spain	Italy	Serbia	Iran	Morocco	Pakistan
Observed dry yield range (t ha^{-1})	4.8–25.8	-	8.5–19.8	4.0–12.0	10.0–25.0	7.3–15.0
Evaluated variables	CC, B, Y	-	Y	CC, B, Y	Y	CC, B, Y
Crop parameters						
Threshold air temperatures						
Base temperature, T$_{base}$ (°C)	3	5	5	5	5	5
Upper temperature, T$_{upper}$ (°C)	25	30	30	30	30	30
Crop transpiration						
Kc$_{Tr,x}$	1.15	1.10	1.10	1.10	1.15	1.10
Production						
Normalized water productivity, WP * (g m^{-2})	18.0	17.0	17.0	18.0	18.0	16.7–18.6
Reference harvest index, HIo (%)	75	70	70	60	70	68–73
Water stress response						
Top soil thickness (cm) [1]	20	10	10	10	10	10
Canopy expansion p$_{upper}$	0.20	0.20	0.20	0.25	0.20	0.10–0.20
Canopy expansion p$_{lower}$	0.60	0.60	0.60	0.70	0.60	0.45–0.55
Canopy expansion shape factor	3.0	3.0	3.0	4.0	3.0	0.5–3.2
Stomatal closure p$_{upper}$	0.65	0.65	0.65	0.65	0.57	0.45–0.65
Stomatal closure shape factor	3.0	3.0	3.0	2.5	2.5	2.5–2.8
Canopy senescence p$_{upper}$	0.75	0.75	0.75	0.75	0.75	0.45–0.55
Canopy senescence shape factor	3.0	3.0	3.0	2.5	2.5	1.2–3.0
HIo adjustment-Before yield formation (+)	None	None	None	None	None	None
HIo adjustment-During yield formation (+)	6	4	4	4	4	4
HIo adjustment-During yield formation (−)	None	None	None	None	1	None

[1] Top soil thickness considered for the estimation of soil water depletion (Program settings: Crop parameters). -: Information is not available.

Table 3. Statistics (root mean square error–RMSE- and index of agreement–*d*-) for the comparisons between observed and simulated values of green canopy cover (CC), total dry biomass (B), and root dry yield (Y) for the calibration and validation of the AquaCrop model. Slope, intercept, and r^2 are for the linear regression of observed against simulated values. n is the number of observed data used in the calibration and validation process.

Variable	n	Observations Range	Simulations Range	RMSE	*d*	Slope	Intercept	r^2
Calibration								
CC (%)	60	2–100	0–100	11.39	0.999	1.038	−2.748	0.924
B (t ha^{-1})	18	1.56–34.0	4.50–30.21	2.10	0.983	0.857	2.172	0.957
Y (t ha^{-1})	4	4.83–25.80	3.49–25.18	0.85	0.999	1.056	−1.256	0.994
Validation								
Y (t ha^{-1})	8	16.18–27.00	16.26–27.63	1.17	0.998	0.945	1.278	0.908

The satisfactory performance in the simulation of CC led to a reasonable fit in biomass accumulation for all treatments, as indicated by the high value of *d* and moderate RMSE (Table 3). Figure 2 depicts

the comparison between observed and simulated seasonal course of biomass accumulation for all
the treatments of the AIMCRA experiment. A slight tendency to over-predict (in all treatments) and
under-predict (in control and rainfed treatments) at the beginning and end of the season, respectively,
can be observed (also see slope and intercept in Table 3). These results are more than satisfactory when
taking into account the uncertainty in the measurement errors in biomass sampling, which indicates a
proper adjustment of the crop parameters related to transpiration, as well as the normalized water
productivity (WP *) (Table 2).

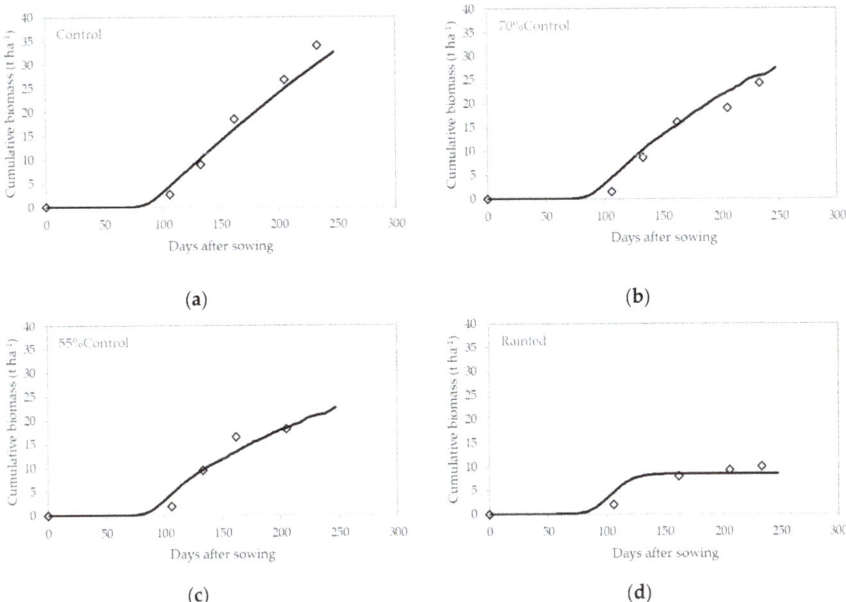

Figure 2. Observed and simulated seasonal course of biomass accumulation in dry terms for the
irrigation treatments of the AIMCRA experiment used for model calibration: (**a**) control, (**b**) 70%
control, (**c**) 55% control, and (**d**) rainfed.

Figure 3 presents the comparison between observed and simulated dry root yield at harvest
using the calibrated and the default crop parameters. The calibration improved yield prediction
relative to the use of the default parameters, where simulated yields were below the observed yields,
particularly near the maximum (Figure 3). After calibration, the simulation of yield was good for
all treatments (deviation of less than 5%), with the exception of the rainfed treatment (deviation of
28%). The under-prediction of yield in this treatment must be mostly related to the differences between
observed and simulated CC and biomass at the end of the season (Figures 1d and 2d). The premature
canopy senescence stopped the build-up of the simulated harvest index (HI), which reached a value
of only 41%. The simulation of HI under terminal drought conditions, where accelerated canopy
senescence has a major impact, is particularly difficult [17]. Nevertheless, the overall yield predictions
in the range of 4.8 to 25.8 t ha^{-1} are good, as shown in Table 3 (RMSE = 0.85 t ha^{-1}, d = 0.999). It is also
important to highlight how well the model simulates the positive impact of water stress on the HI
during yield formation, as observed in the 70% control and 55% control treatments (Figure 3).

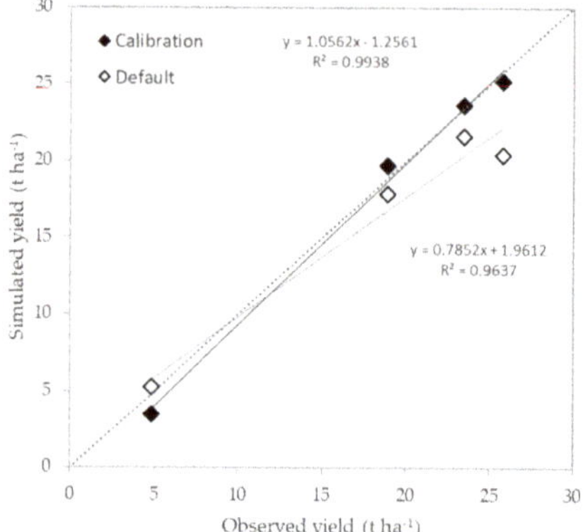

Figure 3. Simulated versus observed dry root yield at harvest for the AIMCRA experiment used for model calibration with the calibrated and default crop parameters.

Validation of the model was performed with the calibrated crop parameters using the dataset from commercial farms described above. The agreement between observed and simulated dry root yield at harvest is presented in Figure 4 for a large spectrum of production levels (from 16 to 27 t ha^{-1}). This comparison shows that all simulated yields have a deviation of less than 10%, without an apparent trend for over-prediction or under-prediction (Slope = 0.945, Intercept = 1.278, and r^2 = 0.908, Table 3). Thus, there was a very good fit, with a high value of *d* and moderate RMSE (Table 3).

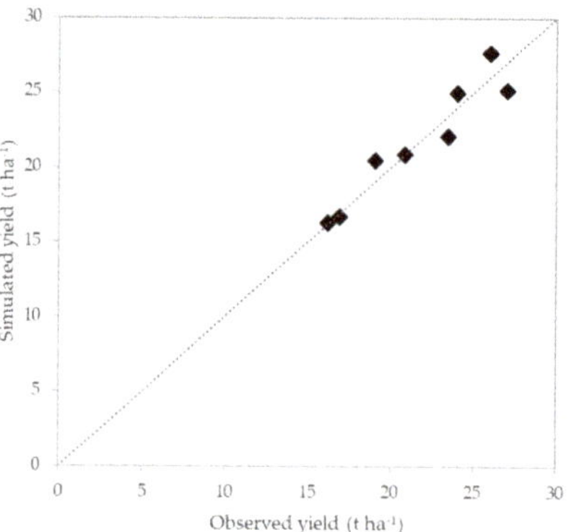

Figure 4. Simulated versus observed dry root yield at harvest for the FARM1-8 datasets used for the model validation.

3.2. Crop Response to Different Irrigation Water Allocations

The calibrated and validated AquaCrop model for sugar beet was used to simulate the crop response to different IWA under optimum irrigation schedules in the two main production areas of Spain with climatic datasets of 30 years generated by CLIMAGEN software. Figure 5 depicts the range of cumulative rainfall and ETo values just for the growing cycle of sugar beet in Valladolid (North area) and in Sevilla (South area) for the 30 simulated seasons. The North area presents much higher ETo (around 1030 mm) and lower rainfall (around 200 mm) than the South area (around 630 and 500 mm of ETo and rainfall, respectively) ($p < 0.05$, Tukey test), although a higher inter-annual variability is observed in the South, particularly in rainfall. Note that there are different sowing times, with spring and autumn for North and South areas, respectively. This is reflected in the current irrigation water allocation of 600 mm in the North versus 400 mm in the South.

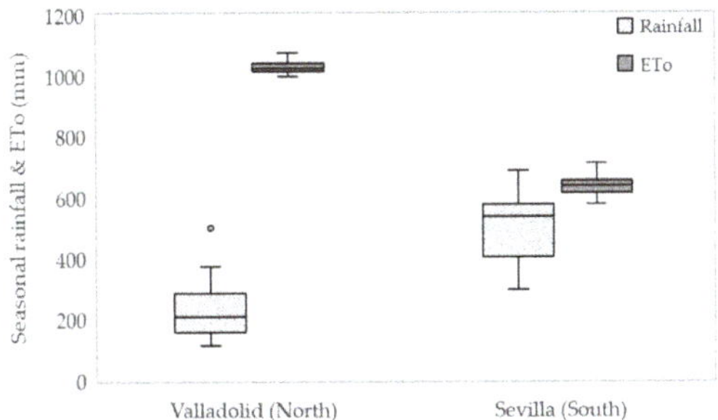

Figure 5. Cumulative rainfall and reference evapotranspiration (ETo) for the sugar beet growing cycle in Valladolid (North productive area) and Sevilla (South productive area), and for the 30 simulated seasons. The box represents 50% of the data, where the box's upper boundary represents the upper quartile (Q3) of the data, the lower boundary represents the lower quartile (Q1) and the line inside the box represents the median (Q2). The whiskers show the largest and smallest values. The circle denotes outlier values.

The potential dry root yields under the different irrigation water allocation scenarios for the 30 simulated seasons in the two productive areas were simulated with AquaCrop and are presented in Figure 6. The average potential yield for the reference IWA scenario is higher in the South, with an average of 24 versus 22 t ha^{-1} in the North. Yields similar to those in the South were achieved in the North under the IWA scenario of 700 mm (Figure 6a). As a consequence, the irrigation WP is considerably higher in the South than the North area (Figure 7), with average values of 6.1 Kg m^{-3} and 3.7 Kg m^{-3} under the IWA reference scenario, respectively. Similarly, evapotranspiration WP (defined as the ratio of the dry root yield to the water consumed–ET-) followed the same pattern, with average values of 4.1 Kg m^{-3} and 2.7 Kg m^{-3} for the North and South areas, respectively. As might be expected, the increase in IWA led to a downward trend in irrigation WP in both areas ($p < 0.05$, Tukey test) (Figure 7). Regarding the year-to-year variability, the highest variability in yield occurred in the North under a 25% reduction in IWA (Figure 6), where a greater dependency on seasonal rainfall under this deficit IWA scenario increased the inter-annual variability. On the contrary, the maximum variability in irrigation WP was found in the South for the highest IWA (Figure 7). For this location, in this scenario, an increase in IWA did not enhance yields in up to 33% of the years, due to the rainfall patterns, which explained the high variability observed.

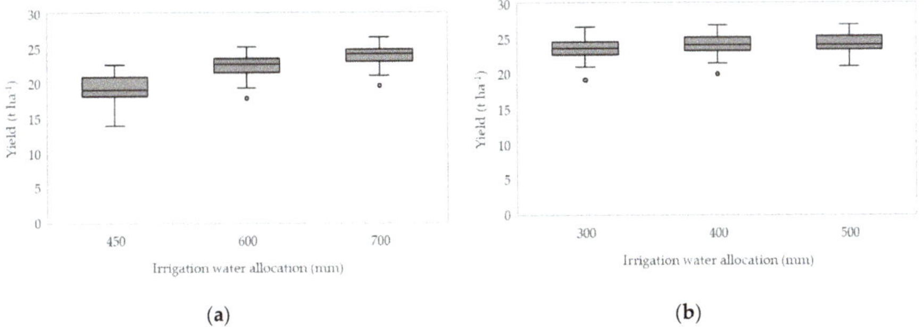

(a)

(b)

Figure 6. Potential dry root yield at harvest under three different irrigation water allocation scenarios for the 30 simulated seasons in the two main productive areas: (**a**) North (Valladolid), and (**b**) South (Sevilla). The box represents 50% of the data, where the box's upper boundary represents the upper quartile (Q3) of the data, the lower boundary represents the lower quartile (Q1), and the line inside the box represents the median (Q2). The whiskers show the largest and smallest values. The circles denote outlier values.

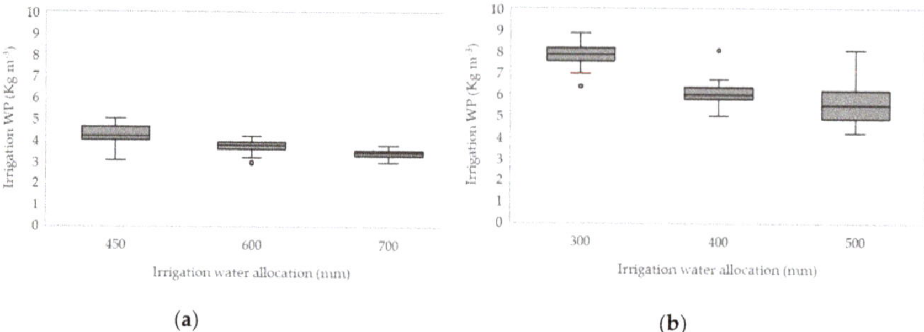

(a)　　　　　　　　　　　　　　　　　　　(b)

Figure 7. Simulated irrigation water productivity (WP) under three different irrigation water allocation scenarios for the 30 simulated seasons in the two main productive areas: (**a**) North (Valladolid), and (**b**) South (Sevilla). The box represents 50% of the data, where the box's upper boundary represents the upper quartile (Q3) of the data, the lower boundary represents the lower quartile (Q1), and the line inside the box represents the median (Q2). The whiskers show the largest and smallest values. The circles denote outlier values.

3.3. Yield and Water Productivity Gaps

The average yield and irrigation WP gaps for each scenario are presented in Figures 8 and 9, respectively. In the South, the potential yields were not affected by increasing the IWA in 100 mm, unlike the Northern area where the yields rose almost 1.5 t ha^{-1} (Figure 8). A similar situation was observed under a 25% reduction in IWA, with a yield gap of around 3 t ha^{-1} in the North in comparison with only 0.5 t ha^{-1} in the South. The greater seasonal rainfall in the South leads to negligible yield gaps (Figure 8). In the case of the irrigation WP gaps (Figure 9), the pattern is the opposite of that observed in the yield gaps. The highest irrigation WP gaps were simulated in the South, with an average increase of 1.8 Kg m^{-3} by increasing the IWA in 100 mm.

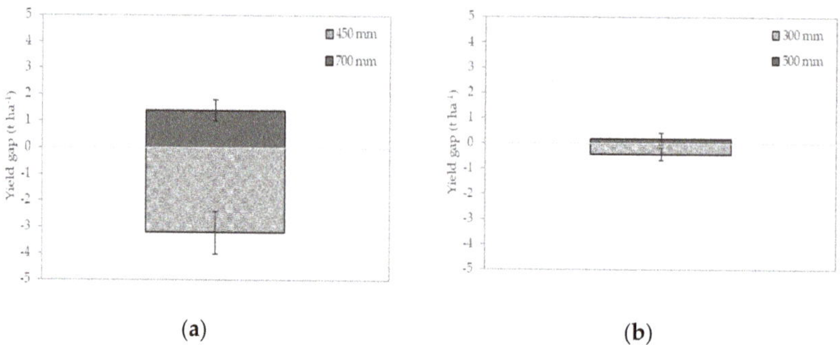

(a)　　　　　　　　　　　　　　　　　　　(b)

Figure 8. Average potential dry root yield gap of the different irrigation water allocation scenarios with respect to the normal irrigation water allocation (reference scenario) for the 30 simulated seasons in the two main productive areas: (**a**) North (Valladolid), and (**b**) South (Sevilla). Vertical bars represent the standard deviation.

The detailed year-to-year simulated yields comparison in the three IWA scenarios is presented in Figure 10. In the North, yields under low IWA were always less than the reference yields and quite variable (Figure 10a). By contrast, in the South, there was very little impact of a reduced IWA given that the inter-annual rainfall variability is corrected with optimal irrigation (Figure 10b).

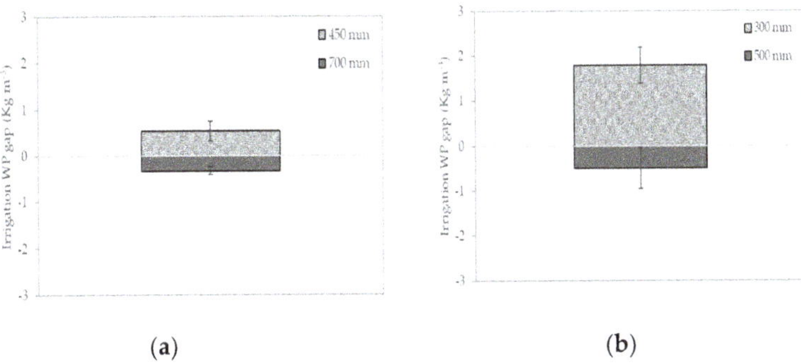

Figure 9. Average simulated irrigation water productivity (WP) gap of the different irrigation water allocation scenarios with respect to the normal irrigation water allocation (reference scenario) for the 30 simulated seasons in the two main productive areas: (**a**) North (Valladolid), and (**b**) South (Sevilla). Vertical bars represent the standard deviation.

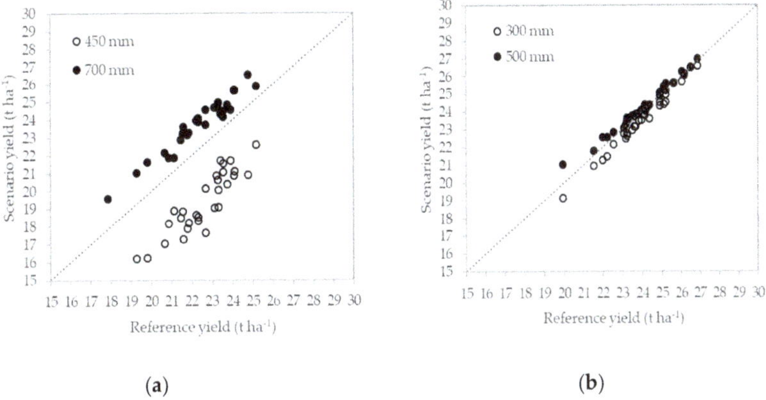

Figure 10. Comparison of potential dry root yield at harvest under different irrigation water allocation scenarios (scenario yield) with potential dry root yield at harvest under the normal irrigation water allocation (reference yield) for the 30 simulated seasons in the two main productive areas: (**a**) North (Valladolid), and (**b**) South (Sevilla).

4. Discussion

In the view of the calibration and validation results (Figures 1–4), AquaCrop effectively reproduced (Table 3) the main features of yield responses of sugar beet to water deficits. It is important to acknowledge that AquaCrop is focused mainly to simulate water-limited yields. Calibration was carried out with the results of a single deficit irrigation experiment, while the model validation was carried out with data from eight farms differing in location, years, varieties, sowing dates, and irrigation (Table 1), with an observed yield range between 16 to 27 t ha^{-1} (Figure 4). This high variability in environmental conditions and management provides robustness to the simulated results. No previous studies of calibration/parameterization and validation of AquaCrop for sugar beet have covered this broad range of production levels (Table 2). Only Reference [20] used a comparable yield range (by introducing differences in sowing dates and irrigation) from experimental fields located in the Tadla irrigation scheme (Morocco), but neither B or CC were experimentally assessed, and agreement statistics were not reported. These shortcomings make it difficult to evaluate the goodness of the adjustment they proposed for the crop parameters (Table 2). The same applies to Reference [18], where only the model performance to predict yield is assessed. In the case of Reference [19], the calibration process (based on irrigation and fertilization experiments) was performed using only a full irrigation treatment (the deficit irrigation treatments were included only in the validation process). This hampers the proper adjustment of the crop water stress parameters. The calibration process in Reference [21] is more unsuitable, since the crop parameters values changed for the different treatments.

Based on the calibration performed here, a set of conservative parameters was adjusted, as reported in Table 2. These crop parameters values do not differ greatly from those in the default crop file for sugar beet within the AquaCrop model database (Table 2). However, there were significant changes in some key crop parameters, such as the normalized WP (WP *), which was increased from 17 g m^{-2} in the default file to 18 g m^{-2} (Table 2), which has an important impact on yield simulation (Figure 3). The new WP * value was already proposed by References [19] and [20] (Table 2). By contrast, Reference [18] considerably under-predicted the yield in the irrigated treatments using the default WP * value of 17 g m^{-2} (Table 2). Another important change was the modification of the base and upper temperature (Table 2) for better simulation of the crop development and the impact of low temperatures on stomatal conductance [29]. Our proposed changes are in line with those recommended by Reference [37] for the base temperature and by References [38] and [39] for the upper temperature. Although the default crop water stress parameters (Table 2) satisfactorily simulated the crop response under deficit irrigation (Figure 1b–d, Figures 2b–d and 3), a small change in the parameter that adjust HIo for water stress during yield formation (Table 2) allowed the correct simulation of the increase in HI due to the water stress suffered during the leaf expansion, as observed by Reference [40]. Furthermore, the validation of the default crop water stress parameters was crucial, since important modifications in the determination of soil water stress were performed in versions 6.0 and 6.1 of AquaCrop [29] in comparison with previous versions. In the 6.1 version, the water depletion in the top soil is compared against whole root zone depletion in order to determine which part of the soil profile controls the water stress. Therefore, the proper definition of the top soil thickness for each crop (Table 2) is also a relevant task for the calibration of the new model versions.

Using these calibrated parameters, AquaCrop accurately simulated the evolution of CC (Figure 1) and biomass (Figure 2), as well as the final root yield (Figure 3). Nevertheless, despite the improvements introduced for better accounting of water stress in the new model versions (v6.0 and 6.1), AquaCrop was not able to simulate the effect of a light rainfall on the water stress reductions (Figure 1d), likely because, in the field, the rainfall triggers new root growth in the surface layers that facilitates rapid extraction of the infiltrated soil water, which quickly reduces water stress, as shown in grain sorghum [41]. The model could not simulate such an adaptive response and this resulted in under-prediction of the final yield for the rainfed treatment (Figure 3). Note that AquaCrop predictions were less accurate for the most deficit irrigated treatments in Reference [19]. This has been observed in other works for different crops, such as in maize [42] and in dry beans [43]. There seems to be a need for improving the AquaCrop simulations under severe water stress. Nevertheless, in view of the good results obtained in the extensive validation test performed in this study with a wide range of production levels (Table 3), it can be concluded that AquaCrop reproduced quite well the main features of sugar beet production as a function of the water supply. Overall, the model performance may be considered satisfactory, particularly when compared with the performance of more complex models such as STICS in Reference [13], but not in Reference [14], or CSM-CERES-Beet [12], even when water stress conditions were not simulated, as in the latter. Therefore, the calibrated crop parameters for sugar beet can be used with confidence in AquaCrop under different environmental conditions and management and for a wide range of current varieties. However, while commercial varieties seem to have similar yield responses to water stress [44,45], the existence of lines with greater drought tolerance [44] could result in new commercial varieties (recent breeding priority) that may require a re-calibration of the crop water stress parameters in the future.

Once AquaCrop has been properly calibrated and validated, it is possible to use it for the simulation of some features that require excessive time and resources in field experimentation. An illustration of this potential is presented here, by simulating the potential yield and the irrigation

WP as a function of IWA in the two main production areas in Spain. The results highlighted that sowing time, associated with location and IWA, significantly impact sugar beet production and its irrigation WP in a given environment. The average potential yield and irrigation WP were higher in the South (autumn sowing) than in the North (spring sowing), with an irrigation water saving of 33%, which is a significant amount for a water-limited environment (Figures 6 and 7). These results can be attributed to the higher evaporative demand and the lower rainfall in spring sowing in the North, which causes a higher irrigation demand (Figure 5). On the other hand, the longer period in autumn sowings during which assimilates are translocated and accumulated in the roots leads to higher root yields. The benefits of autumn sowing have been reported by Reference [46] in a three-year experiment in another Mediterranean environment. Furthermore, the values of irrigation WP reported in Reference [46] for the autumn and spring sowings (5.6 and 4.1 g m^{-3}, respectively) are quite similar to those obtained in this case (6.1 and 3.7 g m^{-3}, respectively). In fact, even the evapotranspiration WP values were nearly identical (2.8 g m^{-3} with respect to 2.7 g m^{-3} simulated in this study). This confers a great reliability on the model performance and in the analysis we have carried out. In view of the results, in environments with low-risk of winter frost and with common summer drought, autumn sowings should be encouraged, both for their high yields and higher irrigation WP. Early plantings of sugar beet in Northern environments have also become a breeding priority, by selecting sugar beet lines with frost resistance [47] and bolting resistance [48].

Water authorities in many parts of the world have proposed to optimize water use through a proper IWA for each crop. AquaCrop can be a useful tool to carry out this undertaking, as has been demonstrated in this study. In the North, the curvilinear relationship between yield and AIW [8,49] was evident in the results ($p < 0.05$, Tukey test) (Figure 6a), increasing the yields almost 1.5 t ha^{-1} by increasing the IWA in 100 mm or with a yield decrease of around 3 t ha^{-1} under a 25% reduction in IWA (Figure 8a). This inversely impacts the irrigation WP (Figures 7a and 9a). The tradeoff between yield and irrigation WP must be taken into account by the water authorities and policymakers in their decisions not to compromise the profitability of the sugar beet production. On the contrary, in the South, an increase or reduction in IWA did not have a significant impact on the yields ($p < 0.05$, Tukey test) (Figures 6b and 8b) due to greater influence of the seasonal rainfall on yields. In this case, however, a high increase in irrigation WP was simulated under a 25% reduction in IWA (Figures 7b and 9b), which could lead to a significant improvement in the water use efficiency without having a significant yield penalty. In the South, under water scarcity situations, restrictions in IWA could be applied to sugar beet without having significant negative impacts when compared with the impact on summer crops or on sugar beet in the Northern area.

Author Contributions: Data curation, formal analysis, investigation, methodology, software, validation and writing – original draft, M.G.-V.; Funding acquisition and writing – review & editing, R.M.-V. and E.F.; Supervision, M.G.-V., R.M.-V. and E.F.

Funding: AIMCRA and the H2020 SHui project (Project number: 773903) funded this research.

Acknowledgments: We thank S. Blanco, J.A. Centeno, and C. Ruz for help with this work.

Conflicts of Interest: The authors declare no conflict of interest.

References

1. FAOSTAT Database. Available online: http://www.fao.org/faostat/en/#home (accessed on 15 July 2019).
2. Rinaldi, M.; Horemans, S. Sugar beet. In *Crop Yield Response to Water*; Steduto, P., Hsiao, T.C., Fereres, E., Raes, D., Eds.; FAO: Rome, Italy, 2012; pp. 202–208.
3. Singels, A.; van der Laan, M. Sugarcane. In *Crop Yield Response to Water*; Steduto, P., Hsiao, T.C., Fereres, E., Raes, D., Eds.; FAO: Rome, Italy, 2012; pp. 174–180.
4. Sadeghian, S.Y.; Yavari, N. Effect of Water-Deficit Stress on Germination and Early Seedling Growth in Sugar Beet. *J. Agron. Crop Sci.* **2004**, *190*, 138–144. [CrossRef]
5. Davidoff, D.; Hanks, R.J. Sugar beet production as influenced by limited irrigation. *Irrig. Sci.* **1989**, *10*, 1–17. [CrossRef]

6.   Pidgeon, J.D.; Werker, A.R.; Jaggard, K.W.; Richter, G.M.; Lister, D.H.; Jones, P.D. Climatic impact on the productivity of sugar beet in Europe, 1961–1995. *Agric. For. Meteorol.* **2001**, *109*, 27–37. [CrossRef]

7.   Jones, P.D.; Lister, D.H.; Jaggard, K.W.; Pidgeon, J.D. Future climate impact on the productivity of sugar beet (*Beta vulgaris* L.) in Europe. *Clim. Chang.* **2003**, *58*, 93–108.

8.   Fereres, E.; Soriano, M.A. Deficit irrigation for reducing agricultural water use. *J. Exp. Bot.* **2007**, *58*, 147–159. [CrossRef] [PubMed]

9.   Werker, A.R.; Jaggard, K.W. Dependence of sugar beet yield on light interception and evapotranspiration. *Agric. For. Meteorol.* **1998**, *89*, 229–240. [CrossRef]

10.   Doorenbos, J.; Kasssam, A.H. *Yield Response to Water*; FAO Irrigation and Drainage Paper No. 33; FAO: Rome, Italy, 1979.

11.   Qi, A.; Kenter, C.; Hoffmann, C.; Jaggard, K.W. The Broom's Barn sugar beet growth model and its adaptation to soils with varied available water content. *Eur. J. Agron.* **2005**, *23*, 108–122. [CrossRef]

12.   Anar, M.J.; Lina, Z.; Hoogenboomb, G.; Sheliab, V.; Batchelord, W.D.; Tebohe, J.M.; Ostliee, M.; Schatze, B.G.; Khanf, M. Modeling growth, development and yield of Sugarbeet using DSSAT. *Agr. Syst.* **2019**, *169*, 58–70. [CrossRef]

13.   Beaudoina, N.; Launay, M.; Saubouac, E.; Ponsardina, G.; Marya, B. Evaluation of the soil crop model STICS over 8 years against the "on farm" database of Bruyeres catchment. *Eur. J. Agron.* **2008**, *29*, 46–57. [CrossRef]

14.   Baey, C.; Didier, A.; Lemaire, S.; Maupas, F.; Cournèdea, P.-H. Parametrization of five classical plant growth models applied to sugar beet and comparison of their predictive capacity on root yield and total biomass. *Ecol. Model.* **2014**, *290*, 11–20. [CrossRef]

15.   Lemaire, S.; Maupas, F.; Cournède, P.-H.; de Reffye, P. A morphogenetic crop model for sugar-beet (*Beta vulgaris* L.). In Proceedings of the International Symposium on Crop Modeling and Decision Support: ISCMDS 2008, Nanjing, China, 19–22 April 2008.

16.   Taky, A. Maîtrise des Excès d'eau Hivernaux et de L'Irrigation et de Leurs Conséquences sur la Productivité de la Betterave Sucrière dans le Périmètre Irrigué du Gharb (Maroc). Analyse Expérimentale et Modélisation. Ph.D. Thesis, AgroParisTech, Paris, France, 2008.

17.   Steduto, P.; Hsiao, T.C.; Raes, D.; Fereres, E. Aquacrop—The FAO Crop Model to Simulate Yield Response to Water: I. Concepts and Underlying Principles. *Agron. J.* **2009**, *101*, 426–437. [CrossRef]

18.   Stricevic, R.; Cosic, M.; Djurovic, N.; Pejic, B.; Maksimovic, L. Assessment of the FAO AquaCrop model in the simulation of rainfed and supplementally irrigated maize, sugar beet and sunflower. *Agric. Water Manag.* **2011**, *98*, 1615–1621. [CrossRef]

19.   Alishiri, R.; Paknejad, F.; Aghayari, F. Simulation of sugar beet growth under different water regimes and nitrogen levels by AquaCrop. *Int. J. Biosci.* **2014**, *4*, 1–9.

20.   van Straaten, J.W. Regional Modelling of Water Stress. Irrigation Water Requirement Meets Water Availability in the oum er Rbia Basin. Master's Thesis, Utrecht University, Utrecht, The Netherlands, 2017.

21.   Malik, A.; Shakir, A.S.; Ajmal, M.; Khan, M.J.; Khan, T.A. Assessment of AquaCrop Model in Simulating Sugar Beet Canopy Cover, Biomass and Root Yield under Different Irrigation and Field Management Practices in Semi-Arid Regions of Pakistan. *Water Resour. Manag.* **2017**, *31*, 4275–4292. [CrossRef]

22.   Araji, H.A.; Wayayok, A.; Khayamim, S.; Teh, C.B.S.; Abdullah, A.F.; Amiri, E.; Bavani, A.M. Calibration of AquaCrop model to simulate sugar beet production and water productivity under different treatments. *Appl. Eng. Agric.* **2019**, *35*, 211–219. [CrossRef]

23.   Vanuytrecht, E.; Raes, D.; Willems, P. Regional and global climate projections increase mid-century yield variability and crop productivity in Belgium. *Reg. Environ. Chang.* **2016**, *16*, 659–672. [CrossRef]

24.   Gobin, A.; Kersebaum, K.C.; Eitzinger, J.; Trnka, M.; Hlavinka, P.; Takác, J.; Kroes, J.; Ventrella, D.; Dalla Marta, A.; Deelstra, J.; et al. Variability in the water footprint of arable crop production across European regions. *Water* **2017**, *9*, 93. [CrossRef]

25.   Geerts, S.; Raes, D.; Garcia, M. Using AquaCrop to derive deficit irrigation schedules. *Agric. Water Manag.* **2010**, *98*, 213–216. [CrossRef]

26.   Garcia-Vila, M.; Fereres, E. Combining the simulation crop model AquaCrop with an economic model for the optimization of irrigation management at farm level. *Eur. J. Agron.* **2012**, *36*, 21–31. [CrossRef]

27.   Carmona, G.; Varela-Ortega, C.; Bromley, J. Supporting decision making under uncertainty: Development of a participatory integrated model for water management in the middle Guadiana river basin. *Environ. Model. Softw.* **2013**, *50*, 144–157. [CrossRef]

28. Liu, J.; Pattey, E. Retrieval of leaf area index from top-of-canopy digital photography over agricultural crops. *Agric. For. Meteorol.* **2010**, *150*, 1485–1490. [CrossRef]

29. Raes, D.; Steduto, P.; Hsiao, T.C.; Fereres, E. Chapter 3. Calculation procedures. In *AquaCrop Version 6.0—6.1. Reference Manual*; Raes, D., Steduto, P., Hsiao, T.C., Fereres, E., Eds.; FAO: Rome, Italy, 2018; pp. 1–141.

30. Raes, D.; Steduto, P.; Hsiao, T.C.; Fereres, E. AquaCrop—The FAO Crop Model to Simulate Yield Response to Water: II. Main Algorithms and Software Description. *Agron. J.* **2009**, *101*, 438–447. [CrossRef]

31. Hsiao, T.C.; Fereres, E.; Steduto, P.; Raes, D. AquaCrop parameterization, calibration, and validation guide. In *Crop Yield Response to Water*; Steduto, P., Fereres, E., Raes, D., Eds.; FAO: Rome, Italy, 2012; pp. 70–87.

32. Vanuytrecht, E.; Raes, D.; Willems, P. Global sensitivity analysis of yield output from the water productivity model. *Environ. Model. Softw.* **2014**, *51*, 323–332. [CrossRef]

33. Saxton, K.E.; Rawls, W.J. Soil water characteristic estimates by texture and organic matter for hydrologic solutions. *Soil Sci. Soc. Am. J.* **2006**, *70*, 1569–1578. [CrossRef]

34. Willmott, C.J. Some comments on the evaluation of model performance. *Bull. Meteorol. Soc.* **1982**, *63*, 1309–1313. [CrossRef]

35. CLIMAGEN. Available online: http://www.uco.es/fitotecnia/contents/index.html (accessed on 15 July 2019).

36. Geng, S.; Auburn, J.S.; Brandstetter, E.; Li, B. *A Program to Simulate Meteorological Variables: Documentation for SIMMETEO*; Agronomy Progress Rep. 204; Dep. of Agronomy and Range Science, Univ. of California: Davis, CA, USA, 1988.

37. Milford, G.F.J.; Pocock, T.O.; Riley, I. An analysis of leaf growth in sugar beet. I. Leaf appearance and expansion in relation to temperature under controlled conditions. *Ann. Appl. Biol.* **1985**, *106*, 163–172. [CrossRef]

38. Terry, N. Developmental Physiology of Sugar Beet: I. The influence of light and temperature on growth. *J. Exp. Bot.* **1968**, *19*, 795–811. [CrossRef]

39. Radke, J.K.; Bauer, R.E. Growth of Sugar Beets as Affected by Root Temperatures. Part I: Greenhouse Studies. *Agron. J.* **1969**, *61*, 860–863. [CrossRef]

40. Werker, A.R.; Jaggard, K.W.; Allison, M.F. Modelling partitioning between structure and storage in sugar beet: Effects of drought and soil nitrogen. *Plant Soil* **1999**, *207*, 97–106. [CrossRef]

41. Fereres, E. Short and Long-Term Effects of Irrigation on the Fertility and Productivity of Soils. In Proceedings of the 17th Colloquium of the International Potash Institute, Bern, Switzerland, 2–6 May 1983; pp. 283–304.

42. Hsiao, T.C.; Heng, L.; Steduto, P.; Rojas-Lara, B.; Raes, D.; Fereres, E. AquaCrop—The FAO crop model to simulate yield response to water: III. Parameterization and testing for maize. *Agron. J.* **2009**, *101*, 448–459. [CrossRef]

43. Espadafor, M.; Couto, L.; Resende, M.; Henderson, D.W.; García-Vila, M.; Fereres, E. Simulation of the responses of dry beans (*Phaseolus vulgaris* L.) to irrigation. *Trans. ASABE* **2017**, *60*, 1983–1994. [CrossRef]

44. Ober, E.S.; Luterbacher, M.C. Genotypic variation for drought tolerance in *Beta vulgaris*. *Ann. Bot.* **2002**, *89*, 917–924. [CrossRef]

45. Hoffmann, C.M.; Huijbregts, T.; Van Swaaij, N.; Jansen, R. Impact of different environments in Europe on yield and quality of sugar beet genotypes. *Eur. J. Agron.* **2009**, *30*, 17–26. [CrossRef]

46. Rinaldi, M.; Vonella, A.V. The response of autumn and spring sown sugar beet (*Beta vulgaris* L.) to irrigation in southern Italy: Water and radiation use efficiency. *Field Crops Res.* **2006**, *95*, 103–114. [CrossRef]

47. Reinsdorf, E.; Koch, H.J.; Märlander, B. Phenotype related differences in frost tolerance of winter sugar beet (*Beta vulgaris* L.). *Field Crops Res.* **2013**, *151*, 27–34. [CrossRef]

48. Hoffmann, C.M.; Kluge-Severin, S. Growth analysis of autumn and spring sown sugar beet. *Eur. J. Agron.* **2011**, *34*, 1–9. [CrossRef]

49. Stewart, J.I.; Hagan, R.M. Functions to predict effects of crop water deficits. *J. Irrig. Drain. Div.* **1973**, *99*, 421–439.

Article

REUTIVAR: Model for Precision Fertigation Scheduling for Olive Orchards Using Reclaimed Water

Carmen Alcaide Zaragoza [1,*], Irene Fernández García [2], Rafael González Perea [3], Emilio Camacho Poyato [1] and Juan Antonio Rodríguez Díaz [1]

[1] Department of Agronomy, University of Córdoba, Campus Rabanales, Edif. da Vinci, 14071 Córdoba, Spain; ecamacho@uco.es (E.C.P.); jarodriguez@uco.es (J.A.R.D.)

[2] Electrical Engineering Department, University of Córdoba, Campus Rabanales, Edif. da Vinci, 14071 Córdoba, Spain; g52fegai@uco.es

[3] Department of Plant Production and Agricultural Technology, School of Advanced Agricultural Engineering, University of Castilla-La Mancha, Campus Universitario, s/n, 0207 Albacete, Spain; Rafael.GonzalezPerea@uclm.es

* Correspondence: g12alzac@uco.es

Received: 31 October 2019; Accepted: 10 December 2019; Published: 13 December 2019

Abstract: Olive orchard is the most representative and iconic crop in Andalusia (Southern Spain). It is also considered one of the major economic activities of this region. However, due to its extensive growing area, olive orchard is also the most water-demanding crop in the Guadalquivir River Basin. In addition, its fertilization is commonly imprecise, which causes over-fertilization, especially nitrogen. This leads to pollution problems in both soil and water, threating the environment and the system sustainability. This concern is further exacerbated by the use of reclaimed water to irrigate since water is already a nutrient carrier. In this work, a model which determines the real-time irrigation and fertilization scheduling for olive orchard, applying treated wastewater, has been developed. The precision fertigation model considers weather information, both historical and forecast data, soil characteristics, hydraulic characteristics of the system, water allocation, tree nutrient status, and irrigation water quality. As a result, daily information about irrigation time and fertilizer quantity, considering the most susceptible crop stage, is provided. The proposed model showed that by using treated wastewater, additional fertilization was not required, leading to significant environmental benefits but also benefits in the total farm financial costs.

Keywords: reclaimed water; fertigation scheduling; precision irrigation; olive orchard

---

1. Introduction

Freshwater resources are mainly used for agricultural irrigation, accounting for more than 70% of all water withdrawals worldwide [1]. This makes agriculture especially vulnerable to drought periods. This problem is emphasized in a context of climate change since alterations in temperature and rainfall patterns and an increase in the occurrence probability of extreme events have been forecasted [2]. Consequently, in arid and semi-arid areas where irrigation agriculture is a major activity, such as Mediterranean countries, extremely high water stress is predicted [3,4]. This will lead to an increase in water demand and potential soil moisture deficit. This situation is particularly critical in Andalusia (Southern Spain), since irrigated agriculture plays a key role in its economy. Olive is the most representative and iconic crop in this region, not only for its importance in the landscape and culture, but also for being considered one of its major economic activities. However, due to its extensive area, olive orchard is the most water demanding crop in the Guadalquivir River Basin, with a water demand higher than 580 hm^3/year for more than one and a half million hectares, which implies about 20% of the total agricultural water demand in that region [5]. On the other hand, olive fertilization is commonly imprecise, which causes over-fertilization, especially nitrogen. This leads to pollution problems in atmosphere, soil, and water, threating the environment and the biosystem sustainability. Thus, irrigated agriculture will need to face the challenge of ensuring production using water and fertilizers in a more sustainable way.

Strategies to improve the efficient use of water and fertilizer, increasing the irrigation districts sustainability, have been proposed. The significant development of information and communication technologies (ICTs) has also encouraged the application of these strategies. For instance, the authors of Reference [6] developed a mobile and desktop application for farmers (IrriFresa App) which incorporated the precision irrigation principles for strawberry crop. Its use in commercial farms led to significant water savings which ranged from 11% to 33%. Other authors developed ICT applications focused on irrigation and fertilization management. Thus, References [7] and [8] proposed methodologies to determine the most economic fertilizer considering the water quality of the irrigation system. However, in both works, the optimal nutrient solution, which is usually unknown by farmers, was needed as a model input. The authors of Reference [9] developed a mobile application with Android Studio IDE (integrated development environment) to determine the fertilizer quantity depending on the crop type and the irrigation system conditions. Nevertheless, this mobile application was only available for greenhouse vegetables. The Institute for Agricultural and Fisheries Research and Training (IFAPA) developed a tool for irrigation and fertigation (fertigation) scheduling for olive orchard [10]. This management tool considers average historic agroclimatic information and determines a monthly fertigation scheduling for olive orchard. However, it is only available online which, sometimes, is not functional for farmers in the day to day irrigation management. It neither considers real-time weather information nor weather forecast. This could cause an inaccurate recommendation because of climate variability.

On the other hand, in order to reduce the pressure on water resources, the use of treated water as an alternative to conventional water sources has also been addressed. Treated wastewater is defined as water arising from any combination of domestic, municipal, or industrial origin that has been processed in a wastewater treatment plant [11]. Reclaimed or reused water is formerly treated wastewater with an additional treatment which makes it suitable for reusing in different purposes, such as agriculture, landscape irrigation, or recharge of groundwater aquifers, among others [12,13]. References [14] and [15] assured that the use of this water constitutes a strategy both for the water scarcity problem and the sustainability of the irrigated agriculture improvement. After long-term research about citrus trees and horticultural crops irrigated with reclaimed water, References [16] and [17] determined that the use of reclaimed water for irrigating citrus trees and horticultural crops generates significant fertilizer savings without posing a risk for human health. Due to its importance and extensive cultivation in Mediterranean areas, different authors have studied the impacts of using reclaimed water as irrigation water in olive orchards on soil pollution [18,19], tree development [20,21], and oil quality [21,22]. They concluded that negative impacts were not found when reclaimed water was properly controlled and managed. The authors of Reference [23] compared olive crop irrigation using reclaimed water and freshwater for eight irrigation seasons. Their work showed that fertilizer applications were not required when reclaimed water was used as irrigation water since the water already satisfied the olive nutrient requirements. Their study also highlighted the importance of a regular water quality control to consider nutrients provided with the water. Therefore, the use of reclaimed water in irrigated agriculture could reduce both the intensive use of fertilizer and the associated energy and economic costs. However, special attention must be paid when reclaimed water is used for irrigation because water is already a nutrient carrier and its nutrient content is variable along the year. Nevertheless, when reclaimed water is properly managed and supported by new advances in the ICT, this non-conventional water source can become a strategic solution to the problem raised [24].

In this work, a new model which determines the precision fertigation scheduling in real-time for olive orchard, considering the particularities of applying reclaimed water, has been developed. The innovation of this model was the integration of the irrigation and fertilization olive management techniques combined with weather forecast and agroclimatic records, updated in real-time, and adapted for the specific case of irrigation with reclaimed water. This model aims to improve the irrigation system sustainability, applying nutrients according to the crop needs alone and concentrating the water application on the most critical crop stages to water stress. This model is intended for farmers and technicians to manage water and fertilizer in the most efficient way. The model has been tested in an olive orchard commercial farm located in Córdoba (Southern Spain).

2. Materials and Methods

The methodology focused on the development of a precision fertigation model for olive orchard, called REUTIVAR, using reclaimed water as water source. The model was developed considering both historical and forecast weather information, soil characteristics, hydraulic characteristics of the irrigation system, irrigation water quality, water allocation, and soil and tree nutrient status. A detailed description of the model, the required inputs, and the case study are provided below.

2.1. Model Description

The model aims to provide the optimal real-time fertigation scheduling in an easy and simple way. It is made up of four independent but interconnected modules: (1) farm characteristics, (2) climate data, (3) irrigation scheduling, and (4) fertilization scheduling (Figure 1). The model was implemented in MATLAB^{TM} (MathWorks Inc., MA, USA) [25].

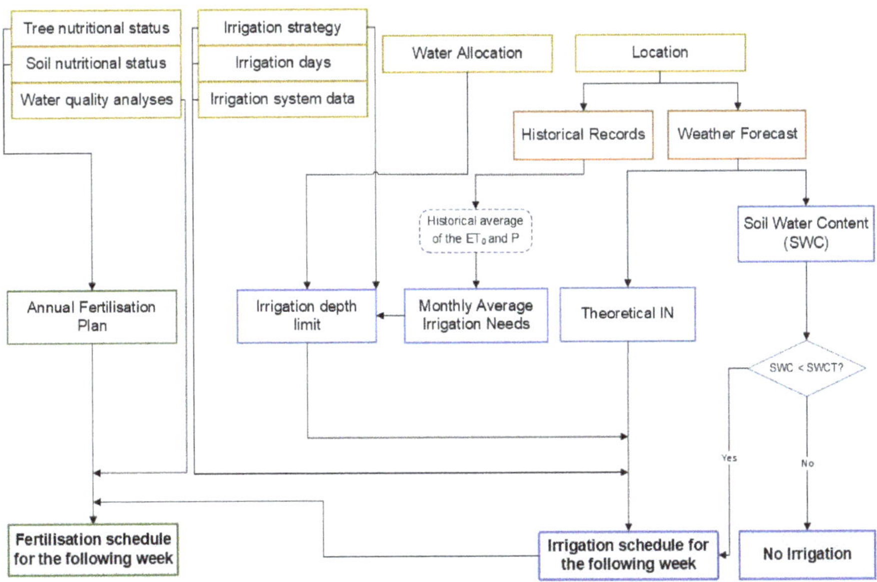

Figure 1. REUTIVAR Flow chart.

2.1.1. Farm Characteristics

Data of the farm is required in this module: location, area, planting pattern, soil texture, water allocation, irrigation system (flow and spacing of emitters), irrigation scheduling, and crop nutritional status. Soil texture was obtained using the USDA (US Department of Agriculture) soil texture triangle. Subsequently, using the ROSETTA model [26], the soil moisture retention curves were calculated.

Growing cycle, crop coefficient, and crop nutrient uptake were also needed, although they were embedded in the model. Water quality analyses are vital when reclaimed water is used for irrigation since the nutrient water content can be different depending on the case. Hence, water quality analyses were carried out regularly. Finally, soil nutritional status was also evaluated to improve the precision of the fertigation scheduling. This input variable was considered as an optional variable to take into account those cases in which soil nutritional status data are not available.

2.1.2. Climate Data

This module considered two aspects: historical agroclimatic data and weather forecasting data, both related to farm location. Firstly, the nearest agroclimatic station to the farm, selected from the available stations in the Agroclimatic Stations Network of the Regional Government of Andalusia [27], was determined. From this agroclimatic station, using web scraping techniques (i.e., an automated process to extract data from websites), the daily values of precipitation (P) and reference evapotranspiration (ET_0) of the available entire time series were obtained. Then, the daily average values of P and ET_0 were computed as well as the monthly average irrigation needs ($IN_{historical}$). $IN_{historical}$ were calculated as the difference between the crop evapotranspiration (ET_c) and the effective precipitation (P_{eff}) according to Allen [28]. ET_c and P_{eff} are defined later in Section 2.1.3.

The daily weather prediction of the study area one-week forward was obtained using AEMET (Agencia Estatal de Meteorología) OpenData. AEMET OpenData is the API REST (application programming interface representational state transfer) of the Spanish State Meteorological Agency [29]. The climate data obtained were: mean temperature (°C), maximum temperature (°C), minimum temperature (°C), maximum relative humidity (%), minimum relative humidity (%), wind speed (km/h), and cloudiness index (%). From these climate variables, the value of ET_0 was calculated by the FAO (Food and Agriculture Organization) Penman–Monteith equation [28]. Finally, by web scraping techniques, forecasted precipitation was obtained from eltiempo.es [30]. Both weather forecasting and historical data were used to schedule the irrigation events.

2.1.3. Irrigation Scheduling

Irrigation scheduling was determined based on theoretical daily irrigation requirements for the following week (IN_d). However, parameters such as monthly water thresholds, soil water content, irrigation system, and irrigation scheduling options were also considered.

Firstly, because of water scarcity problems, monthly water thresholds for irrigation (IWT) were established to ensure water availability at the most critical crop stage. IWT were determined considering the irrigation strategy and $IN_{historical}$, and the total water allocation according to the volume of reclaimed water in the treatment plant and the water allocation established by the water agency [5]. However, although the initial approach was based on historical records, the weekly irrigation recommendations were determined according to weather forecast. Therefore, fortnightly, REUTIVAR checked if the water initially scheduled, considering historical data, had been consumed, i.e., if the irrigation water applied from the beginning of the irrigation season matched to the IWT for that period. Otherwise, IWT was recalculated for the following months.

Both $IN_{historical}$ and IN_d were calculated as the difference of ET_c and P_{eff}. The main difference between both variables is that $IN_{historical}$ was calculated using historical agroclimatic data and IN_d using the weather forecasting. P_{eff}, the amount of rainfall actually stored in the soil, was calculated using a fixed percentage of P [31]. In this case, it was considered a value of the 80% of P. The crop irrigation needs were calculated to refill the daily ET_c, obtained by the methodology proposed by FAO [32]:

$$ET_c = ET_0 \cdot k_c \cdot k_r \tag{1}$$

where ET_0 (mm) was calculated from the weather forecast (see Section 2.1.2), k_c is the crop coefficient (in this work the values proposed by Reference [33] were used, Table 1), and k_r is a parameter related to the tree canopy. k_r is equal to 1 for crops with more than 60% of soil cover and ranges from 0 to 1 otherwise. In that case, k_r is obtained by using Equation (2), proposed by Reference [34].

$$k_r = 2 \cdot S_c / 100 \quad \text{for } S_c \leq 60\%$$
$$k_r = 1 \quad \text{for } S_c > 60\% \tag{2}$$

where S_c (%) is the percentage of the soil covered by the canopy at midday and it is calculated as a function of the average canopy diameter, D (m), and plant density (N(tree/ha)):

$$S_c = \frac{\pi \cdot D^2 \cdot N}{400} \tag{3}$$

Table 1. Olive crop coefficient (k_c) in Córdoba (Spain).

Jan	Feb	Mar	Apr	May	Jun	Jul	Aug	Sept	Oct	Nov	Dec
0.65	0.65	0.65	0.60	0.55	0.55	0.50	0.50	0.55	0.55	0.60	0.65

A soil water balance was used to establish the occurrence of irrigation events. This happened when soil water content was less than 25% of the field capacity value (FC). This is called soil water content threshold (SWCT) in this work. The daily soil water content (SWC_d) provides information about the amount of water that the crop can extract from its root zone (Equation (4)). Daily, REUTIVAR simulated the soil water balance until that date by using historical data. Then, a soil water content prediction for the following week was conducted considering the weather forecast.

$$SWC_d = SWC_{d-1} + P_{eff,d} + IN_d - ET_{c,adj,d} - R_d - D_d \tag{4}$$

where d is the irrigation day, SWC_{d-1} is the soil water content in the day d-1 (mm), IN_d is the applied irrigation depth (mm), $ET_{c,adj}$ is the crop adjusted evapotranspiration to take into account the crop difficulty to extract water when the soil water content diminishes (mm), R_d is the runoff (mm), and D_d is the deep percolation (mm). Both R_d and D_d are deemed null in this work, since REUTIVAR is designed for a drip irrigation system. To compute the SWC, information about soil type and rooting depth were also required.

However, the amount of water that the crop can extract from the soil is not uniformly distributed along the soil drying period and changes depending on the soil moisture. Thus, when the soil is wet, the resistance to water extraction is low, and the crop water uptake can satisfy the atmospheric water demand, i.e., water uptake equals ET_c. In contrast, when the water uptake does not reach ET_c, this term must be adjusted, as shown in Equation (5) [28].

$$ET_{c,adj} = ET_c \cdot \frac{TAW - D_r}{TAW - RAW} \tag{5}$$

where TAW is the total available water (mm) calculated from Equation (6), D_r is the root zone depletion (mm), calculated as the difference between TAW and SWC_{d-1}, and RAW is the readily available soil water in the root zone (Equation (7)).

$$TAW = 1000 \cdot (\theta_{FC} - \theta_{PWP}) \cdot Z_r \tag{6}$$

where θ_{FC} is the water content at field capacity (m^3/m^3), θ_{PWP} is the water content at permanent wilting point (m^3/m^3), and Z_r is the rooting depth (m), which was considered as 1 m for the olive case.

$$RAW = p \cdot TAW \tag{7}$$

where p is the fraction of TAW in which the crop can extract water without suffering water stress, and its value is also obtained from Allen [28]. In the olive case, p is considered 0.75 [33].

In addition, the model considered the days of the week the user chose to undertake irrigation to establish the irrigation volume. To do this, REUTIVAR calculated the irrigation needs for the whole week using the weather forecast. In case of internet connection or data source failure, the daily average values were considered temporally. Once connectivity was recovered, the model distributed the irrigation recommendations amongst the selected days by assigning to each of those days its correspondent volume plus the one of the following days, until the next day of irrigation (Equation (8)).

$$
\begin{aligned}
IN_d &= \sum_{d=1}^{n-1} IN_d \cdot k_{strategy} \ \text{ for } IN_d < \left(\tfrac{IWT}{dm}\right)\cdot id \\
IN_d &= \left(\tfrac{IWT}{dm}\right)\cdot id \ \text{ for } IN_d > \left(\tfrac{IWT}{dm}\right)\cdot id
\end{aligned}
\tag{8}
$$

where n is the next irrigation day chosen by the user, $k_{strategy}$ is the applied coefficient depending on the irrigation scheduling options, dm is the number of days of the month, and id is the number of days from the current day to the next irrigation event.

Three irrigation scheduling options were included in the model: full irrigation (FI), sustained deficit irrigation (SDI), and regulated deficit irrigation (RDI). These strategies are currently the most widespread options, as shown in Reference [35] for olive orchard. These strategies are explained in detail below.

- FI: in this strategy, the irrigation events are scheduled to cover the total irrigation olive needs, i.e., to fully refill daily ET_c. Therefore, in this case, $k_{strategy}$ equals 1.
- SDI: a percentage of the total olive crop irrigation needs is applied equally along the irrigation period. This percentage can be selected and modified manually, and it corresponds to the value of $k_{strategy}$ in the previous equation.
- RDI: a percentage of the FI is also applied but with varying the irrigation volume according to the crop phenological phase. This strategy concentrates the water stress on the least critical stage to oil production. Specifically, this period is the pit hardening, which ranges from the ending of the fruit set to the beginning of the fruit growth [36,37]. In addition, the pit hardening stage matches the summer, when transpiration efficiency is also minimal. Therefore, $k_{strategy}$ is variable based on these variations.

Finally, the irrigation time (t) was determined according to Equation (9).

$$
t_d = \frac{IN_d \cdot A \cdot 10^4}{IE \cdot q_e \cdot n_e}
\tag{9}
$$

where t_d is the irrigation time for the day d (h/day), A is the sector area (ha), IE is the irrigation efficiency, which was considered 0.95 for the model, q_e is the emitter flow (L/h), n_e is the emitter number, and 10^4 is the unit conversion factor. There is the possibility of limiting t_d according to the off-peak energy tariff hours.

2.1.4. Fertilization Scheduling

The fertilization schedule was established according to an annual plan, which varied depending on the crop uptake and the previous year nutritional status, as shown in Reference [38] for olive trees. According to this author, the nutritional status of the tree is determined by using foliar diagnosis [39]. The samples must be taken in July since from that date, the foliar nutrient content is stable. In addition, for that period, the critical nutrient level in leaves for olive are tabulated [40] (Table 2). The analysis results were compared with the leaf nutrient levels and, only in case of deficiency, fertilization applications were scheduled.

Table 2. Interpretation of nutrient content level of olive leaves taken in July, expressed as dry matter percentage according to Reference [40].

Element (%)	Deficient	Adequate	Toxic
Nitrogen (N)	1.20	1.30–1.70	>1.70
Phosphorus (P)	0.05	0.10–0.30	-
Potassium (K)	0.40	>0.80	-

If fertilizer application was required, the total fertilizer amount was calculated according to an estimation of the annual nutrient uptake for olive crop [38]. The following year, foliar analyses were conducted again to increase or decrease the nutrient dose. A flowchart decision tree of the methodology to establish the annual fertilization plan is shown in Figure 2. For nitrogen (N), if leaf nutrient content was lower than 1.3%, the estimation of N requirements was 0.5 kg/tree. However, total N application could not exceed 100 kg/ha. For phosphorus (P), in case P leaf content was less than 0.08%, then 0.5 kg/tree were needed. Finally, for potassium (K), the estimated application was 1 kg/tree if K leaf content was less than 0.7%. The total fertilizer amount was distributed along the irrigation season depending on the crop grow cycle, following the recommendations of Reference [41] (Table 3). The nutrient requirements were met by applying N, P_2O_5, and K_2O, which are the main components of commercial fertilizers.

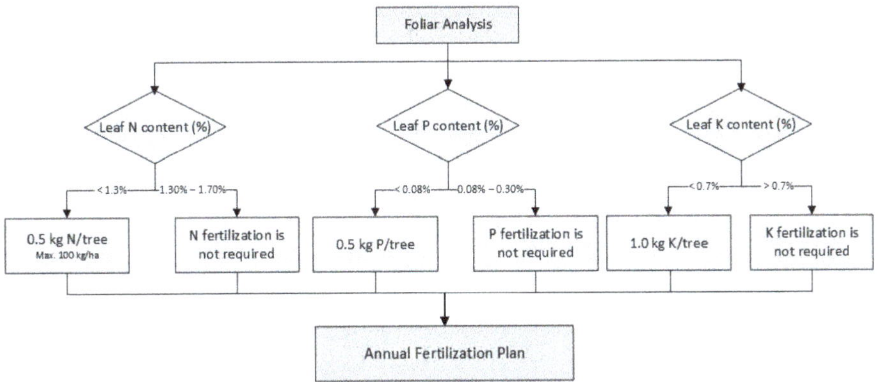

Figure 2. Fertilization flowchart decision tree.

Table 3. Monthly distribution of nutrient applications on fertigation (%).

Month	N	P_2O_5	K_2O
April	5	4	4
May	28	25	17
June	28	25	17
July	25	23	31
August	14	23	31

The final nutrient quantity to be applied was calculated as that estimated in the annual plan minus the nutrient content determined from water quality samples. This quantity was adjusted to the irrigation schedule, i.e., the fertilizer application days always coincided with the irrigation days. Finally, the user can select between fertilizing once a week or in each irrigation event.

2.2. Graphical Interface

Finally, the model previously described was integrated in a desktop application to make its use easier. The desktop application was developed by the graphical interface development environment

(App Designer) of MATLAB^{TM} version R2018a [42]. Then, it was compiled using MATLAB Compiler^{TM} with MATLAB Runtime 9.4.

2.3. Case Study

The developed model was tested in a commercial olive orchard located in the Tintín Irrigation District (TID), Córdoba (Southern Spain). In this region, the climate is typically Mediterranean, with annual average rainfall of 590 mm, mainly in spring and autumn. The annual average temperature is 16.9 °C and the ET_0 is 3 mm/day. Olive grove is the main crop (*Olea europea* L., cv Hojiblanca and *Olea europea* L., cv Nevadillo azul) spaced at 8 × 8 m and devoted to oil production. The average production of the farm is around 7000 kg/ha. The water applied in TID comes from the village sewage treatment plant. This water is stored in a reservoir where chemical and physical treatments are carried out to obtain the required water quality for reuse. The water is then distributed by subsurface drip irrigation (2.2 L/h pressure compensating drippers spaced at 1 m and installed at a depth of 40 cm). A water allocation of 1500 m^3/(ha·year) is applied according to the irrigation district manager criteria. Fertilizers are applied by fertigation and also based on the farmer's experience.

2.4. Management Scenarios

In order to test the model as well as to highlight its usefulness, five scenarios have been assessed:

1. Actual management scenario (AM). This scenario took into account the fertilization and irrigation scheduling established by the irrigation district manager. It consisted of irrigation applications during 8 h on Tuesdays, Thursdays, Saturdays, and Sundays for the entire irrigation season. Regarding fertilization, they applied 10 kg/ha of N, 6 kg/ha of P_2O_5, and 12 kg/ha of K_2O.
2. Optimal water and fertilization management scenario (OWAF). This scenario is obtained by applying the REUTIVAR model. It considered the RDI strategy taking into account soil, water, and crop analysis to establish an optimal fertigation scheduling using historical and forecasted climatic data for the actual conditions. The irrigation days were selected according to the TID criteria, but the irrigation events' duration was variable during the irrigation season, according to the most critical crop stages to water stress.
3. Optimal water, fertilization, and electricity cost management scenario (OWAFE). As in the previous scenario, this option considered the RDI strategy to determine the optimal irrigation and fertilization scheduling, but the electricity tariff was included. In this scenario, the optimal fertigation was scheduled only during off-peak energy tariff hours. Therefore, the irrigation days were similar to previous scenarios. The irrigation event duration was also adjusted to the crop stages. However, in contrast to OWAF, there was also a daily time limit for irrigation. This limit was 8 h during the weekdays. No limits were established on weekends and in August.
4. Actual management scenario considering a hypothetical olive nutrient deficiency (AM-ND). This scenario was the same as AM, i.e., it considered the fertigation scheduling established by the irrigation district manager, but it also simulated a hypothetical olive N, P, and K deficiency. This scenario intended to evaluate the efficiency and adequacy of the TID fertilization application strategy in case of nutrient deficiencies.
5. Optimal water, fertilization, and electricity cost management scenario considering hypothetical olive nutrient deficiency (OWAFE-ND). As in the previous scenario, this option considered a hypothetical N, P, and K deficiency to determine the optimal irrigation and fertilization scheduling in that case but considering the management of the OWAFE scenario.

3. Results and Discussion

3.1. Analysis of Soil and Nutrients in the Study Area

In the study farm, the texture and soil nutrient content, water quality, and tree nutritional status were analyzed to run the model. Regarding the soil, tests at different sites in the farm were carried out to assess soil texture, electrical conductivity, pH, and nutrient content. The soil samples were taken considering changes in the morphology, color, and slope to cover as much soil diversity as possible, as recommended by Parra [43]. Four places were selected for this purpose: two of them were located under the tree (U) and the remaining two were located between trees (B). At each place, two samples were taken at different depths, according to the type of crop. For the olive tree case, the samples were taken between 0 and 15 cm (P1) and between 15 and 30 cm (P2). In terms of assessing the nutritional status of the tree, foliar diagnosis was carried out [39], two samples of 100 leaves from 50 trees each were analyzed in July 2018 and 2019. Finally, water quality analyses were carried out monthly. The water samples were taken in the pipe between the reservoir and the pumping station. Thus, the nutrient content of the water in the model was updated fortnightly using these results.

Several analyses were carried out during the 2018 and 2019 irrigation seasons in the olive orchard farm located in TID. Table 4, Table 5, and Table 6 show the soil, foliar, and water quality analysis results, respectively.

Table 4. Soil analyses undertaken in September 2018 in the case study farm.

Name Sample	Depth (cm)	Texture			P (mg/kg)	K (mg/kg)
		Clay (%)	Silt (%)	Sand (%)		
U1P1	0–15	37.8	26.8	35.4	14.1	563
U1P2	15–30	38.1	28.0	33.9	7.7	454
U2P1	0–15	34.1	23.5	42.4	9.9	454
U2P2	15–30	37.4	20.0	42.6	5.8	317
B1P1	0–15	37.2	28.2	34.6	23.6	872
B1P2	15–30	34.6	28.9	36.5	25.7	794
B2P1	0–15	32.4	23.7	43.9	22.8	978
B2P2	15–30	36.3	21.3	42.4	27.0	598

Table 5. Olive foliar analyses conducted in July 2018 and 2019 in the studied areas (values expressed as dry matter percentage).

Element (%)	'Hojiblanca'		'Nevadillo azul'	
	2018	2019	2018	2019
N	2.02	1.80	1.35	1.81
P	0.11	0.11	0.13	0.14
K	0.92	0.82	0.88	0.84

Table 6. Water nutrient analysis undertaken in the case study farm during the 2019 irrigation season.

Date	$N\text{-}NH_4$ (mg/L)	$N\text{-}NO_3$ (mg/L)	$P\text{-}PO_4$ (mg/L)	K (mg/L)
21 August 2018	1.0	1.5	1.7	-
25 September 2018	1.5	1.3	1.2	-
23 May 2019	10.7	1.8	0.5	-
13 June 2019	13.0	1.9	0.4	-
29 June 2019	12.2	1.7	0.2	32
9 August 2019	2.3	2	0.2	-
9 September 2019	3.3	1	0.2	-

The farm soil was defined as clay loam. From the moisture retention curves, the FC value was determined as 0.28 cm^3/cm^3 and the permanent wilting point (PWP) as 0.14 cm^3/cm^3. Finally, as for the macronutrient amount, the phosphorus (P) amount between trees was higher than under the tree. Nevertheless, the P amount contained in all the locations suggests that a P fertilization response was unlikely, especially for olive orchard. This is because P extractions are low, and this nutrient is easily reusable for olive trees. In addition, high P content could cause zinc (Zn) blockages, originating Zn deficiency in the tree. Likewise, considering the potassium (K) analysis, all the samples showed high levels of K and, consequently, the expected fertilizer response was unlikely. The nitrogen (N) levels in soil were not analyzed due to the high short-term mobility of this element in the soil.

As shown in Table 5, the amount of P and K in olive leaf were within the recommended range in the 2018 irrigation season and the values were adequate in the following year. The P leaf level remained steady for the two seasons. The K leaf content was slightly lower in the 2019 irrigation season, although it stayed within the correct range. However, for the N level, except in 'Nevadillo azul' variety in the 2018 irrigation season, all the leaf samples showed mildly high nitrogen content. According to Molina-Soria and Fernández-Escobar [44], a nitrogen leaf content higher than 1.7% causes impacts on flower and a decrease in oil quality. Other authors also agree with the damages that an excess of nitrogen can produce in olive, such as a decrease in the olive frost tolerance, a delay in the fruit ripening, leading to a reduction in fat yield, and also soil pollution produced by nitrogen leaching [45–47].

As shown in Table 6, the nitrogen content varied along the 2019 irrigation season, especially the ammoniacal nitrogen (N-NH4). The highest nitrogen concentrations occurred in May, June, and July, when nitrogen needs increase. Additionally, in May and June, the irrigation needs for olive are also larger. Therefore, when a RDI strategy is considered, the total nitrogen application must be higher. The nitrogen fluctuation highlighted the importance of a regular water quality control since the N fertilizer amount to be applied could be over or underestimated. Regarding phosphates, the amount contained in water was higher in 2018 than in 2019. During the 2019 irrigation season, the variations were not significant. However, the concentration difference between years also emphasized the relevance of water quality controls. Finally, about potassium concentration, it was only possible to take one sample because of technical problems. The K concentration in that sample was considerably high, which indicated substantial nutrient application throughout the irrigation season. The K concentration was assumed constant for the whole irrigation season.

3.2. Analysis of Management Scenarios

The model was applied for 2019 irrigation season data and the five proposed scenarios. Figure 3 shows the AM, OWAF, OWAFE, AM-ND, and OWAFE-ND irrigation schedules along the 2019 irrigation season. For AM and AM-ND, the irrigation schedule distribution was the same and, for that reason, both are represented in a single figure (Figure 3b). It also occurred for OWAFE and OWAFE-ND scenarios, which are represented in Figure 3d.

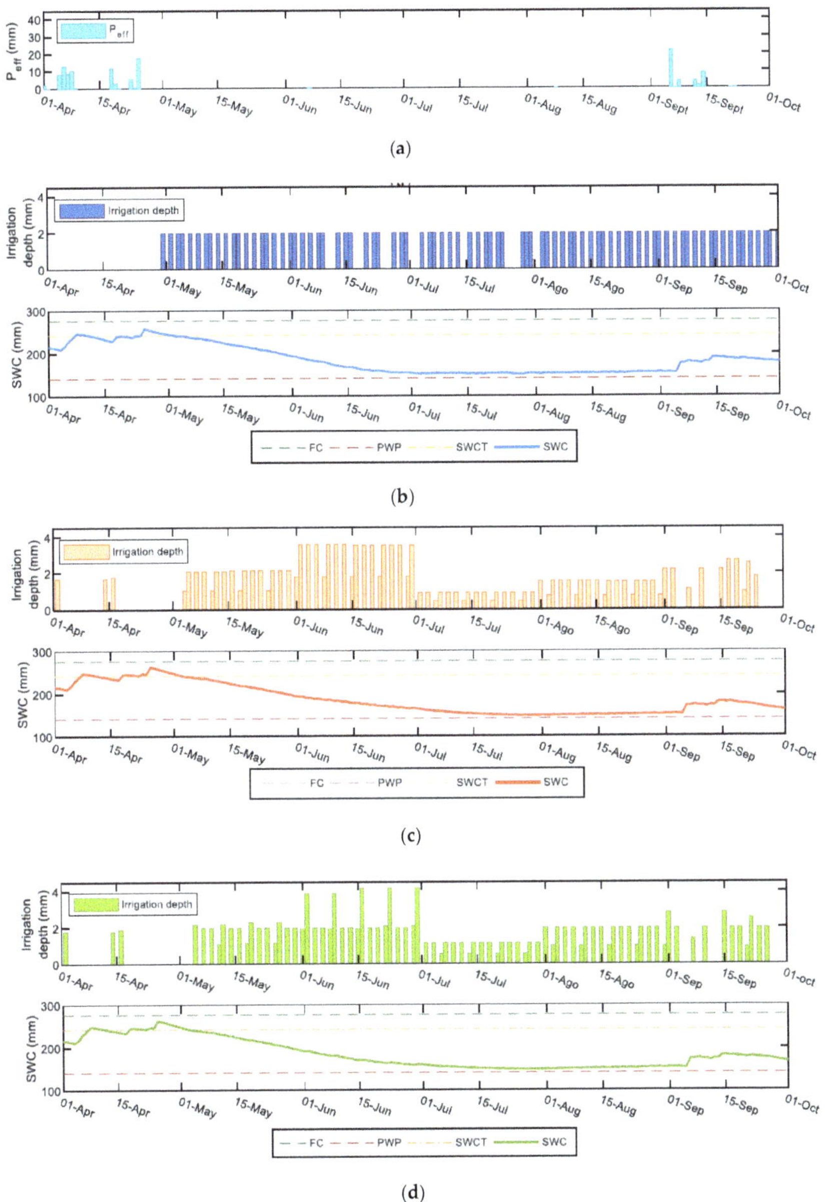

Figure 3. (**a**) Seasonal distribution of daily effective precipitation for the 2019 irrigation season. (**b**) Seasonal distribution of daily soil water content and irrigation scheduling in the actual management (AM) and actual management scenario considering a hypothetical olive nutrient deficiency (AM-ND) scenarios. (**c**) Seasonal distribution of daily soil water content and irrigation scheduling in the optimal water and fertilization management (OWAF) scenario. (**d**) Seasonal distribution of soil water content and irrigation scheduling in the optimal water, fertilization, and electricity cost management (OWAFE) and the optimal water, fertilization, and electricity cost management scenario considering hypothetical olive nutrient deficiency (OWAFE-ND) scenarios.

The total P_{eff} from April to September was 123 mm (Figure 3a) and the average ET_0 during the irrigation season about 4.5 mm·day^{-1}, which can be considered as a regular year. The total water applied in the AM scenario was 1623 m^3/ha for the 2019 irrigation season. It did not adjust the total water applied to the water allocation according to the irrigation district manager criteria (1500 m^3/ha). This could affect the profitability of the studied irrigation district since the water use in that region is limited by the water authority, causing financial penalties in case of exceeding the established limit. However, OWAF, OWAFE, and OWAFE-ND adjusted the total water applied to this water allocation limit. In particular, OWAF applied a total of 1492 m^3/ha, and OWAFE and OWAFE-ND, a total of 1497 m^3/ha. The three of them concentrated the irrigation events in May, June, and September, when olive is more sensitive to water stress [36,48]. However, for the OWAF scenario, the water application distribution could be steadier since time restrictions were not considered. In contrast, OWAFE and OWAFE-ND took into account the electricity tariff existing in TID. In these scenarios, the optimal fertigation was scheduled only during off-peak hours, when power and energy are cheaper. For the TID electricity tariff, the off-peak hours included from 12 am to 8 am every day, as well as 24 h for Saturdays, Sundays, and bank holidays. This effect reduced the water application in the most water-demanding months according to the RDI strategy. Thus, the volume of water applied in May and June was lower in OWAF than in the OWAFE and OWAFE-ND scenarios.

In terms of soil water content, in all the scenarios, the water content never fell below the PWP. This could be because the soil is extremely dry and the resistance to water extraction is considerably higher when water content in soil is lower than the RAW. It was also noticed that, despite applying the scheduled irrigation in all the cases, the soil water content decreased from May to the first precipitation event in September. However, this decrease was slightly lower in OWAFE and OWAFE-ND and noticeably lower in OWAF, since more water was applied in June. This allowed the soil not to be depleted abruptly and for a shorter period of time. Despite this, in these three scenarios, the soil water content diminished considerably. This is because of the limit of water allocation, noticeably lower than the olive orchard irrigation needs. Therefore, the soil water content only increased after the first precipitation events.

Regarding fertilization needs in both OWAF and OWAFE scenarios, fertilizer application was not required according to REUTIVAR, as recommended in Reference [38]. This is because the nutrient leaf levels of the olive orchard were in the correct range, i.e., there was no nutrient deficiency. In the AM scenario, although fertilizers were not needed either, some were applied. Then, olive nutrient deficiency was forced for the AM-ND and OWFE-ND scenarios to evaluate how REUTIVAR would operate in that hypothetical situation. Table 7 shows the nutrient amount applied for all the scenarios.

Table 7. N, P, and K applications through water and fertilizers for AM, AM-ND, OWAF, OWAFE, and OWAFE-ND scenarios.

Scenario	AM/AM-ND			OWAF			OWAFE			OWAFE-ND		
Nutrient	N	P	K	N	P	K	N	P	K	N	P	K
Applied with water (kg/ha)	15.8	0.5	52	18.3	0.5	48.0	15.5	0.4	48.0	15.5	0.4	48.0
Applied with fertilizer (kg/ha)	9.9	2.7	10.2	0	0	0	0	0	0	59.9	73.7	108.0
TOTAL	**25.6**	**3.2**	**62.2**	**18.3**	**0.5**	**48.0**	**15.5**	**0.4**	**48.0**	**75.0**	**74.1**	**156.0**

The use of reclaimed water as a water source for irrigation involves the nutrient application within the water, as shown in Table 7. For this reason, regular water quality controls are essential for the proper irrigation system performance. In both OWAF and OWAFE scenarios, considering the average production of the farm, the nitrogen removal by harvest and pruning was compensated with the applied nitrogen within water irrigation in addition to the rainwater nitrogen and the organic matter mineralization [49]. More nitrogen was applied in the OWAF scenario compared to the OWAFE scenario because in June, when more water was applied in OWAF, the nitrogen content in the water was also higher. As for phosphorus, olive trees show low P extractions by harvest, and they can also reuse it. For that reason, olive orchards in that region do not usually present P deficiency problems and they do not frequently respond to P applications. Therefore, the P application within the water was enough to cover olive needs. Finally, the potassium element showed the highest concentration within the irrigation water. However, the extraction of this element is significantly high in olive trees. Hence, potassium deficiency is the major nutritional problem in olive orchard in Andalusia, playing a key role in the olive tree nutrition. Nevertheless, the K amount applied in the irrigation water covered all potassium removals for harvest. Thus, the fertilizer application was not required, which led to a reduction in the total annual farm cost and a decrease in pollution keeping the olive oil production.

In the hypothetical high nutrient deficiency situation, i.e., the AM-ND and OWAFE-ND scenarios, considering the farm area and the tree spacing, the total nutrient needs according to Reference [38] recommendations, were: 78 kg/ha of N, 78 kg/ha of P, and 156 kg/ha of K. Based on these recommendations, in the AM-ND scenario, a fertilizer deficiency was determined. In both AM and AM-ND fertilization, decisions were not based on the tree nutritional status, which can cause fatal damages to the farm, its production, and profitability. On the other hand, in the OWAFE-ND scenario, the nutrient needs were covered, which would lead to an increase in farm yield. Under this scenario, the use of reclaimed water as irrigation water would entail savings of 21%, 1%, and 31% of the N, P, and K needs, respectively.

In summary, the use of the fertigation recommendations (OWAF and OWAFE scenarios) provided by the REUTIVAR model entailed a better water distribution along the irrigation season compared to conventional practices (AM) in which water is applied without considering the most critical olive phenological stages. As for nutrient applications, an excess of nutrients was applied in the AM compared to OWAF and OWAFE scenarios, in which the nutrient content of reclaimed water is considered. Furthermore, in the hypothetical case of nutrient deficiency, since the conventional practice is not based on nutrient tree status, not enough fertilizers were applied in the AM-ND scenario. In contrast, if the model recommendation would be followed, all the nutrient requirements would be covered.

3.3. Graphical Interface

Finally, the developed model was integrated in a user-friendly graphical interface to make the daily fertigation scheduling management easier (Figure 4).

In the screen (a) of the graphical interface (Figure 4a), the user must introduce the required data to run the model: farm location, type of soil, water allocation, irrigation system characteristics, irrigation strategy, foliar analysis results, and water quality results. In this screen, the user can select between different irrigation strategy options: RDI, SDI, and FI. They can also select the days of the week they wish to irrigate and if they wish to fertilize whenever irrigation is applied or weekly. Then, all this information is stored in an internal database through the save button. The fertigation schedule for the following week is provided in the screen (b) (Figure 4b). Firstly, in this screen, a precipitation and temperature forecast for the following week is provided. Then, a simulation of the soil water content for that week is shown. At the bottom of the screen, irrigation and fertilization scheduling recommendations are provided. The irrigation scheduling is given in volume (mm) and in time units (hours), which is more functional. In addition, information about total water applied from the beginning of the irrigation season and water expected to be applied for that week are also shown. Finally, in this screen, two buttons can also be observed: the refresh button and the edit data button. The refresh button allows to manually update the real-time information of screen (b). By pushing the edit button, screen (a) reappears and the user can modify any of the data previously introduced.

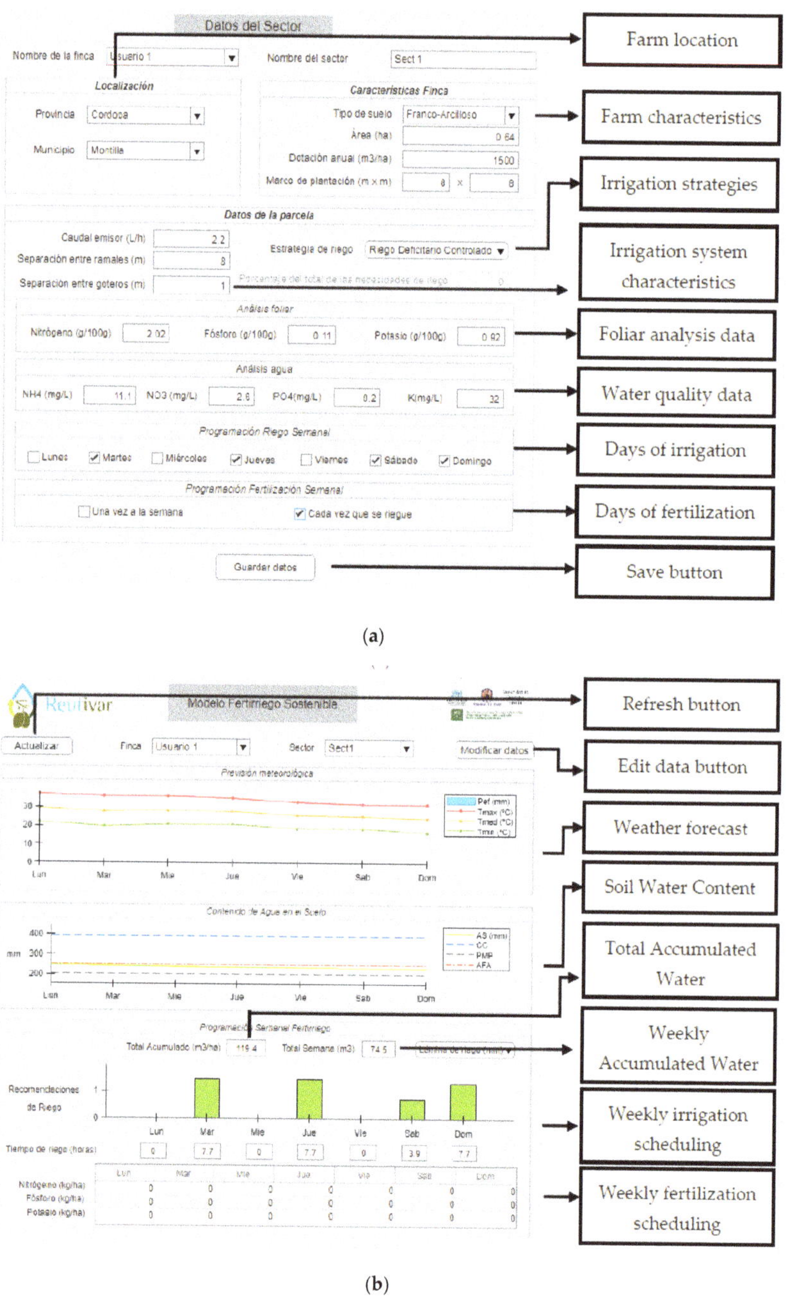

(a)

(b)

Figure 4. Graphical interface of the REUTIVAR model for fertigation schedule recommendations from 3 October 2019 to 9 October 2019. (**a**) Required data to run the precision fertigation model, (**b**) Irrigation and figure recommendations of the model.

4. Conclusions

A precision fertigation model using reclaimed water, REUTIVAR, was developed. REUTIVAR was validated in the 2019 irrigation season in a commercial olive orchard. To do this, five scenarios were simulated. This validation proved that REUTIVAR adjusts the established water allocation along the irrigation season, preventing additional costs to the farm. However, it also showed that this water allocation was much lower than the olive orchard irrigation needs. The scenario simulations also indicated that with the current farm characteristics, the RDI strategy could be applied, even if electricity tariff limitations were considered. This could involve important improvements in both quality and quantity of final oil production. This validation also proved that, thanks to the nutrients which the water carry, additional fertilization was not required. This implies large benefits in the environment but also in the total financial costs of the farm. On the other hand, if the hypothetical nutrient situation is assumed, the current fertigation management of the farm would not cover the nutrient needs either. This shows the importance of the nutritional crop status diagnosis to establish the fertilization decisions.

This work confirms that irrigation with reclaimed water should be managed in a more sustainable way since famers tend to overfertilize. However, following REUTIVAR's recommendations, it is possible to save fertilizer costs with positive effects on the environment and farmer's incomes. In addition, although the model was validated for olive orchard, REUTIVAR could be adapted to other crops, offering a useful tool to manage reclaimed water in an efficient and sustainable way.

Author Contributions: Conceptualization, J.A.R.D. and E.C.P.; methodology, C.A.Z., I.F.G., R.G.P. and J.A.R.D.; investigation, C.A.Z., J.A.R.D. and E.C.P.; resources, J.A.R.D. and E.C.P.; data curation, J.A.R.D. and E.C.P.; writing—original draft preparation, C.A.Z.; writing—review and editing, C.A.Z., I.F.G., R.G.P., J.A.R.D. and E.C.P.; visualization, C.A.Z.; supervision, J.A.R.D. and I.F.G.; project administration, J.A.R.D. and E.C.P.; funding acquisition, J.A.R.D. and E.C.P.

Funding: This work is part of the REUTIVAR (Modelo Sostenible del Olivar Mediante el Uso Aguas Regeneradas) project, co-funded by the Regional Government of Andalusia and the European Union through EARF 2014–2020.

Conflicts of Interest: The authors declare no conflict of interest.

References

1. FAO. *The Future of Food and Agriculture: Trends and Challenges*; FAO: Rome, Italy, 2017; ISBN 9789251095515.
2. Beniston, M.; Stephenson, D.B.; Christensen, O.B.; Ferro, C.A.T.; Frei, C.; Goyette, S.; Halsnaes, K.; Holt, T.; Jylhä, K.; Koffi, B.; et al. Future extreme events in European climate: An exploration of regional climate model projections. *Clim. Chang.* **2007**, *81*, 71–95. [CrossRef]
3. Bisselink, B.; Bernhard, J.; Gelati, E.; Adamovic, M.; Guenther, S.; Mentaschi, L.; De Roo, A. *Impact of a Changing Climate, Land Use, and Water Usage on Europe's Water Resources: A Model Simulation Study*; Publications Office of the European Union: Brussels, Belgium, 2018; ISBN 9789279802874.
4. Maddock, A.; Samuel Young, R.; Reig, P. *Ranking the World's Most Water-Stressed Countries in 2040*; World Resources Institute: Washington, DC, USA, 2015.
5. CHG. *Confederación Hidrográfica del Guadalquivir (Spanish Government) Plan Hidrológico de la Demarcación Hidrográfica del Guadalquivir*; Ciclo de Planificación Hidrológica 2015–2021; Ministerio de Agricultura, Alimentación y Medio Ambiente: Madrid, Spain, 2016; pp. 2015–2021.
6. González Perea, R.; Fernández García, I.; Martin Arroyo, M.; Rodríguez Díaz, J.A.; Camacho Poyato, E.; Montesinos, P. Multiplatform application for precision irrigation scheduling in strawberries. *Agric. Water Manag.* **2017**, *183*, 194–201. [CrossRef]
7. Bueno-Delgado, M.V.; Molina-Martínez, J.M.; Correoso-Campillo, R.; Pavón-Mariño, P. Ecofert: An Android application for the optimization of fertilizer cost in fertigation. *Comput. Electron. Agric.* **2016**, *121*, 32–42. [CrossRef]
8. Pagán, F.J.; Ferrández-Villena, M.; Fernández-Pacheco, D.G.; Rosillo, J.J.; Molina-Martínez, J.M. Optifer: An application to optimize fertiliser costs in fertigation. *Agric. Water Manag.* **2015**, *151*, 19–29. [CrossRef]

9.   Pérez-Castro, A.; Sánchez-Molina, J.A.; Castilla, M.; Sánchez-Moreno, J.; Moreno-Úbeda, J.C.; Magán, J.J. cFertigUAL: A fertigation management app for greenhouse vegetable crops. *Agric. Water Manag.* **2017**, *183*, 186–193. [CrossRef]

10. Institute for Agricultural and Fisheries Research and Training (IFAPA) SERVIFAPA—Programación del Riego y la Fertilización del Olivar. Available online: https://www.juntadeandalucia.es/agriculturaypesca/ ifapa/servifapa/recomendador-olivar/ (accessed on 11 October 2019).

11. ISO 20670:2018, Water Reuse—Vocabulary. Available online: https://www.iso.org/obp/ui/#iso:std:iso:20670: ed-1:v1:en (accessed on 15 April 2019).

12. Arizona Department of Environmental Quality Glosary of Environmental Terms. Available online: https: //legacy.azdeq.gov/function/help/glossary.html (accessed on 15 April 2019).

13. Raschid-sally, L.; Jayakody, P. *Drivers and Characteristics of Wastewater Agriculture in Developing Countries: Results From a Global Assessment*; IWMI: Colombo, Sri Lanka, 2009; ISBN 9789290906988.

14. Chen, W.; Lu, S.; Jiao, W.; Wang, M.; Chang, A.C. Reclaimed water: A safe irrigation water source? *Environ. Dev.* **2013**, *8*, 74–83. [CrossRef]

15. Ródenas, M.A.; Albacete, M. The River Segura: Reclaimed water, recovered river. *J. Water Reuse Desalin.* **2014**, *4*, 50–57. [CrossRef]

16. Pedrero, F.; Alarcón, J.J.; Nicolás, E.; Intriago, J.C.; Kujawa-Roeleveld, K.; Camposeo, S.; Vivaldi, G.A. Multidisciplinary Approach on Reclaimed Water Use Projects in Agriculture. In Proceedings of the III Jornada Agua y Sostenibilidad: La Reutilización de Aguas en España y Europa. Pasado, Presente y Futuro, Murcia, Spain, 15 December 2016; pp. 1–13.

17. Maestre-Valero, J.F.; González-Ortega, M.J.; Martínez-Álvarez, V.; Martin-Gorriz, B. The role of reclaimed water for crop irrigation in southeast Spain. *Water Supply* **2019**, *19*, 1555–1562. [CrossRef]

18. Petousi, I.; Fountoulakis, M.S.; Saru, M.L.; Nikolaidis, N.; Fletcher, L.; Stentiford, E.I.; Manios, T. Effects of reclaimed wastewater irrigation on olive (Olea europaea L. cv. 'Koroneiki') trees. *Agric. Water Manag.* **2015**, *160*, 33–40. [CrossRef]

19. Segal, E.; Dag, A.; Ben-Gal, A.; Zipori, I.; Erel, R.; Suryano, S.; Yermiyahu, U. Olive orchard irrigation with reclaimed wastewater: Agronomic and environmental considerations. *Agric. Ecosyst. Environ.* **2011**, *140*, 454–461. [CrossRef]

20. Ayoub, S.; Al-Shdiefat, S.; Rawashdeh, H.; Bashabsheh, I. Utilization of reclaimed wastewater for olive irrigation: Effect on soil properties, tree growth, yield and oil content. *Agric. Water Manag.* **2016**, *176*, 163–169. [CrossRef]

21. Bedbabis, S.; Ferrara, G.; Ben Rouina, B.; Boukhris, M. Effects of irrigation with treated wastewater on olive tree growth, yield and leaf mineral elements at short term. *Sci. Hortic. (Amst.)* **2010**, *126*, 345–350. [CrossRef]

22. Bourazanis, G.; Roussos, P.A.; Argyrokastritis, I.; Kosmas, C.; Kerkides, P. Evaluation of the use of treated municipal waste water on the yield, oil quality, free fatty acids' profile and nutrient levels in olive trees cv Koroneiki, in Greece. *Agric. Water Manag.* **2016**, *163*, 1–8. [CrossRef]

23. Erel, R.; Eppel, A.; Yermiyahu, U.; Ben-Gal, A.; Levy, G.; Zipori, I.; Schaumann, G.E.; Mayer, O.; Dag, A. Long-term irrigation with reclaimed wastewater: Implications on nutrient management, soil chemistry and olive (Olea europaea L.) performance. *Agric. Water Manag.* **2019**, *213*, 324–335. [CrossRef]

24. Trinh, L.T.; Vu, G.N.H.; Van Der Steen, P.; Lens, P.N.L. Climate Change Adaptation Indicators to Assess Wastewater Management and Reuse Options in the Mekong Delta, Vietnam. *Water Resour. Manag.* **2013**, *27*, 1175–1191. [CrossRef]

25. Pratap, R. *Getting Started with Matlab. A Quick Introduction for Scientist and Engineers*, 7th ed.; Oxford University Press: Oxford, UK, 2017.

26. Schaap, M.G. ROSETTA model. *J. Hydrol.* **1999**, *251*, 1–4.

27. Junta de Andalucía Estaciones Agroclimáticas de Andalucía. Available online: https://www.juntadeandalucia. es/agriculturaypesca/ifapa/ria/servlet/FrontController (accessed on 1 July 2019).

28. Allen, R.G. FAO Irrigation and Drainage Paper Crop by. *Irrig. Drain.* **1998**, *300*, 300.

29. Agencia Estatal de Meteorología (AEMET) AEMET OpenData. Available online: https://opendata.aemet.es/ centrodedescargas/inicio (accessed on 1 May 2019).

30. Pelmorex Corp Pelmorex Corp Weather. Available online: https://www.pelmorex.com/en/weather/ (accessed on 7 June 2019).

31. Smith, M. *Cropwat: A computer Program for Irrigation Planning and Management*; FAO Land and Water Development Division: Rome, Italy, 1992.

32. Doorenbos, J.; Pruit, W.O. Guidelines for predicting crop water requirements. *FAO Irrig. Drain.* **1977**, *24*, 15–20.

33. Orgaz, F.; Fereres, E. El Riego. In *El Cultivo del Olivo*; Junta de Andalucía, Ediciones Mundi-Prensa, Eds.; Ediciones Mundi-Prensa: Madrid, Spain, 2001; pp. 285–306.

34. Fereres, E.; Pruitt, W.O.; Beutel, J.A.; Henderson, D.W.; Holzapfel, E.; Shulbach, H.; Uriu, K. ET and drip irrigation scheduling. In *Drip Irrigation Management*; University of California. Div. of Agric. Sci. No. 21259; University of California: Oakland, CA, USA, 1981; pp. 8–13.

35. Padilla-Díaz, C.M.; Rodriguez-Dominguez, C.M.; Hernandez-Santana, V.; Perez-Martin, A.; Fernández, J.E. Scheduling regulated deficit irrigation in a hedgerow olive orchard from leaf turgor pressure related measurements. *Agric. Water Manag.* **2016**, *164*, 28–37. [CrossRef]

36. Rallo, L.; Cuevas, J. Fructificación y producción. In *El Cultivo del Olivo*; Junta de Andalucía, Ediciones Mundi-Prensa, Eds.; Ediciones Mundi-Prensa: Madrid, Spain, 2017; pp. 145–186.

37. Orgaz, F.; Fereres, E.; Testi, L. El Riego. In *El Cultivo del Olivo*; Junta de Andalucía, Ediciones Mundi-Prensa, Eds.; Ediciones Mundi-Prensa: Madrid, Spain, 2017; pp. 461–490.

38. Fernández-Escobar, R. Fertilización. In *El Cultivo del Olivo*; Junta de Andalucía, Ediciones Mundi-Prensa, Eds.; Ediciones Mundi-Prensa: Madrid, Spain, 2017; pp. 419–460.

39. Fernández-Escobar, R.; Parra, M.A.; Navarro, C.; Arquero, O. Foliar diagnosis as a guide to olive fertilization. *Span. J. Agric. Res.* **2009**, *7*, 212–223. [CrossRef]

40. Fernández-Escobar, R. Trends in olive nutrition. *Acta Hortic.* **2018**, *1199*, 215–223. [CrossRef]

41. García García, C. Abonado del olivar. In *Guía Práctica de la Fertilización Racional de Los Cultivos en España. Parte II. Abonado de Los Principales Cultivos en España*; Ministerio de Medio Ambiente y Medio Rural y Marino: Madrid, Spain, 2009.

42. The MathWorks, Inc. Mathworks App Designer. In *MATLAB App Building*; The MathWorks, Inc.: Natick, MA, USA, 2019.

43. Parra, M.A. Suelo. In *El Cultivo del Olivo*; Junta de Andalucía, Ediciones Mundi-Prensa, Eds.; Ediciones Mundi-Prensa: Madrid, Spain, 2017; pp. 251–287.

44. Molina-Soria, C.; Fernández-Escobar, R. A Proposal of New Critical Leaf Nitrogen Concentrations in Olive. *Acta Hortic.* **2012**, *949*, 283–286. [CrossRef]

45. Fernández-Escobar, R.; Navarro, S.; Melgar, J.C. Effect of Nitrogen Status on Frost Tolerance of Olive Trees. *Acta Hortic.* **2010**, *924*, 41–45. [CrossRef]

46. Fernández-Escobar, R.; Marin, L.; Sánchez-Zamora, M.A.; García-Novelo, J.M.; Molina-Soria, C.; Parra, M.A. Long-term effects of N fertilization on cropping and growth of olive trees and on N accumulation in soil profile. *Eur. J. Agron.* **2009**, *31*, 223–232. [CrossRef]

47. Fernández-Escobar, R.; Antonaya-Baena, M.F.; Sánchez-Zamora, M.A.; Molina-Soria, C. The amount of nitrogen applied and nutritional status of olive plants affect nitrogen uptake efficiency. *Sci. Hortic. (Amst.).* **2014**, *167*, 1–4. [CrossRef]

48. García, J.M.; Cuevas, M.V.; Fernández, J.E. Production and oil quality in 'Arbequina' olive (Olea europaea, L.) trees under two deficit irrigation strategies. *Irrig. Sci.* **2013**, *31*, 359–370. [CrossRef]

49. García-Novelo, J.M. *El Balance de Nitrógeno en el Olivar*; Universidad de Córdoba: Montería, Colombia, 2006.

MDPI

St. Alban-Anlage 66

4052 Basel

Switzerland

Tel. +41 61 683 77 34

Fax +41 61 302 89 18

www.mdpi.com

Water Editorial Office

E-mail: water@mdpi.com

www.mdpi.com/journal/water

9 783039 287901